VOLUME THIRTY THREE

# ADVANCES IN BIOMEMBRANES AND LIPID SELF-ASSEMBLY

# EDITORIAL BOARD

VOLUME THIRTY THREE

# ADVANCES IN BIOMEMBRANES AND LIPID SELF-ASSEMBLY

## Biological Membrane Vesicles: Scientific, Biotechnological and Clinical Considerations Part 2

Edited by


**Antonella Bongiovanni**
*Institute for Biomedical Research and Innovation (IRIB)*
*National Research Council of Italy (CNR), Palermo, Italy*

**Gabriella Pocsfalvi**
*Extracellular Vesicles and Mass Spectrometry Laboratory,*
*Institute of Biosciences and BioResources (IBBR),*
*National Research Council of Italy (CNR), Naples, Italy*

**Mauro Manno**
*National Research Council of Italy (CNR)*
*Institute of Biophysics (IBF)*
*CNR-IBF, Palermo, Italy*

**Veronika Kralj-Iglič**
*University of Ljubljana, Faculty of Health Sciences,*
*Laboratory of Clinical Biophysics, Ljubljana, Slovenia*
*Extracellular Vesicles and Mass Spectrometry Laboratory,*
*Institute of Biosciences and BioResources (IBBR),*
*National Research Council of Italy (CNR), Naples, Italy*



ELSEVIER

ACADEMIC PRESS
An imprint of Elsevier

Academic Press is an imprint of Elsevier
125 London Wall, London EC2Y 5AS, United Kingdom
525 B Street, Suite 1650, San Diego, CA 92101, United States
50 Hampshire Street, 5th Floor, Cambridge, MA 02139, United States
The Boulevard, Langford Lane, Kidlington, Oxford OX5 1GB, United Kingdom

First edition 2021



**Notices**
Knowledge and best practice in this field are constantly changing. As new research and experience broaden our understanding, changes in research methods, professional practices, or medical treatment may become necessary.

Practitioners and researchers must always rely on their own experience and knowledge in evaluating and using any information, methods, compounds, or experiments described herein. In using such information or methods they should be mindful of their own safety and the safety of others, including parties for whom they have a professional responsibility.

To the fullest extent of the law, neither the Publisher nor the authors, contributors, or editors, assume any liability for any injury and/or damage to persons or property as a matter of products liability, negligence or otherwise, or from any use or operation of any methods, products, instructions, or ideas contained in the material herein.

ISBN: 978-0-12-824608-5
ISSN: 2451-9634



*Publisher:* Zoe Kruze
*Acquisitions Editor:* Jason Mitchell
*Developmental Editor:* Tara Nadera
*Production Project Manager:* Abdulla Sait
*Cover Designer:* Greg Harris

Typeset by TNQ Technologies

# Contents

# Contributors

**Saba Battelino**
Department of Otorhinolaryngology and Cervicofacial Surgery, University Medical Centre Ljubljana, Ljubljana, Slovenia; Faculty of Medicine, University of Ljubljana, Ljubljana, Slovenia

**Darja Božič**
Laboratory of Clinical Biophysics, Faculty of Health Sciences, University of Ljubljana, Ljubljana, Slovenia; Laboratory of Physics, Faculty of Electrical Engineering, University of Ljubljana, Ljubljana, Slovenia

**Giovambattista Capasso**
Department of Medical Translational Sciences, University of Campania "L. Vanvitelli", Naples, Italy

**Immacolata Fiume**
Extracellular Vesicles and Mass Spectrometry Group, Institute of Biosciences and BioResources, National Research Council of Italy, Naples, Italy

**Paola Franco**
Institute of Genetics and Biophysics (IGB), National Research Council (CNR) of Italy, Naples, Italy

**Aleš Iglič**
Laboratory of Physics, Faculty of Electrical Engineering, University of Ljubljana, Ljubljana, Slovenia; Laboratory of Clinical Biophysics, Faculty of Medicine, University of Ljubljana, Ljubljana, Slovenia; Extracellular Vesicles and Mass Spectrometry Laboratory, Institute of Biosciences and BioResources, National Research Council of Italy, Naples, Italy

**Zala Jan**
Laboratory of Clinical Biophysics, Faculty of Health Sciences, University of Ljubljana, Ljubljana, Slovenia

**Marko Jeran**
Laboratory of Clinical Biophysics, Faculty of Health Sciences, University of Ljubljana, Ljubljana, Slovenia; Laboratory of Physics, Faculty of Electrical Engineering, University of Ljubljana, Ljubljana, Slovenia

**Annamaria Kisslinger**
Institute of Experimental Endocrinology and Oncology (IEOS), National Research Council (CNR) of Italy, Naples, Italy

**Veronika Kralj-Iglič**
University of Ljubljana, Faculty of Health Sciences, Laboratory of Clinical Biophysics, Ljubljana, Slovenia; Extracellular Vesicles and Mass Spectrometry Laboratory, Institute of Biosciences and BioResources, National Research Council of Italy, Naples, Italy

**Giovanna L. Liguori**
Institute of Genetics and Biophysics (IGB), National Research Council (CNR) of Italy, Naples, Italy

**Francesca Mantile**
Institute of Genetics and Biophysics (IGB), National Research Council (CNR) of Italy, Naples, Italy

**Ljubiša Pađen**
Laboratory of Clinical Biophysics, Faculty of Health Sciences, University of Ljubljana, Ljubljana, Slovenia

**Manca Pajnič**
Laboratory of Clinical Biophysics, Faculty of Health Sciences, University of Ljubljana, Ljubljana, Slovenia

**Žiga Pandur**
University of Ljubljana, Biotechnical Faculty, Department of Food Science and Technology, Laboratory of Microbiology, Ljubljana, Slovenia; University of Ljubljana, Faculty of Mechanical Engineering, Ljubljana, Slovenia

**Gabriella Pocsfalvi**
Extracellular Vesicles and Mass Spectrometry Group, Institute of Biosciences and BioResources, National Research Council of Italy, Naples, Italy

**Marzia Pucci**
Department of Biomedicine, Neuroscience and Advanced Diagnostics (Bi.N.D.), Section of Biology and Genetics, University of Palermo, Palermo, Italy

**Stefania Raimondo**
Department of Biomedicine, Neuroscience and Advanced Diagnostics (Bi.N.D.), Section of Biology and Genetics, University of Palermo, Palermo, Italy; Navhetec s.r.l, Palermo, Italy

**Christopher Stanly**
Extracellular Vesicles and Mass Spectrometry Group, Institute of Biosciences and BioResources, National Research Council of Italy, Naples, Italy

**David Stopar**
University of Ljubljana, Biotechnical Faculty, Department of Food Science and Technology, Laboratory of Microbiology, Ljubljana, Slovenia

**Maria Patrizia Stoppelli**
Institute of Genetics and Biophysics (IGB), National Research Council (CNR) of Italy, Naples, Italy

**Francesco Trepiccione**
Department of Medical Translational Sciences, University of Campania "L. Vanvitelli", Naples, Italy

**Bojana Uršič**
Department of Urology, University Medical Centre Ljubljana, Ljubljana, Slovenia; Laboratory of Clinical Biophysics, Faculty of Health Sciences, University of Ljubljana, Ljubljana, Slovenia

**Domen Vozel**
Department of Otorhinolaryngology and Cervicofacial Surgery, University Medical Centre Ljubljana, Ljubljana, Slovenia; Faculty of Medicine, University of Ljubljana, Ljubljana, Slovenia

# Preface

Throughout the last decade, our understanding of cell-to-cell communication has made an exceptional progress, which in turn has revealed biological membranes as key players in cell life and development. The finding that cell-derived membranous nanoparticles (e.g., extracellular vesicles, EVs) mediate intercellular communication by transferring biomolecules, including proteins, lipids, DNA and RNA has revolutionized cell biology and epigenetic fields, opening new prospects for the cross-talk between cells within the same organ, and among different tissues and even organisms in health and disease and as cross-kingdom communication mean.

Furthermore, the discovery of EVs as carriers of functional small molecules and proteins has raised great interest in the drug delivery field, as it may be possible to harness these vesicles for therapeutic delivery of miRNA, siRNA, mRNA, lncRNA, peptides and synthetic drugs. Native and drug-loaded mammalian cell-derived EVs (e.g., mesenchymal cell-derived EVs) have been recently identified and now constitute a rapidly growing research interest known as "cell-free therapy".

In this special volume, original research articles and reviews on all aspects related to the theme of "Extracellular Vesicles: from basic sciences to therapeutic exploitation" are presented. The volume features different functional and applicative themes revealing a fundamental role of EVs in membrane biology, biochemistry and biophysics, and in the functioning of living organisms including human beings. The contributions are organized into two volumes with 6 chapters in each volume.

In Volume 32, Chapters 1 and 2 (Amyloid β-peptide interaction with GM1 containing model membrane by Rita Carrotta et al., and Propolis flavonoids and terpenes, and their interactions with the model lipid membranes: a review, by Luka Šturm and Nataša Poklar Ulrich, respectively) consider molecular interactions that underlie membrane composition and function. Protein composition, regulation and their relation to the EV biological functions are presented in Chapters 3 and 4 (The Human EV membranome by Ana Paulina Ramos Juarez et al., and Post-translational lipidation in extracellular vesicles: chemical mechanism, biological function and application by Sabrina Picciotto et al., respectively). Close connection between methods and identity of EVs are addressed in Chapters 5 and 6 (Pursuing mechanisms of extracellular vesicle formation. Effects of sample processing, by Darja Božič et al., and Multi-technique analysis of extracellular vesicles: not only size matters by Fabio Perissinotto et al., respectively).

In Volume 33, the clinical aspects of extracellular vesicles are considered in Chapters 1–3 (Single patient analysis for clinical applications by Christopher Stanly et al., Biological role and clinical relevance of extracellular vesicles as key mediators of cell communication in cancer by Francesca Mantile et al., and Treatment with platelet- and extracellular vesicle-rich plasma in otorhinolaryngology. A review and future perspectives by Domen Vozel et al., respectively). Chapters 4 and 5 provide important insights into future development of the field of EVs. Plant EVs as novel and promising systems for industrial application are introduced in Chapter 4 (Plant extracellular vesicles: the safe for bioactive compounds by Stefania Raimondo and Marzia Pucci) while the problem of repeatability and accuracy of measurement that is common to all research in EV field is presented in Chapter 5 (A quality management based approach for exosomes research by Annamaria Kisslinger and Giovanna Liguori). The last chapter completes the volume with insights into the evolution of membrane systems from which we may learn fundamental mechanisms connecting living and non-living systems (Evolution of mechanical stability from lipid layers to complex bacterial envelope structures by Žiga Pandur and David Stopar).

CHAPTER ONE

# Urinary extracellular vesicles: single patient analysis for clinical applications

**Christopher Stanly[a], Immacolata Fiume[a], Bojana Uršič[b], Veronika Kralj-Iglič[d,e], Francesco Trepiccione[c], Giovambattista Capasso[c] and Gabriella Pocsfalvi[a,*]**

[a]Extracellular Vesicles and Mass Spectrometry Group, Institute of Biosciences and BioResources, National Research Council of Italy, Naples, Italy
[b]Laboratory of Clinical Biophysics, Faculty of Health Sciences, University of Ljubljana, Ljubljana, Slovenia
[c]Department of Medical Translational Sciences, University of Campania "L. Vanvitelli", Naples, Italy
[d]University of Ljubljana, Faculty of Health Sciences, Laboratory of Clinical Biophysics, Ljubljana, Slovenia
[e]Extracellular Vesicles and Mass Spectrometry Laboratory, Institute of Biosciences and BioResources, National Research Council of Italy, Naples, Italy
*Corresponding author: E-mail: gabriella.pocsfalvi@ibbr.cnr.it

## Contents



*Advances in Biomembranes and Lipid Self-Assembly,* Volume 33
ISSN 2451-9634
https://doi.org/10.1016/bs.abl.2020.09.004

## Abstract

Urinary extracellular vesicles (uEVs) are a heterogeneous group of membrane-bound vesicles that originate from the urinary system. They are considered to be an informative resource for the non-invasive study of molecular events in health and disease as they carry a variety of molecular constituents from their cells of origin, such as lipids, proteins and RNAs. For early stages of biomarker discovery schemes, untargeted omics-based approaches, which preferably employs uEVs isolated from pooled urine samples, may suit better. However, targeted approaches that preferably uses uEV isolated from individual urine samples are preferred for validation and clinical translation. The volume of urine needed for downstream analytical purposes and the sampling method (i.e. pooled vs. individual samples) are important practical aspects which need to be taken into consideration along with the challenges associated with the different methods of uEVs isolation. Here we provide an overview of the various isolation methods with special focus on the initial sample volumes as well as the sampling method applied. Our original data on the isolation of uEVs by employing salt precipitation and the sucrose/$D_2O$ double cushion ultracentrifugation method is also presented along with the latest trends over the last three years on their clinical applications. In this work, we demonstrate the performance and the reproducibility of the method using small volumes of individual and pooled samples as well as samples of a small group of patients with type 2 diabetes. Our data shows that the isolation method is compatible with both individual and pooled samples-based quantitative proteomics strategies.

## 1. Introduction

Extracellular vesicles (EVs) are released into their surrounding environment by cells and can be isolated from all biofluids including urine [1,2]. They are a heterogeneous group of membrane-enclosed vesicles that can be divided broadly into small extracellular vesicles (sEVs) and large extracellular vesicles (lEVs) based on their size [3]. The sEVs usually have a diameter of 30−150 nm and are believed to originate from the endocytic pathway whereas the lEVs have a typical diameter of 200−1000 nm and are known to directly bud off the plasma membrane of the parent cell [3]. It has also been seen that some vesicles are released by cells undergoing programmed cell death [4]. EVs are packed with proteins, mRNA, miRNAs and DNAs and they are able to deliver them to distant recipient cells. However, the mechanisms of their uptake by recipient cells differ from cell to cell [5].

## 1.1 Methods and practical considerations for isolation of urinary extracellular vesicles

Urinary extracellular vesicles (uEVs) are secreted from the urothelial cells, i.e. the epithelial cells of the urinary tract, including the kidneys, ureters, bladder, and urethra and the male reproductive tract [6]. It should be noted that the vast majority (99.96%) of uEVs originate from the kidney [7]. The methods used for the isolation of uEVs are the same as the methods applied for the isolation of EVs from cell cultures or from other biofluids. However, there are several challenges associated with the isolation of uEVs from the urine matrix. There are a number of optimized methods for the isolation of uEVs which can be grouped based on the principles used for separations as centrifugation, filtration, precipitation and chromatography-based separation.

In 2004, Pisitkun et al., for the first time, isolated low-density urinary membrane vesicles from a pooled urine sample of six individual healthy human subjects (50 mL each) by differential ultracentrifugation (dUC) [8]. Since then dUC is one of the most commonly used isolation methods in the field of uEVs research. dUC involves a series of centrifugation steps at different velocities. It takes advantage of the differences in size and density of the membrane-bound vesicles and separates them from other matrix components, aggregates and nanoparticles [9–11]. Initially, a low-velocity centrifugation step (300–500 × *g*) is performed to remove dead cells and debris. This is followed by centrifugation at speeds of 10,000 to 20,000 × *g* to isolate larger vesicles (microvesicles). Finally, the supernatant is centrifuged at speeds of 100,000 to 200,000 × *g* to pellet the smaller vesicles. One of the major drawbacks of dUC is the co-purifying contaminants, along with the possibility of aggregation and rupturing of vesicles [10]. Therefore, dUC is generally employed as the first purification step during the isolation of uEVs and is often followed by density gradient ultracentrifugation (DGUC). DGUC separates EVs based on their buoyant density. Sucrose and iodixanol (OptiPrep™) are the most frequently used gradient materials for uEVs purification. These compounds are stable and inert toward the vesicles and their densities and viscosities are well known. DGUC may be carried out in continuous [12] or discontinuous [13] gradient modes and both have been applied for the isolation of uEVs.

Filtration can also be used for the enrichment of uEVs as it was shown by Cheruvanky et al. in their original work of 2007 [11]. Microfiltration by using low protein binding, hydrophilized polyvinylidene difluoride membrane was demonstrated to efficiently enrich uEVs [14]. Filtration-based uEVs isolation method was further developed by the introduction of sequential filtration that comprises dead-end filtration, tangential flow filtration, and track-etched membrane filtration; and is ideally suited for a large sample volume of urine [15]. Recently, optimized ultrafiltration (UF), that combines low-velocity centrifugation with a two-step filtration followed by precipitation using the ExoQuick-TC™ exosome precipitation solution, was also used for the isolation of uEVs [1]. The main advantage of filtration is that it is faster and simpler compared to UC-based methods and, also it is easily scalable. However, non-specific binding of abundant soluble proteins and vesicles to the membrane surfaces of the filters and clogging are the frequently reported shortcomings of this method.

Precipitation-based methods are increasingly being used in uEVs research including a method that uses a proprietary precipitation reagent (i.e., ExoQuick-TC: System Biosciences; Mountain View, CA) (Exo-Qick-TC system). The main advantage of precipitation-based methods is that they allow the purification of uEVs using low-velocity centrifugation. Rider et al., in their study adapted the polyethylene glycol (PEG) precipitation protocols for virus precipitation and developed an ExtraPEG method for the isolation of uEVs [16]. In this study, they found that ExtraPEG method yielded uEVs which were highly comparable to the vesicles isolated by dUC and were indeed superior to vesicles isolated through commercially available methods. A novel uEVs-based gene expression assay (ExoDX™ Prostate test) that assesses the risk of prostate cancer also involves the use of polymer-based precipitation for the preparation of uEVs sample.

Chromatography is also used in the isolation of uEVs. Affinity chromatography exploits the presence of specific proteins and carbohydrates on the surface of the EVs. Antibodies against EV specific surface tetraspanins such as CD9, CD63 and CD81, or EV proteins or carbohydrate-binding molecules such as heparin, lectin and heat shock protein-binding peptides can be immobilized to a solid phase, these antibodies then specifically bind specific vesicle sub-populations. Bead-based methods containing three lectins [17] and concanavalin-A [18] specific to uEVs has been developed as a sample preparation step for exosome-based biomarker research. Size-exclusion chromatography (SEC) is becoming a popular method for the isolation of plasma and cell-derived EVs as it is capable of separating EVs from protein complexes and lipoproteins. Besides the commercial columns, Sepharose 2B, Sepharose CL-2B and Sepharose CL-4B gels are the most frequently used column material for SEC-based EV separations. The application of SEC directly for the isolation of uEVs from urine is hampered by the low (i) initial sample volume and (ii) EV concentration in urine. The volume of the sample applied to the SEC column is relatively low (i.e. 0.5%—4% of the total column volume) and the sample gets further diluted after SEC. Nevertheless, SEC was successfully used to isolate uEVs from very small amounts of urine samples (1 mL) of living and deceased kidney donors to profile the organ status before kidney transplantation by next-generation sequencing and proteomics [19]. When a higher quantity and/or purer uEVs sample is required, the initial urine samples have to be concentrated before SEC can be performed. Ultracentrifugation or ultrafiltration is usually the method of choice for this concentration and pre purification step.

In a recent review, Svenningsen et al. compared six standard methods (dUC, DGUC, SEC, UF, precipitation and affinity isolation) that are commonly used for the isolation of uEVs [7]. They retrieved data from EV knowledgebase EV-TRACK [20] on the isolation methods used for the enrichment of human uEVs and identified dUC as the most widely preferred one. Most studies used a combination of isolation methods when high-quality EVs are required for any downstream analysis. Approximately 18% of the published studies employed a combination of dUC and DGUC and 10% opted for a combination of dUC and UF. A combination of three methods dUC + SEC + UF was used in 6% of the published studies. Fewer than 2% of studies used a combination of dUC + affinity isolation, dUC + SEC, precipitation + UF, SEC + UF, affinity + UF or preicipiation + gUC. Each of these methods has its own advantages and disadvantages. DUC method requires specialized equipment while recovery rates of EVs vary. Although DGUC is known to reduce non-EV protein contaminants, the procedure is time-consuming and it requires specialized equipment and expertize. Although SEC is a fast procedure, it does not require specialized equipment and the biological activity of the isolated vesicles is also retained. The main drawback of this technique for the isolation of urinary EVs is that the final sample is diluted. Filtration is simple and rapid but it is greatly hindered by the non-specific binding of soluble proteins and by the frequent clogging of filter pores. The isolation of uEVs by precipitation is one of the most preferred methods because of its simplicity, but the uEV isolates contain high levels of impurities thus rendering it less suitable for Omics studies. High purity uEVs purification can be realized by employing affinity isolation, however, the isolated vesicles may represent only a subtype which is not always in the scope of the study.

Though standard EV isolation methods work for the isolation of uEVs, some peculiarities related to the urine sample should be considered to obtain optimal results. These are the (i) highly diluted nature of the urine sample and (ii) the presence of uromodulin which is the most abundant protein in human urine also known Tamm-Horsfall protein (THP). Uromodulin is a glycoprotein secreted after cleavage of the GPI anchor precursor protein. It is known to form large extracellular filaments, networks and aggregates by the interaction of cysteine-rich domains that co-sediment with uEVs. It was shown that uromodulin is a co-purifying impurity rather than a major component of uEV preparations. The presence of uromodulin usually negatively influences the downstream applications, therefore utmost care must be taken for its efficient removal. A number of methods are used to remove uromodulin, such as treatment with reducing agents like dithiothreitol (DTT) [21], detergent-like CHAPS [14], and salt precipitation [22]. Uromodulin depolymerization in the presence of DTT results in the release of the vesicles trapped in the polymeric network. The major shortcoming of using DTT is that some uromodulin monomers still can remain in the isolated EV fraction which can interfere with the downstream analysis. Salt-precipitation using NaCl was the originally described method for the purification of uromodulin from urine described by Igor Tamm and Frank Horsfall in 1952 [23]. More recently this method was also used to reduce the amount of uromodulin in the urine sample before the isolation of EVs [22]. Protein precipitation by adding $ZnSO_4$ was also shown to efficiently remove uromodulin prior to the vesicle isolation [18].

Another important aspect, especially when the aim of the study is biomarker development is the choice between single patient- or pooled urine sample-based sampling strategies. Most of the early uEV studies use pooled urine samples from patients and healthy donors which has a number of practical advantages like high sample volume, low cost and statistical value. However, sample pooling has different drawbacks too of which the requirement for larger spaces for sample storage and the loss of individual characteristics of the sample are the most notable ones. Quantitative assessment of individual patient samples is preferable especially when the aim is to confirm potential biomarker candidates in validation studies. The "fit-for-purpose" biomarker discovery schemes suggest that untargeted omic approaches may be better suited for early stages of biomarker discovery, while targeted approaches are preferred for validation and implementation [24].

Untargeted omics studies typically involve multiplex analysis that necessities a higher amount of samples and number of biological replicates while non-targeted approaches are extremely sensitive and only requiring small amounts of test samples. These key factors should be taken into account when designing or implementing a uEV-based biomarker study.

## 1.2 Urinary biomarker research relies on the studies of EV cargos

Urinary EVs are gaining importance as a resource for next-generation liquid biopsy [25]. Protein, peptide and RNA biocargo of uEVs are a great source of potential biomarkers for non-invasive renal monitoring as they provide a full representation of the entire urinary system. The molecular cargos of uEVs are not generally present as free soluble molecules in urine, therefore they are a rich source for non-invasively obtaining novel sets of potential biomarkers for various diseases. Mass-spectrometry based proteomics helps to understand the proteome and the change of protein concentration in uEVs [4]. However, the straightforward translation of uEVs proteomics studies into clinical use is complicated if the original sample contains non-exosomal proteins at high and variable concentrations [26]. The vesicular proteome is not only a great source for biomarker discovery but it also provides a wide range of information regarding the biogenesis and pathophysiological functions of EVs. The major proteins enriched in uEVs are cytoskeletal proteins (actins, cofilin-1, ezrin/radixin/moesin, profiling-1 and tubulins), metabolic enzymes (enolases, glyceraldehyde 3-phosphate dehydrogenase, peroxiredoxins and pyruvate kinase), tetraspanins (CD9, CD63, and CD 81), integrins, ribosomal proteins, Rab proteins, cytosolic proteins, plasma membrane proteins, heat shock proteins, annexins, Tsg101, alix, syntenin-1 and proteins involved in vesicle trafficking [6].

There has been an explosion of interest in the RNA content of extracellular vesicles ever since extracellular vesicles mediated transfer of RNA to target cell was first reported by *Valadi* et al. in 2007 [27]. A number of studies recently have identified both mRNA and miRNA in EVs isolated from various cell types and body fluids [28]. The ability of EVs to alter gene expression of neighboring and distant cells by transferring functional mRNA or miRNA is a very exciting finding [27]. Urinary miRNAs originate primarily from the kidney and urinary tract cells, cells infiltrating the renal tissue and shed in the urine, or the systemic circulation [29].

Lipids are important structural and functional constituents of EVs. Urinary lipidomics is an intriguing area of research in biomedical sciences since urine is a rich source of non-invasive biomarkers. The exploration of uEVs lipids as cancer biomarkers has just begun and the field is rapidly gaining attention. In one study Skotland et al. identified 107 lipid species by a high-throughput mass spectrometry lipidomics analysis. They quantified the 36 most abundant lipid species in uEVs of prostate cancer patients and healthy control and identified nine lipid species that were found to be significantly different when the two groups were compared. The highest significance was shown for phosphatidylserine and lactosylceramide [30]. Sphingomyelin, phosphatidylserine, cholesterol and saturated fatty acids are enriched in EVs when compared to total cell membranes [31]. In the second study, Glover et al. analyzed the lipids from uEVs of Hereditary α-Trytasemia(HαT) patients and healthy controls and found a higher number of uEVs in patient samples as compared to healthy control samples. Hence, normalization of uEVs number was performed prior to lipid analysis. They identified 521 lipid species out of which 64 were significantly decreased in HαT patients relative to the control group. The significantly decreased lipid species were diacylglycerols and lysophosphatidylethanolamines [32].

## 1.3 EV-based biomarkers and therapeutics for diabetic nephropathy in diabetes mellitus

Diabetic nephropathy (DN) or diabetic kidney disease is a syndrome characterized by the presence of pathologically increased levels of urine albumin excretion, diabetic glomerular lesions, and decrease of glomerular filtration rate (GFR) of diabetic patients. DN is a significant cause of chronic kidney disease and end-stage renal failure globally and it is also associated with an increased risk of adverse cardiovascular outcomes, morbidity and mortality. Currently, the glomerular filtration rate (GFR), blood urea nitrogen, serum creatinine, proteinuria, and albuminuria are the measures used to assess the presence and progress of diabetic nephropathy. The establishment of prognostic and diagnostic molecular biomarkers for DN has received major interest in recent years in general. Advances in uEVs studies are promising and these membrane-bound vesicles might be useful sources of novel biomarkers for DN.

Biomarker research aims to detect specific feature in the biocargo of uEVs (miRNA and protein) that could serve for the diagnosis and prognosis of diabetic nephropathy (DN). Injury to podocytes is considered a major contributor to diabetic kidney disease: their loss causes proteinuria and progressive glomerulosclerosis [33]. Sakurai et al. in a recent study showed that patients with type 2 diabetes (T2D) overexpress Elf3 protein in uEVs that suggests irreversible injuries in podocytes [34]. De et al. found that C-megalin content in uEVs was found to increase with the progression of the albuminuria in patients with T2D [35].

Significant increase of uromodulin mRNA and miRNA et-7c-5p in uEVs were shown to correlate with both renal function and progression of DN and proposed as potential biomarkers of disease progression [36]. The miRNA profiles of T2D patient at varying stages of kidney involvement in T2D showed that four miRNAs (miR-362-3p, miR-877-3p, and miR-150-5p and miR-15a-5p) were found to be differentially expressed in patients with DN [37].

The use of EVs as an alternative therapeutic agent to stem cell in tissue regeneration, immune modulation and cancer therapy is a rapidly advancing field of research. EVs isolated from stem cells (urinary stem cells (USCs) and adipose stem cells (ASCs)) have been proposed for DN treatment. A Protective effect of human USCs-derived EVs overexpressing microRNA-16-5p (miR-16-5p) injected into diabetic rats was observed against nephropathy [38]. In a mouse model, ASCs-derived EVs were able to attenuate diabetes by reducing the concentration of urine proteins, the levels of serum creatinine and blood urea nitrogen and the rate of podocyte apoptosis [39]. These EV-based options could become an important therapeutic strategy in future.

## 1.4 Recent trends in uEV research

A PubMed search (performed in March 2020) using "urine exosome" term retrieves a total of 438 articles using filters "English" language and "human" species over the period of 2010–2020 (Fig. 1).

A steady increase in the number of papers published from the beginning of the decade until 2018 can be observed showing that uEVs is an important sector in EV research. Analyzing the latest (last three years) trend in uEV research (Table 1) we can see that the majority of the publications are devoted to biomarker research (36 out of 44 original research articles analyzed) and only 7 studies are dedicated to fundamental research like the development of novel methods. The latest trend seems to prefer sampling of individuals (34 studies) rather than pooled urine samples (10 studies) for the isolation of uEVs. Regarding the type of biocargo studied, we can see that proteins (25 studies) and miRNA (16 studies) are still the main targets in the uEV field and the lipids remained highly neglected (no studies was found). We have analyzed the volumes of the urine samples that these studies used for the isolation of uEVs and found that they largely vary

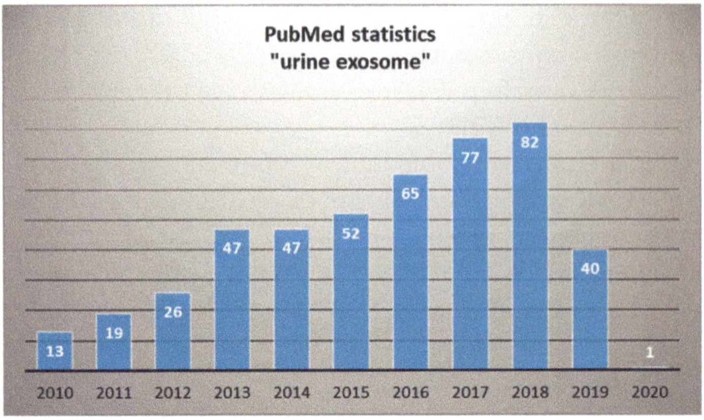


**Fig. 1** PubMed publication statistics: Number of research papers retrieved from the NCBI database PubMed in the last decade when searched with keywords "Urine exosome" on March 30, 2020. Filters applied: in the last 10 years, English, Humans.

ranging from 0.2 mL to 100 mL, although 50 mL was the most popular choice. Moreover, and unfortunately, a high percentage of the published works (25% of the analyzed publications) does not report this information. An interesting observation made through the analysis of publications is that even though there are a variable number of uEVs isolation methods available today, dUC is still a highly employed isolation method (20 studies). DUC isolation has different shortcomings that have been highlighted by a number of studies especially when specific EV biomarker analysis is the goal. Some of the works take special care for the removal of uromodulin before or during the dUC protocol, generally using the DTT treatment. Kit-based methods are getting popular and usually combined with qRT-PCR-based miRNA or western blot-based protein analyses. However, interestingly no LC-MS/MS-based proteomics-based research has been performed with uEVs isolated by kit-based reagents. Most of the studies are based on comparative analysis between healthy and diseased patient samples and the final research goal usually dictates the choice of sample collection method (pooled vs. individual samples), the volume of urine and the number of biological replicates.

In our study, we show that sodium chloride (NaCl) salt-precipitation of uromodulin in combination with the sucrose/$D_2O$ double cushion gradient ultracentrifugation is a suitable method for uEV isolation both from individual and pooled samples using 50 mL of starting urine sample volumes. Shot-gun proteomics on a small cohort of T2D patients was studied to evaluate the quantity and the quality of uEVs for protein biomarker research schemes.

**Table 1** Research papers published in the urinary extracellular research field in the last three years according to NCBI PubMed search using "urine exosome" and filters "10 years, English and human species".

| First author | Year | Volume of urine (mL) | Sample type | uEV isolation method | Physiology | Disease | Type of EV-Isolated | Aim of the work | Cargo studied | Method | DOI |
|---|---|---|---|---|---|---|---|---|---|---|---|
| Huang et al. | 2019 | 0.2 | Individual | HIC | Healthy | No Disease | Exosomes | Fundamental Research | No cargo study | HIC | 10.1007/s00216-019-02022-7 |
| Ku et al. | 2018 | 0.3 | Individual | Acustic trapping | Healthy | No Disease | Exosomes and MVs | Biomarker Research | Protein, miRNA | qRT-PCR, ELISA | 10.1021/acs.analchem.8b00914 |
| Morikawa et al. | 2019 | 0.3 | Individual | kit-based | Healthy and Diseased | Glomerulonephritis | EVs | Biomarker Research | Protein | ELISA | 10.1159/000495217 |
| O Sun et al. | 2018 | 1 | Individual | kit-based | Healthy and Diseased | Hypertension | EVs | Biomarker Research | Protein | Imaging FC | 10.1161/hypertensionaha.118.11766 |
| Wang et al. | 2019 | 2 | Pooled | kit-based | Healthy | No Disease | Exosomes | Biomarker Research | miRNA | PCR, qPCR | 10.1016/j.clinbiochem.2019.03.008 |
| Vicikoa et al. | 2018 | 5 | Individual | kit-based | Healthy and Diseased | Cancer | Exosomes | Biomarker Research | Protein | ELISA | 10.4149/neo_2018_171203n792 |
| Hu et al. | 2018 | 9 | Individual | dUC | Healthy and Diseased | Pre-eclampsia | Evs and Exosomes | Biomarker Research | Protein | WB | 10.1371/journal.pone.0204514 |
| Hasegawa et al. | 2018 | 9 | Pooled | kit-based | Diseased | Prostate Cancer | Exosomes | Biomarker Research | miRNA | qRT-PCR, WB | 10.1158/1541-7786.mcr-17-0321 |
| Srivastava et al. | 2018 | 10 | Individual | DTT/dUC | Healthy and Diseased | Endometrial cancer | Exosomes | Biomarker Research | miRNA | qPCR | 10.1208/s12248-018-0220-y |
| Pathare et al. | 2018 | 20 | Individual | dUC | Healthy and Diseased | Distal renal tubular acidosis | Exosomes | Biomarker Research | Protein | WB | 10.1007/s00424-017-2049-0 |

(*Continued*)

**Table 1** Research papers published in the urinary extracellular research field in the last three years according to NCBI PubMed search using "urine exosome" and filters "10 years, English and human species".—cont'd

| First author | Year | Volume of urine (mL) | Sample type | uEV isolation method | Physiology | Disease | Type of EV-Isolated | Aim of the work | Cargo studied | Method | DOI |
|---|---|---|---|---|---|---|---|---|---|---|---|
| Pathare et al. | 2018 | 20 | Individual | dUC | Healthy and Diseased | Distal renal tubular acidosis | Exosomes | Fundamental Research | Protein | WB | 10.1016/j.kint.2017.10.018 |
| Liu et al. | 2018 | 20 | Individual | ZnSO4/dUC and ConA affinity | Healthy | NA | EVs | Fundamental Research | Protein | LC-MS/MS | 10.1007/s12192-018-0902-5 |
| Chen et al. | 2019 | 25 | Pooled | DTT/dUC and kit-based | Healthy and Diseased | Idiopathic nephrotic syndrome | Exosomes | Biomarker Research | miRNA | qRT-PCR | 10.1016/j.ebiom.2018.11.018 |
| Lv et al. | 2018 | 25 | Individual | PEG precipitation | Healthy and Diseased | Renal fibrosis | Exosomes | Biomarker Research | Protein, miRNA | WB, qRT-PCR | 10.1007/s11255-017-1779-4 |
| Perez-Hernandez et al. | 2018 | 50 | Individual | DTT/dUC | Diseased | Hypertension | Exomes and MVs | Biomarker Research | Protein, miRNA | qRT-PCR, WB | 10.1186/s12967-018-1604-6 |
| Kruglik et al. | 2019 | 50 | Individual | dUC | Healthy | NA | Exosomes | Biomarker Research | Biomolecules | DNA | 10.1039/c8nr04677h |
| Wei et al. | 2019 | 50 | Individual | dUC | Healthy and Diseased | Chronic Hepatitis B | Exosomes | Biomarker Research | miRNA | qRT-PCR | 10.1097/md.0000000000017143 |
| Yu et al. | 2018 | 50 | Individual | dUC | Healthy and Diseased | Renal fibrosis | Exosomes | Biomarker Research | miRNA | WB, qRT-PCR, | 10.1159/000487104 |
| Gheinani et al. | 2018 | 50 | Pooled | dUC/SEC/PEG | Healthy | NA | NVs | Biomarker Research | Protein, miRNA | WB, qRT-PCR | 10.1038/s41598-018-22142-x |
| Chutipongtanate et al. | 2018 | 50 | Individual | dUC/UF | Healthy | NA | EVs | Biomarker Research | Protein | WB, LC-MS/MS | 10.1038/s41598-018-33280-7 |
| He et al. | 2019 | 50 | Individual | dUC/UF | Healthy | NA | Exosomes | Fundamental Research | Protein | WB, qRT-PCR | 10.3892/ijmm.2018.3944 |

| | | | | | | | | | | | |
|---|---|---|---|---|---|---|---|---|---|---|---|
| Lv et al. | 2018 | 50 | Individual | PEG precipitation | Diseased | Renal Fibrosis | Exosomes | Biomarker Research | miRNA | qRT-PCR, WB | 10.1016/j.yexmp.2018.08.004 |
| Hinrichs et al. | 2018 | 70 | Individual and Pooled | dUC | Diseased | Kidney transplant | Exosomes | Fundamental Research | Protein | WB | 10.1152/ajprenal.00545.2017 |
| Raimondo et al. | 2018 | 100 | Pooled | dUC | Healthy and Diseased | Hematuria | Exosomes | Biomarker Research | Protein | WB, LC-MS/MS | 10.1021/acs.jproteome.7b00763 |
| Guan et al. | 2019 | 100 | Individual | dUC/Filtration/SEC | Healthy | NA | Exosomes | Fundamental Research | Protein | LC-MS/MS | 10.1007/s00216-019-01616-5 |
| Vall-Palomar et al. | 2018 | 500 | Pooled | DTT/dUC | Healthy and Diseased | Hypomagnesemia | Exosomes | Biomarker Research | miRNA | WB, qPCR | 10.1186/s12967-018-1651-z |
| Lee et al. | 2018 | 10–20 | Individual | kit-based | Diseased | Urothelial bladder carcinoma | Exosomes | Biomarker Research | DNA | sWGS | 10.1038/s41598-018-32900-6 |
| Gudehithlu et al. | 2019 | 50–100 | Individual | DTT/dUC | Healthy and Diseased | Kidney diseases | Exosomes | Biomarker Research | Protein | ELISA | 10.1007/s10157-019-01734-5 |
| Abe et al. | 2018 | 50–100 | Individual | dUC/GUC | Healthy and Diseased | Diabetic kidney disease | Exosomes | Biomarker Research | mRNA | RT-PCR, qRT-PCR, WB | 10.2152/jmi.65.208 |
| Elsharkawi et al. | 2019 | 5–15 | Individual | kit-based | Healthy and Diseased | Bladder cancer | Exosomes | Biomarker Research | Protein | ELISA | 10.31557/APJCP.2019.20.7.2219 |
| Lozano-Ramos et al. | 2018 | 70–100 | Individual | DTT/UF/SEC | Living and Deceased donors | Kidney transplant | Evs | Biomarker Research | Protein, miRNA | LC-MS/MS, NSG | 10.1186/s12882-018-0985-3 |
| Lim et al. | 2018 | n.m. | Pooled | DTT/dUC | Diseased | Kidney transplant | Exosomes | Biomarker Research | Protein | LC-MS/MS | 10.1371/journal.pone.0204204 |
| Lan et al. | 2018 | n.m. | n.m. | dUC | Healthy | No Disease | Exosomes | Fundamental Research | No cargo study | 2DE | 10.1002/elps.201700388 |

*(Continued)*

**Table 1** Research papers published in the urinary extracellular research field in the last three years according to NCBI PubMed search using "urine exosome" and filters "10 years, English and human species".—cont'd

| First author | Year | Volume of urine (mL) | Sample type | uEV isolation method | Physiology | Disease | Type of EV-Isolated | Aim of the work | Cargo studied | Method | DOI |
|---|---|---|---|---|---|---|---|---|---|---|---|
| Miyazawa et al. | 2018 | n.m. | Pooled | dUC | Healthy | NA | EVs | Biomarker Research | Protein | LC-MS/MS | 10.1007/s10157-018-1538-6 |
| Min et al. | 2018 | n.m. | Pooled | dUC | Healthy and Diseased | IgA nephropathy/glomerulonephritis | Exosomes | Biomarker Research | miRNA | qRT-PCR | 10.1002/jcla.22226 |
| Lv et al. | 2018 | n.m. | Individual | dUC | Healthy and Diseased | Tubulointerstitial inflammation | Exosomes | Biomarker Research | mRNA | qRT-PCR | 10.1681/ASN.2017050523 |
| Feng et al. | 2018 | n.m. | Individual | dUC | Healthy and Diseased | IgA nephropathy | Exosomes | Biomarker Research | mRNA | RT-PCR | 10.1016/j.ajpath.2018.07.017 |
| Keri et al. | 2018 | n.m. | Individual | dUC | Healthy and Diseased | Polycystic kidney disease | Exosomes | Biomarker Research | Protein | WB | 10.1186/s13104-018-3467-6 |
| Dong et al. | 2018 | n.m. | Individual | dUC/GUC | Healthy and Diseased | Dysferlinopathy | Exosomes | Biomarker Research | Protein | WB | 10.7150/thno.22856 |
| Staubach et al. | 2018 | n.m. | Individual | dUC/GUC | Healthy and Diseased | Tubulointerstitial kidney disease | Exosomes | Biomarker Research | Protein | LC-MS/MS | 10.1002/pmic.201700456 |
| Fredsøe et al. | 2019 | n.m. | Individual | kit-based | Diseased | Prostate cancer | Exosomes | Biomarker Research | miRNA | qRT-PCR | 10.1002/ijc.32296 |
| Martelli et al. | 2018 | n.m. | Individual | kit-based | Healthy and Diseased | Nephropathy and cystitis | Exosomes | Fundamental Research | miRNA | qRT-PCR, nested-PCR, WB | 10.3390/v10090466 |
| Sabaratnam et al. | 2019 | spot sample | Individual | | Diseased | Nephrectomy | EVs | Biomarker Research | Protein | WB | 10.1152/ajprenal.00242.2019 |
| Yamamoto et al. | 2018 | | Individual | Filtration | Healthy and Diseased | Diabetic kidney disease | EVs | Biomarker Research | mRNA | qPCR | 10.1159/000489129 |

# 2. Materials and methods

## 2.1 Urine sample collection

The study protocol was approved by the Ethics Committee of the Università della Campania "Luigi Vanvitelli". All subjects have signed written informed consent. Research participants were healthy male volunteers between age 25–45 years old and diabetic patients between age 65–80 years old. The participants had no known history of urinary tract infections. Second-morning urine samples were collected into 120 mL polyethylene urine containers. Protease inhibitors, sodium azide (3.5 mM), phenylmethylsulfonyl fluoride (PMSF) (0.5 mM) and leupeptin (1 μg/mL) were added and samples were stored at -80 °C until processed.

Pooled urine sample preparation: Each healthy volunteer was accepted based on the opinion of the physician that the volunteer has no chronic or major illness. Urine samples were collected two times from a single healthy male in a total amount of about 400 mL. The samples were pooled, protease inhibitors were added, divided into four 100 mL aliquots and stored at -80 °C until use. These samples were used to set up the combined salt-precipitation and sucrose/$D_2O$ double cushion ultracentrifugation method, reproducibility evaluation and interpersonal comparison evaluations.

Individual urine sample preparation: 100 mL of urine sample was collected from a single healthy male, protease inhibitors were added and stored at -80 °C until use. To check the interpersonal variability of 100 mL of second-morning urine, samples from five healthy male subjects we collected and stored at -80 °C until use.

T2D diabetic patient samples: 50 mL of urine samples of three patients suffering from type II diabetes mellitus were collected and stored at -80 °C until use. The uEVs isolation was performed according to the combined salt-precipitation and sucrose/$D_2O$ double cushion method.

## 2.2 Salt-precipitation

The frozen urine samples were thawed and transferred into 50 mL polypropylene centrifuge tubes (Falcon, Fisher Scientific). Each urine sample was centrifuged at 400 × *g* for 20 min at room temperature to remove cells and cellular debris. To the supernatants of each sample, 4.4 M NaCl solution was added to a final concentration of 0.58 M to precipitate urinary mucoproteins, including uromodulin according to the method that was originally described by Tamm and Horsfall [23]. Salt-precipitation also called salting out was carried out for 2 h at room temperature [22]. The sample was clarified by using the below mentioned sequential steps.

### 2.3 Isolation of uEVs by sucrose/$D_2O$ double cushion gradient ultracentrifugation

Urinary EVs were isolated from uromodulin-depleted urine samples using the classical dUC method followed by sucrose/$D_2O$ gradient centrifugation according to Raj et al. [40]. Briefly, sequential centrifugations were performed at 4000 × *g* (50 mL falcon tubes) and 15,000 × *g* (Beckman white cap tubes) at room temperature, followed by high-velocity ultracentrifugation at 200,000 × *g* (Polycarbonate Bottle with Cap Assembly, 26.3 mL, Beckman Coulter, USA) for 1 h at 4 °C using a 70Ti rotor (Beckman Coulter, USA). The resulting pellet was solubilized in a small volume of solubilization buffer composed of 20 mM Tris-HCl pH 8.6. The sample was underlaid with a sucrose/$D_2O$ double cushion consisting of 1 mL of 1 M and 0.8 mL of 2 M sucrose prepared in 20 mM Tris pH 8.6/$D_2O$ in 13.2 mL thin-wall polypropylene tubes (Beckman Coulter, USA) and centrifuged at 110,000 × *g* for 3 h at 4 °C using an SW 41 Ti rotor (Beckman Coulter, USA). The 1 M and 2 M fractions were collected by piercing the bottom of the tube, diluted in solubilization buffer and centrifuged at 110,000 × *g* for 1 h at 4 °C using an SW 41 Ti rotor to remove sucrose. The resulting pellets were resuspended in 10—20 μL solubilization buffers. Protein concentration was determined by the micro-bicinchoninic acid protein assay (Pierce/Thermo Scientific, Rockford, IL, USA) using a nanodrop 200 spectrophotometer (Thermo Fisher Scientific, Wilmington, DE, USA). Samples were stored at -80 °C until used.

AMBIC at room temperature for 30 min in dark. The gel pieces were washed with 100 mM AMBIC, dehydrated with 100% acetonitrile and dried using a vacuum concentrator. The samples were rehydrated in a solution containing 6 ng/μL trypsin (Promega, Madison, WI) in 50 mM AMBIC 10% acetonitrile for 30 min on ice and then 16 h at 37 °C. Peptides were extracted sequentially using 30% acetonitrile/3.5% formic acid (LC-MS grade, Thermo Sci. Rockford, IL, USA), 50% acetonitrile/0.5% formic acid, 100% acetonitrile. The extracted samples were combined and dried using a vacuum concentrator.

The peptides were analyzed by nano-flow reversed-phase liquid chromatography-electrospray ionization tandem mass spectrometry (nano-LC-RP-ESI-MS/MS). 20 μL samples were loaded, purified and concentrated on a reversed-phase monolithic pre-column, 200 μm ID × 5 mm length (LC Packings, Sunnyvale, CA USA) at 25 μL/min flow rate. Peptides were separated at a flow rate of 300 nL/min on a PepSwift Monolithic column, 100 μm ID × 5 cm (LC Packings, Sunnyvale, CA USA) using the following gradient: (solvent A: 2% acetonitrile, 0.1% formic acid solvent B: 98% acetonitrile, 0.1% formic acid) 5%—50% B in 60 min, 5%—98% B in 6 s for 10 min. Eluted peptides were analyzed in information-dependent analysis (IDA) mode using the QSTAR Elite (Applied Biosystems Foster City, CA, USA) mass spectrometer equipped with a nanoflow electrospray ion source.

### 2.3.1 In-solution-based proteomic analysis

Urinary EVs samples (5 μg protein content according to the BCA assay) were solubilized in 0.1% RapiGest SF (Waters Corp, Milford, MA, USA) surfactant in 50 mM AMBIC and lyzed by three freeze-thaw cycles in liquid nitrogen under sonication. Proteins were reduced with 5 mM Tris(2-carboxyethyl)phosphine hydrochloride) (Sigma-Aldrich, Monza, Italy) and alkylated with 10 mM iodoacetamide (Sigma-Aldrich, Saint Louis, MO, USA). The proteins were digested with 1:50 trypsin (Promega, Madison, WI): protein ration at 37 °C for 45 min. After digestion, the sample was acidified with trifluoroacetic acid to a final concentration of 0.5% and centrifuged at 17000 *xg* for 10 min to remove Rapigest. The supernatant was removed and dried in a vacuum concentrator. Peptides were analyzed by nano-RP-LC-ESI-MS/MS as described above. The IDA experiment was set-up to exclude the known molecular ions of uromodulin from the data acquisition. Ion intensities were measured by Analyst QS software and proteins were identified using the Mascot Server (Matrix Science, UK).

### 2.3.2 Data analysis

Analyst QS 2.0 software was used with default parameters to generate and to analyze peak lists extracted from the mass spectra acquired in the information-dependent analysis (IDA) mode. Mascot 2.2 was used to search the peak list against the SwissProt 2013—2014 database using trypsin with one possible missed cleavage, carbamidomethylation of cysteine as fixed and oxidation of methionine were considered variable modifications. An analysis false-positive rate of the protein identifications was performed by searching all tandem mass spectra from the nano-HPLC-ESI-MS/MS analyses against an in-house curated decoy SwissProt human protein database containing forward and reversed sequences. Contaminants such as human keratins and porcine trypsin etc. were included in this database.

Scaffold 4 program packages (Proteome software, Oregon USA), were used to validate MS/MS proteomics experiments and to evaluate the relative expression levels of the proteins commonly identified in IDA experiments.

## 2.4 SDS-PAGE and Western blot analysis

Protein samples (20 µg or as stated), were loaded on a precast Novex 4-12% Bis-Tris NuPAGE gel (Invitrogen, Carlsbad, CA, USA) using MOPS running buffer, according to the manufacturer's instructions and stained with colloidal Coomassie blue.

The western blot analysis was performed on proteins separated by SDS-PAGE and stained with colloidal coomassie blue. The stained gel was then de-stained in water, soaked in 1% SDS pH 8.5 for 30 min, washed twice in water and once using the transfer buffer (Invitrogen, Carlsbad, CA, USA) for 10 min. Proteins were transferred onto PVDF membranes (Invitrogen, Carlsbad, CA, USA), blocked according to manufacture instruction and probed with the primary antibodies-mouse monoclonal antibodies: tumor susceptibility gene 101 [TSG 101, (Y16J) Ref: no sc-101254], apoptosis-linked gene-2 interacting protein X [ALIX, (1A12), Ref: no sc-53540] (Santa Cruz Biotechnology, Santa Cruz, CA). The blots were incubated with alkaline phosphatase-conjugated secondary antibodies and the immune complexes were detected using Western Light chemiluminescent detection system (Applied Bio-systems, Bedford, MA), according to the manufacturer's protocol.

## 2.5 Proteomics of urinary extracellular vesicles

### *2.5.1 In-gel based proteomic analysis*

Urinary EVs samples separated SDS-PAGE and stained with colloidal Coomassie blue were used for in-gel proteomics according to the procedure described by Shevchenko *et al* [41]. The entire gel lane was cut into 1.5 mm bands. The gel pieces were washed using 100 mM ammonium bicarbonate (AMBIC) in 50% acetonitrile thrice for 10 min, once with acetonitrile for 10 min and then dried using a vacuum concentrator. Proteins were reduced using 10 mM 1,4-dithiothreitol (Sigma-Aldrich, Saint Louis, MO, USA in 100 mM AMBIC in 5% acetonitrile at 55° C for 1 hour and alkylated with 55 mM iodoacetamide (Sigma-Aldrich, Saint Louis, MO, USA) in 100 mM.

# 3. Results and discussion

## 3.1 Sucrose/$D_2O$ gradient ultracentrifugation method for low-volume of urine samples

We aimed to set-up a method for the isolation of uEVs from initial urine sample volumes that are feasible to collect easily at the single individual level without pooling in hospital settings. The sucrose/$D_2O$ density gradient ultracentrifugation method published by Raj et al. [40] was chosen as a starting point for scaling down experiments. The original method describes the use of 500—1000 mL pooled urine starting samples for the isolation of about 50—100 μg of highly pure uEVs (expressed in protein content) that were shown to be suitable for downstream proteomics studies. We attempted to scale down the original sample volume to 100 mL. This involved the replacement of the 38.5 mL open thin-walled ultracentrifuge tubes by the 13.2 mL volume tubes in the GUC step of the isolation protocol. The double cushion gradient UC was performed using four individual healthy samples. As it was expected a visible cloudy layer containing the vesicles was observed just above the 1 M fraction (data not shown). This layer was carefully removed, processed for washing steps to remove sucrose. SDS-PAGE image shows the presence of a high amount of uromodulin in the 4 uEV samples (Fig. 2).

Also, the total protein content (approximately 15—20 μg as deduced from the signal intensity in the SDS-PAGE) was higher than expected. Since uEVs isolates from high initial volume sample did not show the presence of such a high concentration of uromodulin over different urine samples including, healthy control, male, female [13], proteinuria and polycystic kidney disease samples [42], our result was unexpected. We deduced that dealing with a small sample volume is troublesome. Especially, the gradient ultracentrifugation step is crucial in the efficient removal of uromodulin from the uEVs isolates which is achieved by the accurate separation of the uEV containing layer from the 2 M fraction and the soluble proteins (above the 1 M fraction) fractions. Reasoning on the observed result, we presumed that there might have been a considerable amount of cross-contamination from the soluble uromodulin-containing fraction (that is just above the 1 M uEV fraction) since collecting a small volume of vesicle-containing cushion layer (0.5—1 mL) from a smaller sized UC tube (by both piercing and pipetting) is rather difficult. Additionally, it seems that even small changes in the sample: buffer ratio with respect to the published method could have a negative effect on the co-purification of uromodulin.

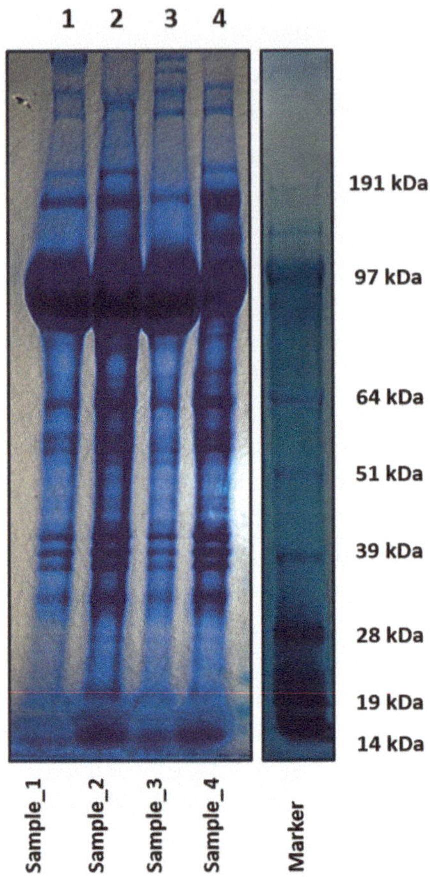


**Fig. 2** SDS-PAGE image of the protein profiles of urinary extracellular vesicles isolated from a pooled urine sample of a healthy male individual. The pooled sample was divided into 4*100 mL Aliquotes and processed for uEVs isolation by the sucrose/$D_2O$ double cushion gradient ultracentrifugation method [37]) in a single experiment.

## 3.2 uEV isolation by the combination of salt-precipitation and sucrose/$D_2O$ gradient ultracentrifugation

To overcome the problem of the co-purification of uromodulin with uEVs during the small volume sucrose/$D_2O$ gradient centrifugation procedure, we decided to add a precipitation step before performing the uEVs isolation (Fig. 3). The precipitation step aimed at reducing the amount of uromodulin in the urine sample before the ultracentrifugation. We have applied the classical procedures that employ salt precipitation as described earlier for the

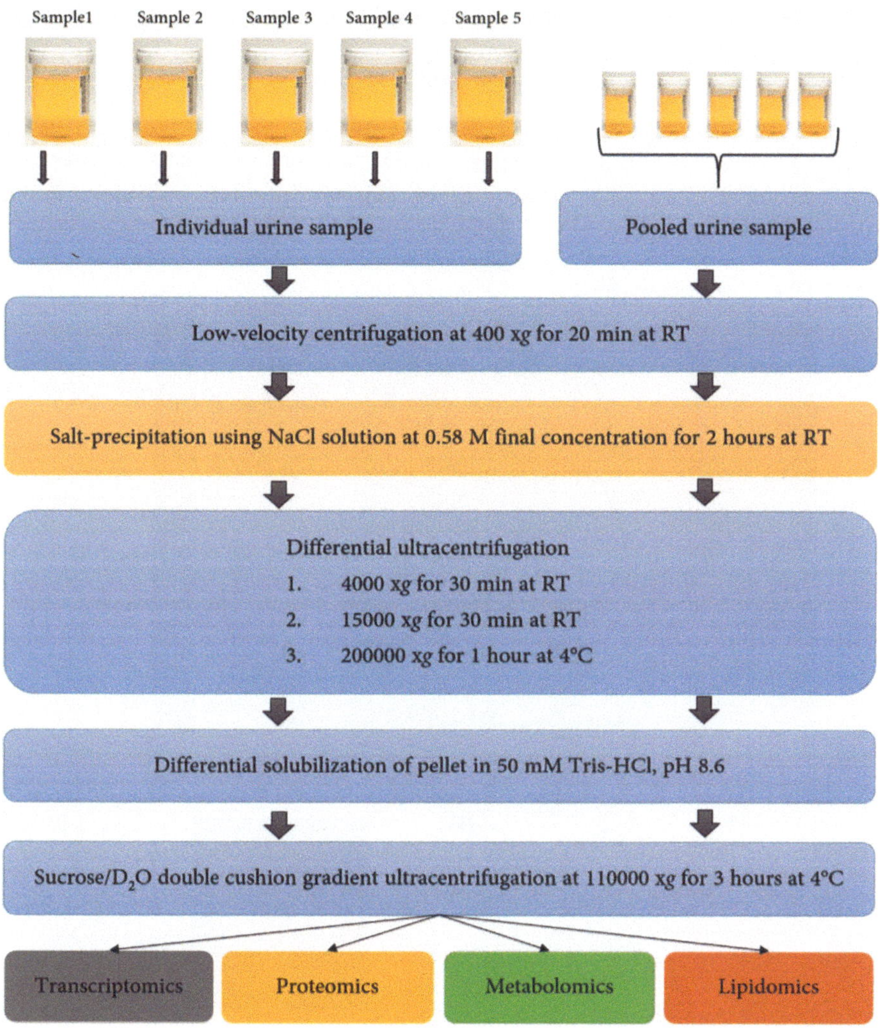


**Fig. 3** Workflow for the isolation of urinary extracellular vesicles (uEVs) using the combined salt precipitation and sucrose/$D_2O$ double cushion gradient ultracentrifugation method from individual and pooled samples for downstream analysis.

isolation of uromodulin itself [23] and observed that it was efficient in the resolving of the uromodulin problem in the uEV samples.

To better evaluate the isolation/purification method as well to test the reproducibility of the combined two-step procedure (Fig. 3) we have purified uEVs from a healthy male individual in three experiments conducted in parallel. The supernatant (S) and pellet samples (P) were analyzed throughout the uEV isolation protocol at each centrifugation step (400 × *g*, 4000 × *g*, 15,000 × *g* and 200,000 × *g*), together with the crude uEVs isolated from the same sample by dUC and the final uEVs (collected from 1 M sucrose cushion) The SDS-PAGE protein profiles of the supernatant samples (Fig. 4A) were according to what was expected very similar to the precipitated proteins from total urine and contained the main soluble urinary

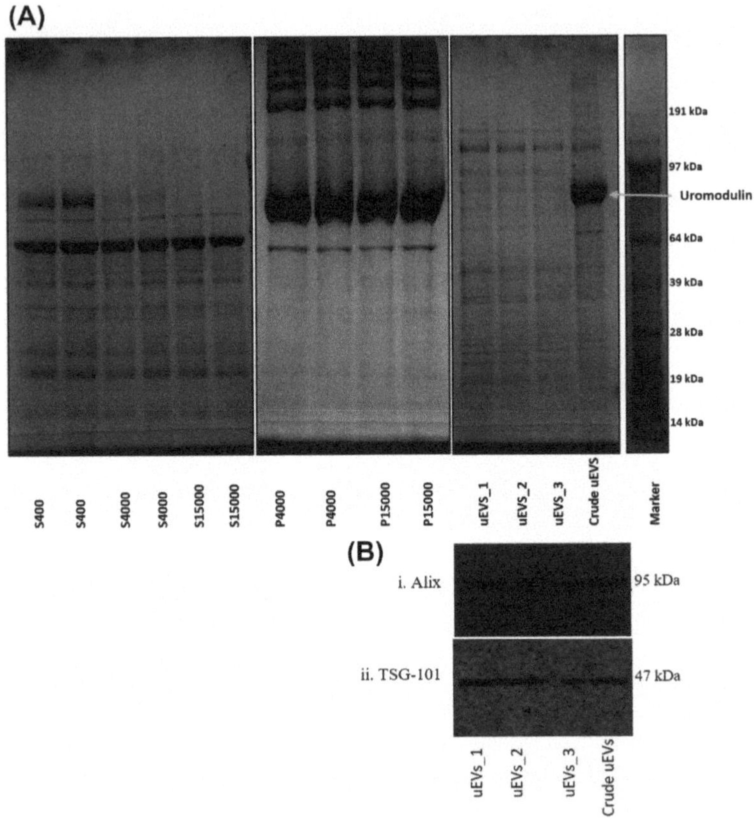


**Fig. 4** (A) SDS-PAGE image of the protein profiles obtained at different stages of the urinary extracellular vesicles (uEVs) isolation procedure based on combined salt precipitation and sucrose/$D_2O$ double cushion gradient ultracentrifugation method and using 100 mL initial urine sample from a healthy male individual in three experiments. The experiments were performed in parallel. P4000, P15000 and S400, S4000, S15000 indicate the pellets (P) and the supernatants (S) obtained at 400 × *g*, 4000 × *g* and 15,000 × *g* centrifugation steps. The crude uEVs sample was obtained by differential ultracentrifugation. The strong bands around 85 kDa correspond to uromodulin. (B) Western blot images showing the detection of two uEVs markers (I) Alix and (ii) TSG-101.

proteins, like uromodulin, albumin, transferrin, immunoglobulins and $\alpha$1 antitrypsin. Fig. 4A shows that the amount of soluble uromodulin in the supernatants was decreasing during the process and it was almost invisible in the S15000 samples. On the other hand, in the pellets (P400, P4000 and P15000), high amounts of uromodulin with different polymerization states were observed (Fig. 4A, middle panel, high-intensity bands). This was confirmed by *in-gel* digestion proteomic analysis of the bands (data not shown). This shows that the buffer used in the protocol efficiently solubilized the uromodulin traces that were left in the sample after salt precipitation. Selected gel bands were analyzed by in-gel proteomic analysis. More importantly, the comparison of the pellets (Fig. 4A, left panel) collected at different centrifugation forces (4000 × *g* and 15,000 × *g* and 200,000 × *g*) with uEVs collected after the GUC step above the 1 M sucrose/$D_2O$ cushion (Fig. 4A, right panel) shows that the final uEVs isolated by the combined scaled-down method showed almost no uromodulin contamination and the protein profile is very similar to the uEVs profiles published previously [40]. Western Blot images of the most commonly used exosome markers, Alix and TSG101 (Fig. 4B) confirmed that the uEVs isolated by the combined salt precipitation and sucrose double cushion method are indeed enriched in membrane vesicles; and the supernatant and pellet samples obtained in the same experiment were absent of these bands. On the other hand, the uEVs isolated by the dUC method (crude uEVs), in spite of undergoing the salt precipitation step, still showed high uromodulin contamination. Hence, the importance of the gradient centrifugation step in obtaining samples suitable for downstream analysis is indicated.

To further understand the effectiveness of the salting-out method for the isolation for uEVs, urine samples were collected from 5 different healthy male individuals were analyzed. uEVs were isolated using the new method (Fig. 3) and the obtained uEVs sample was split into two parts, 2/3 was used for SDS-PAGE and WB and 1/3 for the shot-gun proteomics. The purpose of SDS-PAGE analysis was to monitor the presence of eventually copurified uromodulin. Fig. 5 shows that there is no visible uromodulin copurification with the uEV samples which again proves the efficiency of its removal by the salting out and gradient ultracentrifugation. We can also observe variability in the quantity of uEVs isolated from different individuals' samples. The estimated uEV protein yield from 50 mL of urine was as follows: in Sample 1 it was about 5 μg, in Sample 4 it was about 10 μg while Samples 2, 3 and 5 were similar with an approximate yield of 18 μg. The individual variability

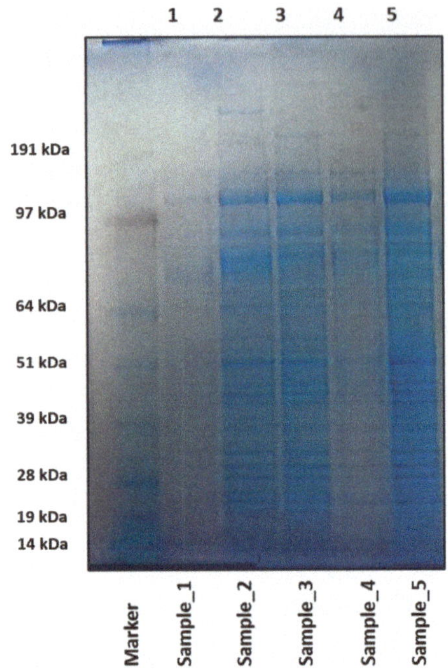


**Fig. 5** SDS-PAGE image of the protein profiles of urinary extracellular vesicles isolated from 100 mL urine samples of five healthy male individuals by the combined salt precipitation and sucrose/$D_2O$ double cushion gradient ultracentrifugation methods scaled down for small initial sample volumes. The uEVs were isolated in five parallel experiments.

of the secreted uEVs is well known and reflects the different water load of the individuals as well as several other factors.

## 3.3 Proteomic analysis of uEVs isolated from healthy individuals and type 2 diabetic patients

### *3.3.1 Healthy control samples*

For the quantitative proteomics experiment of uEVs isolated from healthy individuals, a representative control uEV sample was created by combining in equal portions four uEV samples (1 prepared by dUC and 3 by the combined salt precipitation gUC method), isolated from a single healthy male (Fig. 4). The four uEV samples were compared to the control to study the reproducibility and the efficiency of uromodulin removal in the two isolation methods. Mass spectrometry-based label-free quantitative proteomics was performed. 72 proteins were identified and quantified in these samples. The ten top-ranking proteins were as follow: Aminopeptidase N, dipeptidase-1, Dipeptidyl peptidase 4, Neprilysin, IST1 homolog, Annexin A11, Ezrin, BRO1 domain-containing protein BROX, Gamma-glutamyl transpeptidase 1 and Actin, cytoplasmic 1 (Table 3). These are known to be highly expressed characteristic proteins in uEV isolates and represent a set of marker proteins for them. The relative fold-change in the 3 samples prepared by the novel method were very similar. The result shows that the combined method using 50 mL urine efficiently removes uromodulin and the uEV samples are suitable for downstream proteomic analysis. We also noted that some of the impurities like albumin and uromodulin though in low concentrations but are still present, however they did not influence the protein fold changes considerably.

### *3.3.2 T2D patient's samples*

In a preliminary study, we aimed to test the performance of the novel scaled-down method on patients diagnosed with T2D. T2D is known to cause renal impairment at some stage. Here we analyzed T2D patients who have not yet encountered renal complications. The salt-precipitation/sucrose/$D_2O$ double cushion gUC method was used to isolate uEVs from the urine samples of diabetic patients. Fig. 6 shows the SDS-PAGE profiles of the isolated uEVs and confirms the efficiency of the method in removing uromodulin also in patients' samples.

uEVs samples isolated from the urine of T2D patients were lyzed, in-solution digested and analyzed according to the shot-gun proteomics procedure set-up for the characterization of healthy individual samples. For this experiment, a control sample was created by pooling in equal portions 5 uEVs samples isolated from healthy age-matched individuals (CTRL1_5 merged). The identified proteins and their ranking list (Table 3) were found to be very similar to those obtained for healthy individuals (Table 2). Fold changes respect to healthy control were calculated (Table 3). The relative expression changes are highlighted: in green are the proteins which had relative intensities measured to be higher than in the control sample (up-regulated) and in red are the proteins which had a relative intensity lower than in the control sample (down-regulated). The proteins which were not identified in the control samples are indicated as #Div/0 and the proteins with 0 values are those which were identified only in the control samples. Most significantly, we have found a substantial increase in the expression of intestinal maltase-glucoamylase (MGA-Human) 210 kDa glycoprotein

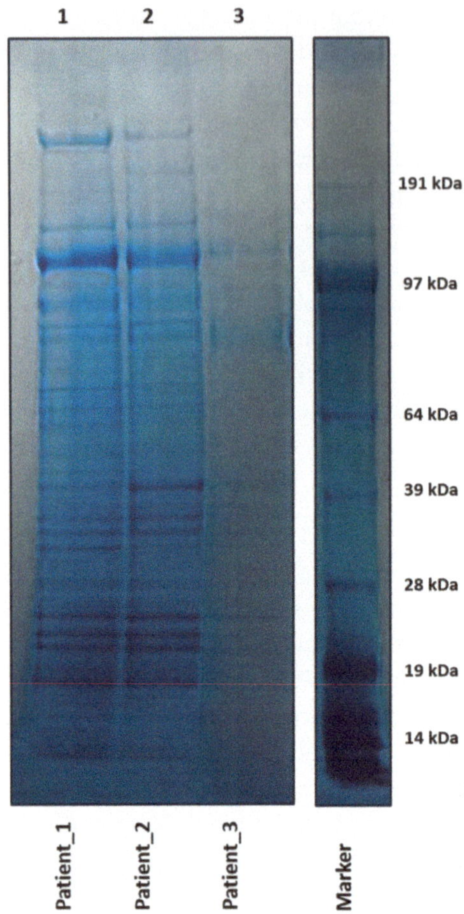


**Fig. 6** SDS-PAGE image of the protein profiles of urinary extracellular vesicles (uEVs) isolated from 50 mL urine samples of three type-2-diabetes patients by the combined salt precipitation and sucrose/$D_2O$ double cushion gradient ultracentrifugation scaled down for small initial sample volume. The uEVs were isolated in three parallel experiments.

in two patients (marked yellow), especially high in one patient. Maltase-glucoamylase is an alpha-glucosidase digestive enzyme, which works in synergy with sucrase-isomaltase and alpha-amylase to digest the full range of dietary starches. MGA-human plays a crucial role in the production of glucose in the human lumen and acts as an efficient drug target for type-2 diabetes [43]. Research focuses on MGA inhibitors which would delay glucose absorption into the bloodstream of the diabetic patient [44]. The presence

**Table 2** List of proteins identified by shot-gun proteomics in the uEVs samples isolated by the salt precipitation and sucrose/D2O double cushion gradient ultracentrifugation method from healthy male individuals.

| N. | Name of the protein | Acc. Number | Identification number from EVPedia |
|---|---|---|---|
| 1 | Aminopeptidase N | AMPN_HUMAN | 2 |
| 2 | Dipeptidase 1 | DPEP1_HUMAN | 65 |
| 3 | Dipeptidyl peptidase 4 | DPP4_HUMAN | 158 |
| 4 | Neprilysin | NEP_HUMAN | 120 |
| 5 | IST1 homolog | IST1_HUMAN | 4 |
| 6 | Annexin A11 | ANX11_HUMAN | 2 |
| 7 | Ezrin | EZRI_HUMAN | 324 |
| 8 | BRO1 domain-containing protein BROX | BROX_HUMAN | 194 |
| 9 | Gamma-glutamyltranspeptidase 1 | GGT1_HUMAN | 109 |
| 10 | Actin, cytoplasmic 1 | ACTB_HUMAN | 17 |
| 11 | Keratin, type II cytoskeletal 1 | K2C1_HUMAN | 2 |
| 12 | Ubiquitin-60S ribosomal protein L40 | RL40_HUMAN | 96 |
| 13 | Solute carrier family 2, facilitated glucose transporter member 5 | GTR5_HUMAN | 40 |
| 14 | Peptidyl-prolyl cis-trans isomerase A | PPIA_HUMAN | 3 |
| 15 | Prominin-1 | PROM1_HUMAN | 90 |
| 16 | Syntenin-1 | SDCB1_HUMAN | 11 |
| 17 | Calmodulin | CALM_HUMAN | 379 |
| 18 | Programmed cell death protein 6 | PDCD6_HUMAN | 3 |
| 19 | Angiotensin-converting enzyme | ACE_HUMAN | 83 |
| 20 | Programmed cell death 6-interacting protein | PDC6I_HUMAN | 3 |
| 21 | Charged multivesicular body protein 5 | CHMP5_HUMAN | 159 |
| 22 | Keratin, type I cytoskeletal 9 | K1C9_HUMAN | 2 |
| 23 | Mucin-1 | MUC1_HUMAN | 70 |
| 24 | Xaa-Pro aminopeptidase 2 | XPP2_HUMAN | 42 |
| 25 | Alpha-enolase | ENOA_HUMAN | 4 |
| 26 | Calpain-7 | CAN7_HUMAN | 76 |
| 27 | Olfactomedin-4 | OLFM4_HUMAN | 60 |
| 28 | Aquaporin-1 | AQP1_HUMAN | 46 |

(*Continued*)

**Table 2** List of proteins identified by shot-gun proteomics in the uEVs samples isolated by the salt precipitation and sucrose/D2O double cushion gradient ultracentrifugation method from healthy male individuals.—cont'd

| N. | Name of the protein | Acc. Number | Identification number from EVPedia |
|---|---|---|---|
| 29 | Guanine nucleotide-binding protein G(I)/G(S)/G(T) subunit beta-2 | GBB2_HUMAN | 3 |
| 30 | Annexin A5 | ANXA5_HUMAN | 2 |
| 31 | 14-3-3 protein zeta/delta | 1433Z_HUMAN | 1 |
| 32 | Keratin, type I cytoskeletal 10 | K1C10_HUMAN | 266 |
| 33 | Charged multivesicular body protein 2a | CHM2A_HUMAN | 163 |
| 34 | Heat shock cognate 71 KDa protein | HSP7C_HUMAN | 392 |
| 35 | Uromodulin | UROM_HUMAN | 52 |
| 36 | Glyceraldehyde-3-phosphate dehydrogenase | G3P_HUMAN | 3 |
| 37 | Unconventional myosin-Ic | MYO1C_HUMAN | 2 |
| 38 | Vacuolar protein sorting-associated protein VTA1 homolog | VTA1_HUMAN | 2 |
| 39 | Serum Albumin | ALBU_HUMAN | 306 |
| 40 | Galectin-3-binding protein | LG3BP_HUMAN | 2 |
| 41 | CD59 glycoprotein | CD59_HUMAN | 240 |
| 42 | Ras-related protein Rab-10 | RAB10_HUMAN | 2 |
| 43 | Phosphatidylethanolamine-binding protein 1 | PEBP1_HUMAN | 238 |
| 44 | Copine-3 | CPNE3_HUMAN | 223 |
| 45 | Annexin A7 | ANXA7_HUMAN | 2 |
| 46 | Guanine nucleotide-binding protein G(k) subunit alpha | GNAI3_HUMAN | 248 |
| 47 | G-protein coupled receptor family C group 5 member C | GPC5C_HUMAN | 94 |
| 48 | Thy-1 membrane glycoprotein | THY1_HUMAN | 84 |
| 49 | ADP-ribosylation factor1 | ARF1_HUMAN | 2 |
| 50 | Phospoglycerate kinase 1 | PGK1_HUMAN | 2 |
| 51 | Rab GDP dissociation inhibitor beta | GDIB_HUMAN | 279 |
| 52 | G-protein coupled receptor family C group 5 member B | GPC5B_HUMAN | 73 |

**Table 2** List of proteins identified by shot-gun proteomics in the uEVs samples isolated by the salt precipitation and sucrose/D2O double cushion gradient ultracentrifugation method from healthy male individuals.—cont'd

| N. | Name of the protein | Acc. Number | Identification number from EVPedia |
|---|---|---|---|
| 53 | Keratin, type II cytoskeletal 2 epidermal | K22E_HUMAN | 2 |
| 54 | Vacuolar protein sorting-associated protein 4A | VPS4A_HUMAN | 140 |
| 55 | Peflin | PEF1_HUMAN | 1 |
| 56 | Ig kappa chain C region | IGKC_HUMAN | 149 |
| 57 | Ig gamma-2 chain C region | IGHG2_HUMAN | 103 |
| 58 | Brain acid soluble protein 1 | BASP1_HUMAN | 205 |
| 59 | Na(+)/H(+) exchange regulatory cofactor NHE-RF1 | NHRF1_HUMAN | 170 |
| 60 | Heat shock protein beta-1 | HSPB1_HUMAN | 2 |
| 61 | Annexin A4 | ANXA4_HUMAN | 261 |
| 62 | Ig alpha-1 chain C region | IGHA1_HUMAN | 136 |
| 63 | Creatine kinase B-type | KCRB_HUMAN | 1 |
| 64 | Ectonucleotide pyro phosphatase | ENPP6_HUMAN | 25 |
| 65 | Betaine–homocysteine S-methyltransferase 1 | BHMT1_HUMAN | 62 |
| 66 | Mannan-binding lectin serine protease 2 | MASP2_HUMAN | 69 |
| 67 | Profilin-1 | PROF1_HUMAN | 361 |
| 68 | Ammonium transporter Rh type C | RHCG_HUMAN | 17 |
| 69 | Prostate stem cell antigen | PSCA_HUMAN | 54 |
| 70 | Immunoglobulin lambda-like polypeptide 5 | IGLL5_HUMAN | 96 |
| 71 | Apolipoprotein D | APOD_HUMAN | 112 |
| 72 | Aminocyclase-1 | ACY1_HUMAN | 62 |

of MGA in uEVs is novel and seems to be an integral characteristic of the patient sample; exceptionally high expression level (more than 39 fold) of MGA was found in sample 1 (Table 3). Another important characteristic of these T2D samples were the high expression levels of two proteins: programmed cell death protein 6 (PDCD6-Human) and its interacting protein (PDCD6I-human). The relevance of PDCD6 and PDCD6I to T2D requires further investigation. Alteration of Mannan-binding lectin-associated

**Table 3** Results of the Scaffold Analysis obtained on the uEVs isolated from samples of Type-2 Diabetes patients and healthy individuals.

| N. | Name of the proteins | Acc. Number | Mol. Mass | T2D_S1 | T2D_S2 | T2D_S3 | CTRL | Fold-change T2D_S1 | Fold-change T2D_S2 | Fold-change T2D_S3 |
|---|---|---|---|---|---|---|---|---|---|---|
| 1 | Aminopeptidase N | AMPN_HUMAN | 110KDa | 0.57 | 0.68 | 0.32 | 0.41 | 1.4 | 1.7 | 0.8 |
| 2 | Dipeptidase 1 | DPEP1_HUMAN | 46KDa | 0.88 | 0.76 | 0.89 | 0.67 | 1.3 | 1.1 | 1.3 |
| 3 | IST1 homolog | IST1_HUMAN | 40KDa | 0.36 | 0.25 | 0.38 | 0.38 | 0.9 | 0.7 | 1.0 |
| 4 | Dipeptidyl peptidase 4 | DPP4_HUMAN | 88KDa | 0.16 | 0.00 | 0.11 | 0.37 | 0.4 | 0.0 | 0.3 |
| 5 | Ezrin | EZRI_HUMAN | 69KDa | 0.42 | 0.21 | 0.29 | 0.13 | 3.2 | 1.6 | 2.2 |
| 6 | Annexin A11 | ANX11_HUMAN | 54KDa | 0.13 | 0.49 | 0.37 | 0.31 | 0.4 | 1.6 | 1.2 |
| 7 | Actin, cytoplasmic 1 | ACTB_HUMAN | 42KDa | 0.54 | 0.24 | 0.36 | 0.36 | 1.5 | 0.7 | 1.0 |
| 8 | Neprilysin | NEP_HUMAN | 86KDa | 0.43 | 0.11 | 0.11 | 0.15 | 2.8 | 0.7 | 0.7 |
| 9 | putative Gamma-glutamyltranspeptidase 3 | GGT3_HUMAN | 62KDa | 0.11 | 0.24 | 0.42 | 0.22 | 0.5 | 1.1 | 1.9 |
| 10 | BRO1 domain-containing protein BROX | BROX_HUMAN | 46KDa | 0.15 | 0.21 | 0.32 | 0.26 | 0.6 | 0.8 | 1.2 |
| 11 | Prominin-1 | PROM1_HUMAN | 97KDa | 0.00 | 0.37 | 0.15 | 0.11 | 0.0 | 3.3 | 1.3 |
| 12 | Solute carrier family 2, facilitated glucose transporter member 5 | GTR5_HUMAN | 55KDa | 0.00 | 0.09 | 0.08 | 0.17 | 0.0 | 0.5 | 0.5 |
| 13 | Mucin-1 | MUC1_HUMAN | 122KDa | 0.00 | 0.00 | 0.16 | 0.07 | 0.0 | 0.0 | 2.2 |
| 14 | Maltase-glucoamylase,intestinal | MGA_HUMAN | 210KDa | 0.31 | 0.02 | 0.00 | 0.01 | 39.1 | 2.8 | 0.0 |
| 15 | Programmed cell death protein 6 | PDCD6_HUMAN | 22KDa | 0.70 | 0.78 | 0.75 | 0.27 | 2.6 | 2.9 | 2.8 |
| 16 | Peptidyl-prolyl cis-trans isomerase A | PPIA_HUMAN | 18KDa | 0.00 | 0.00 | 0.26 | 0.50 | 0.0 | 0.0 | 0.5 |
| 17 | Programmed cell death 6-interacting protein | PDC6I_HUMAN | 96KDa | 0.22 | 0.20 | 0.00 | 0.02 | 12.7 | 11.8 | 0.0 |
| 18 | Calmodulin | CALM_HUMAN | 17KDa | 0.00 | 0.29 | 0.28 | 0.55 | 0.0 | 0.5 | 0.5 |
| 19 | Lys | LAMP2_HUMAN | 45KDa | 0.32 | 0.10 | 0.10 | 0.08 | 4.1 | 1.3 | 1.3 |
| 20 | Olfactomedin-4 | OLFM4_HUMAN | 57KDa | 0.00 | 0.00 | 0.08 | 0.20 | 0.0 | 0.0 | 0.4 |
| 21 | Annexin A5 | ANXA5_HUMAN | 36KDa | 0.00 | 0.44 | 0.43 | 0.10 | 0.0 | 4.4 | 4.3 |
| 22 | Heat shock cognate 71KDa protein | HSP7C_HUMAN | 71KDa | 0.10 | 0.13 | 0.00 | 0.05 | 2.0 | 2.8 | 0.0 |
| 23 | CD59 glycoprotein | CD59_HUMAN | 14KDa | 0.00 | 0.00 | 0.76 | 0.46 | 0.0 | 0.0 | 1.6 |
| 24 | Angiotensin-converting enzyme | ACE_HUMAN | 150KDa | 0.14 | 0.06 | 0.00 | 0.02 | 6.2 | 2.8 | 0.0 |
| 25 | 14-3-3 protein zeta/delta | 1433Z_HUMAN | 28KDa | 0.25 | 0.17 | 0.00 | 0.13 | 1.9 | 1.3 | 0.0 |
| 26 | Calpain-7 | CAN7_HUMAN | 93KDa | 0.00 | 0.05 | 0.00 | 0.06 | 0.0 | 0.9 | 0.0 |
| 27 | Multidrug resistance protein 1 | MDR1_HUMAN | 141KDa | 0.10 | 0.07 | 0.03 | 0.00 | #DIV/0! | #DIV/0! | #DIV/0! |
| 28 | Fruct | ALDOB_HUMAN | 39KDa | 0.37 | 0.12 | 0.00 | 0.04 | 8.5 | 2.8 | 0.0 |
| 29 | Na(+)/H(+) exchange regulatory cofactor NHE-RF1 | NHRF1_HUMAN | 39KDa | 0.00 | 0.12 | 0.00 | 0.09 | 0.0 | 1.3 | 0.0 |
| 30 | Mannan-binding lectin serine protease 2 | MASP2_HUMAN | 76KDa | 0.00 | 0.13 | 0.00 | 0.02 | 0.0 | 5.7 | 0.0 |
| 31 | Podocalyxin | PODXL_HUMAN | 59 KDa | 0.00 | 0.00 | 0.00 | 0.06 | 0.0 | 0.0 | 0.0 |
| 32 | Creatine kinase B-type | KCRB_HUMAN | 43KDa | 0.00 | 0.11 | 0.00 | 0.13 | 0.0 | 0.9 | 0.0 |
| 33 | Glutamyl aminopeptidase | AMPE_HUMAN | 109KDa | 0.13 | 0.00 | 0.00 | 0.02 | 8.3 | 0.0 | 0.0 |
| 34 | Brain acid soluble protein 1 | BASP1_HUMAN | 23KDa | 0.00 | 0.21 | 0.00 | 0.16 | 0.0 | 1.3 | 0.0 |
| 35 | Lys | LAMP1_HUMAN | 45KDa | 0.15 | 0.00 | 0.00 | 0.08 | 2.0 | 0.0 | 0.0 |
| 36 | Angiotensin-converting enzyme 2 | ACE2_HUMAN | 92KDa | 0.15 | 0.00 | 0.00 | 0.00 | #DIV/0! | #DIV/0! | #DIV/0! |

serine protease 2 (MASP-2-human) expression was observed in sample 2. There is increasing evidence that the complement system plays an important role in diabetes and in the development of diabetic vascular complications [45]. In particular, serum and urinary Mannan-binding lectin (MBL) levels are elevated in T2D patients with and without DN and seems to correlate with the renal impairment [45]. Therefore, uEVs MBL expression could be of value for the prediction of renal complications. However, further studies in larger patient cohorts are necessary to evaluate the utility of these potential biomarkers. The analysis also highlights the variability of protein expressions at the individual level and thus supports the utility of single-patient-based validation and clinical studies.

## 4. Conclusions

The development of mass spectrometry-based strategies for protein quantification offers a great tool for biomarker discovery. The non-invasive nature, the biological significance of uEVs and unbiased quantitative analysis of its molecular constituents is a step forward in clinical biomarker research. In summary, we have set-up an effective method for the isolation of highly pure human EVs from urine samples using the combination of NaCl salt precipitation and sucrose/$D_2O$ double cushion gradient ultracentrifugation. The method is suitable for both pooled and individual samples and eliminates most of the uromodulin contamination. Additionally, the method isolates and purifies uEVs that are suitable for downstream "omics" analysis, and in particular for shotgun proteomics workflow. The method was tested on individual and pooled samples of healthy individuals and a small cohort of type 2 diabetes patients for the identification of potential protein biomarkers in uEVs. The results obtained show that the improved uEVs isolation workflow has the potential to accelerate the translation of urinary exosomal biomarkers from bench to bedside for the diagnosis, classification and prognosis of renal diseases.

## Acknowledgements

Authors acknowledge the funding from the European Union's Horizon 2020 research and innovation programme under grant agreements No 801338.

## References

[1] L. He, D. Zhu, J. Wang, X. Wu, A highly efficient method for isolating urinary exosomes, Int. J. Mol. Med. 43 (2019) 83–90, https://doi.org/10.3892/ijmm.2018.3944.

[2] A. Øverbye, T. Skotland, C.J. Koehler, B. Thiede, T. Seierstad, V. Berge, K. Sandvig, A. Llorente, Identification of prostate cancer biomarkers in urinary exosomes, Oncotarget 6 (2015) 30357–30376, https://doi.org/10.18632/oncotarget.4851.
[3] C. Théry, K.W. Witwer, E. Aikawa, et al., Minimal information for studies of extracellular vesicles 2018 (MISEV2018): a position statement of the International Society for Extracellular Vesicles and update of the MISEV2014 guidelines, J. Extracell. Vesicles 7 (1) (2018), https://doi.org/10.1080/20013078.2018.1535750.
[4] G. Pocsfalvi, C. Stanly, A. Vilasi, I. Fiume, G. Capasso, L. Turiák, E.I. Buzas, K. Vékey, Mass spectrometry of extracellular vesicles, Mass Spectrom. Rev. 35 (2016) 3–21, https://doi.org/10.1002/mas.21457.
[5] P. Motamedinia, A.N. Scott, K.L. Bate, N. Sadeghi, G. Salazar, E. Shapiro, J. Ahn, M. Lipsky, J. Lin, G.W. Hruby, K.K. Badani, D.P. Petrylak, M.C. Benson, M.J. Donovan, W.D. Comper, J.M. McKiernan, L.M. Russo, Urine exosomes for non-invasive assessment of gene expression and mutations of prostate cancer, PloS One 11 (2016), https://doi.org/10.1371/journal.pone.0154507 e0154507–e0154507.
[6] P.A. Gonzales, T. Pisitkun, J.D. Hoffert, D. Tchapyjnikov, R.A. Star, R. Kleta, N.S. Wang, M.A. Knepper, Large-scale proteomics and phosphoproteomics of urinary exosomes, J. Am. Soc. Nephrol. 20 (2009) 363 LP–379, https://doi.org/10.1681/ASN.2008040406.
[7] P. Svenningsen, R. Sabaratnam, B.L. Jensen, Urinary extracellular vesicles: origin, role as intercellular messengers and biomarkers; efficient sorting and potential treatment options, Acta Physiol. 228 (2020) e13346, https://doi.org/10.1111/apha.13346.
[8] T. Pisitkun, R.-F. Shen, M.A. Knepper, Identification and proteomic profiling of exosomes in human urine, Proc. Natl. Acad. Sci. U.S.A. 101 (2004) 13368 LP–13373, https://doi.org/10.1073/pnas.0403453101.
[9] C. Stanly, I. Fiume, G. Capasso, G. Pocsfalvi, Isolation of exosome-like vesicles from plants by ultracentrifugation on sucrose/deuterium oxide (D2O) density cushions, Methods Mol. Biol. (2016) 259–269, https://doi.org/10.1007/978-1-4939-3804-9_18.
[10] K. Mizutani, K. Kawakami, K. Horie, Y. Fujita, K. Kameyama, T. Kato, K. Nakane, T. Tsuchiya, M. Yasuda, K. Masunaga, Y. Kasuya, Y. Masuda, T. Deguchi, T. Koie, M. Ito, Urinary exosome as a potential biomarker for urinary tract infection, Cell Microbiol. 21 (2019) e13020, https://doi.org/10.1111/cmi.13020.
[11] A. Cheruvanky, H. Zhou, T. Pisitkun, J.B. Kopp, M.A. Knepper, P.S.T. Yuen, R.A. Star, Rapid isolation of urinary exosomal biomarkers using a nanomembrane ultrafiltration concentrator, Am. J. Physiol. Physiol. 292 (2007) F1657–F1661, https://doi.org/10.1152/ajprenal.00434.2006.
[12] C.Y. Chen, M.C. Hogan, C.J. Ward, Chapter thirteen - purification of exosome-like vesicles from urine, W.F.B.T.-M, in: E. Marshall (Ed.), Cilia, Part A, Academic Press, 2013, pp. 225–241, https://doi.org/10.1016/B978-0-12-397945-2.00013-5.
[13] D.A.A. Raj, in: I. Fiume (Ed.), Urinary Exosomes for Protein Biomarker Research, IntechOpen, Rijeka, 2012, https://doi.org/10.5772/30094 p. Ch. 3.
[14] M.L. Merchant, D.W. Powell, D.W. Wilkey, T.D. Cummins, J.K. Deegens, I.M. Rood, K.J. McAfee, C. Fleischer, E. Klein, J.B. Klein, Microfiltration isolation of human urinary exosomes for characterization by MS, PROTEOMICS – Clin. Appl. 4 (2010) 84–96, https://doi.org/10.1002/prca.200800093.
[15] M.L. Heinemann, J. Vykoukal, in: W.P. Kuo, S. Jia (Eds.), Sequential Filtration: A Gentle Method for the Isolation of Functional Extracellular Vesicles BT - Extracellular Vesicles: Methods and Protocols, Springer New York, New York, NY, 2017, pp. 33–41, https://doi.org/10.1007/978-1-4939-7253-1_4.

[16] M.A. Rider, S.N. Hurwitz, D.G. Meckes Jr., ExtraPEG: a polyethylene glycol-based method for enrichment of extracellular vesicles, Sci. Rep. 6 (2016) 23978, https://doi.org/10.1038/srep23978.
[17] J. Echevarria, F. Royo, R. Pazos, L. Salazar, J.M. Falcon-Perez, N.-C. Reichardt, Microarray-based identification of lectins for the purification of human urinary extracellular vesicles directly from urine samples, Chembiochem 15 (2014) 1621–1626, https://doi.org/10.1002/cbic.201402058.
[18] Z. Liu, D.M. Cauvi, E.M.A. Bernardino, B. Lara, R.E. Lizardo, D. Hawisher, S. Bickler, A. De Maio, Isolation and characterization of human urine extracellular vesicles, Cell Stress Chaperones 23 (2018) 943–953, https://doi.org/10.1007/s12192-018-0902-5.
[19] S.I. Lozano-Ramos, I. Bancu, L. Carreras-Planella, M. Monguió-Tortajada, L. Cañas, J. Juega, J. Bonet, M.P. Armengol, R. Lauzurica, F.E. Borràs, Molecular profile of urine extracellular vesicles from normo-functional kidneys reveal minimal differences between living and deceased donors, BMC Nephrol. 19 (2018) 189, https://doi.org/10.1186/s12882-018-0985-3.
[20] J. Van Deun, P. Mestdagh, P. Agostinis, et al., EV-TRACK: transparent reporting and centralizing knowledge in extracellular vesicle research, Nat. Methods 14 (2017) 228–232, https://doi.org/10.1038/nmeth.4185.
[21] M. Salih, R. Zietse, E.J. Hoorn, Urinary extracellular vesicles and the kidney: biomarkers and beyond, Am. J. Physiol. Physiol. 306 (2014) F1251–F1259, https://doi.org/10.1152/ajprenal.00128.2014.
[22] M. Kosanović, M. Janković, Isolation of urinary extracellular vesicles from Tamm-Horsfall protein–depleted urine and their application in the development of a lectin-exosome-binding assay, Biotechniques 57 (2014) 143–149, https://doi.org/10.2144/000114208.
[23] I. Tamm, F.L. Horsfall Jr., A mucoprotein derived from human urine which reacts with influenza, mumps, and newcastle disease viruses, J. Exp. Med. 95 (1952) 71–97, https://doi.org/10.1084/jem.95.1.71.
[24] C.A. Sobsey, S. Ibrahim, V.R. Richard, V. Gaspar, G. Mitsa, V. Lacasse, R.P. Zahedi, G. Batist, C.H. Borchers, Targeted and untargeted proteomics approaches in biomarker development, Proteomics (2019) 1900029, https://doi.org/10.1002/pmic.201900029, n/a.
[25] B. Pang, Y. Zhu, J. Ni, J. Thompson, D. Malouf, J. Bucci, P. Graham, Y. Li, Extracellular vesicles: the next generation of biomarkers for liquid biopsy-based prostate cancer diagnosis, Theranostics 10 (2020) 2309–2326, https://doi.org/10.7150/thno.39486.
[26] J.L. Welton, S. Khanna, P.J. Giles, P. Brennan, I.A. Brewis, J. Staffurth, M.D. Mason, A. Clayton, Proteomics analysis of bladder cancer exosomes, Mol. & Cell. Proteomics. 9 (2010) 1324 LP–1338, https://doi.org/10.1074/mcp.M000063-MCP201.
[27] H. Valadi, K. Ekström, A. Bossios, M. Sjöstrand, J.J. Lee, J.O. Lötvall, Exosome-mediated transfer of mRNAs and microRNAs is a novel mechanism of genetic exchange between cells, Nat. Cell Biol. 9 (2007) 654–659, https://doi.org/10.1038/ncb1596.
[28] A.H. Gheinani, M. Vögeli, U. Baumgartner, E. Vassella, A. Draeger, F.C. Burkhard, K. Monastyrskaya, Improved isolation strategies to increase the yield and purity of human urinary exosomes for biomarker discovery, Sci. Rep. 8 (2018) 3945, https://doi.org/10.1038/s41598-018-22142-x.
[29] G. De Palma, V.F. Di Lorenzo, S. Krol, A.V. Paradiso, Urinary exosomal shuttle RNA: promising cancer diagnosis biomarkers of lower urinary tract, Int. J. Biol. Markers 34 (2019) 101–107, https://doi.org/10.1177/1724600819827023.
[30] T. Skotland, K. Ekroos, D. Kauhanen, H. Simolin, T. Seierstad, V. Berge, K. Sandvig, A. Llorente, Molecular lipid species in urinary exosomes as potential prostate cancer

biomarkers, Eur. J. Canc. 70 (2017) 122–132, https://doi.org/10.1016/j.ejca.2016.10.011.

[31] N. Singhto, A. Vinaiphat, V. Thongboonkerd, Discrimination of urinary exosomes from microvesicles by lipidomics using thin layer liquid chromatography (TLC) coupled with MALDI-TOF mass spectrometry, Sci. Rep. 9 (2019) 13834, https://doi.org/10.1038/s41598-019-50195-z.

[32] S.C. Glover, M.-Z. Nouri, K.M. Tuna, L.B. Mendoza Alvarez, L.K. Ryan, J.F. Shirley, Y. Tang, N.D. Denslow, A.A. Alli, Lipidomic analysis of urinary exosomes from hereditary α-tryptasemia patients and healthy volunteers, FASEB BioAdv. 1 (2019) 624–638, https://doi.org/10.1096/fba.2019-00030.

[33] D. V Ilatovskaya, V. Levchenko, A. Lowing, L.S. Shuyskiy, O. Palygin, A. Staruschenko, Podocyte injury in diabetic nephropathy: implications of angiotensin II – dependent activation of TRPC channels, Sci. Rep. 5 (2015) 17637, https://doi.org/10.1038/srep17637.

[34] H. Abe, A. Sakurai, H. Ono, S. Hayashi, S. Yoshimoto, A. Ochi, S. Ueda, K. Nishimura, E. Shibata, M. Tamaki, F. Kishi, S. Kishi, T. Murakami, K. Nagai, T. Doi, Urinary exosomal mRNA of WT1 as diagnostic and prognostic biomarker for diabetic nephropathy, J. Med. Invest. 65 (2018) 208–215, https://doi.org/10.2152/jmi.65.208.

[35] S. De, S. Kuwahara, M. Hosojima, T. Ishikawa, R. Kaseda, P. Sarkar, Y. Yoshioka, H. Kabasawa, T. Iida, S. Goto, K. Toba, Y. Higuchi, Y. Suzuki, M. Hara, H. Kurosawa, I. Narita, Y. Hirayama, T. Ochiya, A. Saito, Exocytosis-mediated urinary full-length megalin excretion is linked with the pathogenesis of diabetic nephropathy, Diabetes 66 (2017) 1391 LP–1404, https://doi.org/10.2337/db16-1031.

[36] C.M. Yamamoto, T. Murakami, M.L. Oakes, M. Mitsuhashi, C. Kelly, R.R. Henry, K. Sharma, Uromodulin mRNA from urinary extracellular vesicles correlate to kidney function decline in type 2 diabetes mellitus, Am. J. Nephrol. 47 (2018) 283–291, https://doi.org/10.1159/000489129.

[37] Y. Xie, Y. Jia, X. Cuihua, F. Hu, M. Xue, Y. Xue, Urinary exosomal MicroRNA profiling in incipient type 2 diabetic kidney disease, J. Diabetes Res. 2017 (2017) 6978984, https://doi.org/10.1155/2017/6978984.

[38] Y.-R. Duan, B.-P. Chen, F. Chen, S.-X. Yang, C.-Y. Zhu, Y.-L. Ma, Y. Li, J. Shi, Exosomal microRNA-16-5p from human urine-derived stem cells ameliorates diabetic nephropathy through protection of podocyte, J. Cell Mol. Med. (2019), https://doi.org/10.1111/jcmm.14558. n/a.

[39] J. Jin, Y. Shi, J. Gong, L. Zhao, Y. Li, Q. He, H. Huang, Exosome secreted from adipose-derived stem cells attenuates diabetic nephropathy by promoting autophagy flux and inhibiting apoptosis in podocyte, Stem Cell Res. Ther. 10 (2019) 95, https://doi.org/10.1186/s13287-019-1177-1.

[40] D.A.A. Raj, I. Fiume, G. Capasso, G. Pocsfalvi, A multiplex quantitative proteomics strategy for protein biomarker studies in urinary exosomes, Kidney Int. 81 (2012) 1263–1272, https://doi.org/10.1038/ki.2012.25.

[41] A. Shevchenko, H. Tomas, J. Havli, J. V Olsen, M. Mann, In-gel digestion for mass spectrometric characterization of proteins and proteomes, Nat. Protoc. 1 (2006) 2856–2860, https://doi.org/10.1038/nprot.2006.468.

[42] G. Pocsfalvi, D.A.A. Raj, I. Fiume, A. Vilasi, F. Trepiccione, G. Capasso, Urinary extracellular vesicles as reservoirs of altered proteins during the pathogenesis of polycystic kidney disease, PROTEOMICS – Clin. Appl. 9 (2015) 552–567, https://doi.org/10.1002/prca.201400199.

[43] L. Ren, X. Qin, X. Cao, L. Wang, F. Bai, G. Bai, Y. Shen, Structural insight into substrate specificity of human intestinal maltase-glucoamylase, Protein Cell 2 (2011) 827–836, https://doi.org/10.1007/s13238-011-1105-3.
[44] E.J. Rossi, L. Sim, D.A. Kuntz, D. Hahn, B.D. Johnston, A. Ghavami, M.G. Szczepina, N.S. Kumar, E.E. Sterchi, B.L. Nichols, B.M. Pinto, D.R. Rose, Inhibition of recombinant human maltase glucoamylase by salacinol and derivatives, FEBS J. 273 (2006) 2673–2683, https://doi.org/10.1111/j.1742-4658.2006.05283.x.
[45] L. Jenny, R. Ajjan, R. King, S. Thiel, V. Schroeder, Plasma levels of mannan-binding lectin-associated serine proteases MASP-1 and MASP-2 are elevated in type 1 diabetes and correlate with glycaemic control, Clin. Exp. Immunol. 180 (2015) 227–232, https://doi.org/10.1111/cei.12574.

CHAPTER TWO

# Biological role and clinical relevance of extracellular vesicles as key mediators of cell communication in cancer

**Francesca Mantile, Paola Franco, Maria Patrizia Stoppelli and Giovanna L. Liguori***
Institute of Genetics and Biophysics (IGB), National Research Council (CNR) of Italy, Naples, Italy
*Corresponding author: E-mail: giovanna.liguori@igb.cnr.it

## Contents



*Advances in Biomembranes and Lipid Self-Assembly*, Volume 33
ISSN 2451-9634
https://doi.org/10.1016/bs.abl.2020.05.006

## Abstract

Cells of multicellular organisms constantly communicate with each other during development as well as in adult life to regulate body homeostasis and physiology. The sender cells may either signal to themselves or to specific recipient ones, thus influencing many cell processes such as metabolism, division, differentiation and migration. Cell-to-cell communication is extremely relevant in cancer. Tumor cells communicate among them and with several non tumoral cell types, such as stromal, endothelial, immune and epithelial cells, located both in the tumor microenvironment and at distance. The resultant is to sustain or adverse tumor growth and invasion. A great breakthrough in the comprehension of how cells communicate was achieved with the identification of extracellular vesicles (EVs). EVs are lipid membrane encapsulated structures able to transport and deliver in the cells a great variety of molecules, including proteins, metabolites, nucleic acids, lipids and even entire organelles. Most cells, from unicellular prokaryotes to complex multicellular organisms, are able to vesiculate. EVs are very heterogenous in size, content, function and biochemical/biophysical properties. Cancer cells have been shown to release increasing amount of EVs, with a different content respect to normal counterparts and able to elicit several functional responses in target cells. In this chapter we will present the main aspects of EV-based cancer cell communication, highlighting how EVs have been emerging as an extremely powerful tool to better understand tumor biology as well as for clinical applications in neoplastic diseases.

## 1. Introduction

Cell-cell communication is crucial during development and adult life of pluricellular organisms, in both physiological and pathological conditions. Tumors themselves are all outside isolated entities, but rather complex systemic networks involving communication between tumoral cells with each other as well as with different types of non-transformed cells which compose the tumor microenvironment (TME). The effect of this communication can counteract or support tumor growth and metastasis [1−3].

Communication between cells can occur by direct cell-cell contact or through the secretion of signaling factors. Among the first type of mechanisms there is the formation of junctions between adjacent cells, such as in the epithelial and endothelial cells, otherwise of thin membranous, cytoplasmic connections as cytonemes and tunneling microtubules [4,5]. Cytonemes are signaling filopodia that can span over several cell diameters and contact recipient cells, mediating unilateral communication [6,7]. Tunneling microtubules assure a bilateral and stable communication between two cells, that can exchange several molecules and also entire organelles, depending on the diameter [5,8]. On the other hand, factors secreted in the extracellular space are able to signal to the producing cell itself, the adjacent cells and also to cells located far apart from each other, up to reach the bloodstream and then cells belonging to other tissues/organs.

Over the past 10 years a new paradigm has emerged involving extracellular vesicle (EV) trafficking as an important mechanism of cell-cell communication. EVs are lipid membrane encapsulated particles filled of cellular content released by cells either constitutively or following cell stimulation, stress, transformation or death, processes that EVs both reflect and influence. Among the specific factors that induce EV release, there are changes of pH, hypoxia, irradiation, injury, cellular stress and oncogenic transformation [9–12]. Cells from different organisms, including all eukaryotes, both animal (from yeast, *Drosophila*, *Caenorhabditis elegans*, and parasites to mammals) and plants, but also prokaryotic cells, have been demonstrated to release vesicles into the extracellular environment. EVs have also been isolated from diverse body fluids, including blood, urine, saliva, breast milk, amniotic fluid, cerebrospinal fluid, bile, and semen [13]. These findings point to a high evolutionary conserved cell mechanism involving EVs. EV-mediated cell communication combines the advantages of the microtubules to protect the exchanging molecules from the extracellular environment to the ability of secreted factors to travel long distances and interact with several cells.

As a part of his pangenesis theory in "The Variation of Animals and Plants under Domistication" (1868), Charles Darwin hypothesized that every cell type in the body may generate minute size gemmules that contain molecules and serve to communicate them to other cell types [14]. This seminal intuition fell unnoticed for more than 150 years until contemporary scientists may recognize EVs in Darwin's gemmules [15]. For decades membrane-enclosed vesicles have been shown to be present outside cells both in solid tissues, such as cartilage [16] and in biological fluids, such as blood [17] or semen [18]. Moreover, mammalian cells [19], especially tumor cells [20] and platelets [21] were shown to shed membrane vesicles. Terms as microvesicles (MVs), microparticles or ectosomes have been used to refer to EVs released by outward budding of the cell plasma membrane. MVs released by platelets, monocytes, or neutrophils were shown to have a functional significance and to play a role in blood coagulation [22,23] or binding of opsonized bacteria [24]. In the early 1980s, a more complex EV production pathway was described during reticulocytes maturation, in which vesicles initially formed intracellularly within so-called multivesicular bodies

(MVBs) and then were secreted outside [25–28]. These small EVs formed by vesiculation of intracellular endosomes and released by exocytosis were called exosomes and were initially viewed simply as a means for maturing reticulocytes to discard unwanted components [27]. In the late 1990s a renewed interest in the field of exosome biology came thanks to the pioneering studies of Raposo and coworkers, suggesting that exosomes might be important mediators of intercellular communication during the immune response [29,30]. In the meantime, shedding MVs were also implicated in mediating intercellular communication. MVs have been shown to influence lipid metabolism [23], release of inflammatory mediators [31,32], survival and proliferation [33] of immune or endothelial cells exposed to them. Since then an increasing number of studies implicated EVs as important mediators of cell-cell communication in both physiological and pathological conditions. EV-mediated bilateral communication is involved in immunology [34,35], in reproductive [36] and developmental [37] biology, angiogenesis, tissue remodeling, and growth of the placenta [38,39]. Recent studies of EVs in Drosophila have also demonstrated that EVs may help establish the long-range gradients of Wnt and Hedgehog required for proper anatomic axes and limb development [40–42]. Among pathological conditions in which EVs have been implicated are metabolic, cardiovascular and nervous system pathologies together with several type of cancers [42,43]. Interestingly, EVs can act both promoting or counteracting the disease [44–47]. The study of EVs in cancer has progressed at a rapid pace compared with research into their role in other diseases [48], and EVs have been associated with several hallmark features of cancer [43,49]. EVs influence tumor growth and metastasis, bilateral communication with TME, immunoresistance and resistance to chemotherapy. The role of EVs in cancer progression is likely dynamic and specific to cancer type, genetics, and stage. Moreover, due to their intrinsic characteristics, such as biocompatibility, immunotolerance, cargo stability and targeting abilities, EVs represent a very promising tool for clinical approaches. In this chapter we address the functional significance of EVs in cancer development and progression as well as their clinical implications. In the first part of the chapter, we introduce EVs, giving an overview of their biogenesis mechanism, cargo content and biological role. In the second part, we describe how EVs mediate cancer cell interactions during tumor growth, epithelial-mesenchymal transition (EMT) and metastasis as well as their interactions with TME and immune system. Finally, we focus on the exploitation of EVs for cancer diagnostic and therapeutic approaches.

# 2. Main features of extracellular vesicles

## 2.1 Classification and nomenclature

Due to the huge number of cells producing vesicles in the extracellular space, EVs were described independently in several unrelated research areas, and under various names, and only recently collected together in a unique field. Therefore, a complex and for some extent misleading nomenclature has been produced. Different terms were used to indicate different types of vesicles, based mainly on size and biogenesis (e.g ectosomes, MVs, microparticles, exosomes) but also cell origin or functional aspects (e.g. oncosomes, prostasomes, platelet dusts, apoptotic bodies, migrasomes). Among them, the best studied and characterized classes, in which also the other categories may fall, were ectosomes/MVs and exosomes [43,50,51]. As mentioned before, ectosomes/MVs have been referred to vesicles generated by the direct outward budding and fission of the plasma membrane, whereas the term exosomes was firstly used to indicate vesicles having an endocytic origin, being formed as intraluminal vesicles (ILVs) of the late endosomes or MVBs [50,51]. Due to their different biogenesis (endocytosis or shedding), exosomes and MVs should display different morphological, biochemical and biophysical features. The diameter of exosomes should be close to the diameter of ILVs from which they derive, therefore ranging from 30 to 100−150 nm, depending also by the type of observation used (after fixation and contrasting by electronical microscopy or in a close-to-native state by means of cryo-EM or nanoparticle tracking analysis) [28]. MVs budding from plasmatic membrane escape from this size restriction. At the beginning they were believed to range between 100−150 and 1000 nm, but recently they were shown to be even more heterogeneous, spanning in size between 30−50 nm (notably, as small as, or even smaller, than exosomes) to 2−10 μm [28,43,50−52]. Interestingly, the size of MVs can be also affected by storage conditions [53]. The larger shedding vesicles have been found to be released by cancer cells and for this reason are named oncosomes or large oncosomes. Because of their atypical size (>1000 times larger than exosomes by volume), large oncosomes may have unique properties *in vivo* and may present new opportunities for tumor profiling [52]. As another consequence of their biogenesis, the molecular composition of exosomes might be closer to the composition of endosomal ILVs, whereas MVs should have a composition resembling the one of the cell membrane [28]. In reality, both classes of EVs, actually overlapping at various extent, are also highly heterogeneous within in size, composition, biophysical/biochemical properties and function [43]. The currently available protocols to recover and purify EVs from cell culture supernatants or liquid biopsy samples usually do not reach high yield and result in a heterogeneous population of EVs of unknown origin [54]. Selectively distinguishing one EV population from another to fully study them individually and possibly identify specific characteristics and markers is still a great challenge. Since consensus has not yet emerged on specific markers of EV subtypes, assigning an EV to a particular cell type or biogenesis pathway remains extremely difficult unless, e.g. the EV is caught in the act of being released by live imaging techniques [55]. To complicate the scenario, the term exosome has seen a progressive increasing in popularity and has been often used in a less restrictive but not univocal manner respect to the original definition. It became a "buzz-term", often used to indicate EVs of small size, but arriving also to designate any type of EV [56,57]. A great effort has been made recently by many scientists and by the International Society for Extracellular Vesicles (ISEV) itself, founded in 2012, to make order in the nomenclature, identify the minimal information for studies of EVs and define guidelines regarding their isolation, purification and characterization procedures [13,55–58]. "ISEV endorses EV as the generic term for particles naturally released from the cell that are delimited by a lipid bilayer and cannot replicate, i.e. do not contain a functional nucleus" [55]. Moreover, ISEV invites to consider use of operational terms for EV subtypes instead of terms such as exosomes and MVs/ectosomes that are historically burdened by contradictory definitions. Operational terms can refer to physical characteristics, such as size (small or medium/large EVs), or density (low, middle, high) with each precisely range defined; biochemical composition (positivity for specific markers) or descriptions of conditions or cell of origin [55]. However, despite these position papers, exosomes and MVs/ectosomes are still used by many researchers. Apart from still divergent nomenclature, all the scientific community working in EV field is strongly committed in developing and standardizing new methodological approaches to isolate and characterize more and more precisely the different EV subtypes and assess their specific functions.

## 2.2 Biogenesis and cargo

The functional properties of EVs are dictated by their cargo and by their release and uptake dynamics. Identifying the components responsible for the pro-tumorigenic effects of cancer EVs and the pathways leading to their incorporation into vesicles is currently one of the major challenges in the field [59]. EVs contain bioactive molecules, such as nucleic acids (DNA, mRNA, microRNA, and other non-coding RNAs), proteins (receptors, transcription factors, enzymes, extracellular matrix proteins), metabolites and lipids that can redirect the function of a recipient cell [43,60]. The nature and abundance of EV cargoes are cell-type-specific and are often influenced by the physiological or pathological state of the donor cell, the stimuli that modulate their production and release and the molecular mechanisms that lead to their biogenesis [61]. Interestingly, the cargo itself is the first regulator of EV formation. The ectopic expression of a particular cargo, such as the expression in glioma cells of the mutant *EGFRvIII* gene, coding for a truncated and oncogenic form of the epidermal growth factor receptor, can induce extracellular and systemic release of vesicles containing the cargo itself [62]. Although EV content generally reflects the nature and status of the cell of origin, enrichment of specific proteins and nucleic acids suggests at least a degree of specific cellular sorting into EVs [59,61,63]. It has also been suggested that EVs could be enriched in cargo content (RNA) not found in the producing cells [53]. All these data point to a complex, fascinating, and fine-tuned regulation of EV production as a key process through which cells communicate among each other.

### 2.2.1 Biogenesis

Among the different mechanisms of EV production identified so far, the secretion pathway through MVB and ILV formation is the most studied (Fig. 1). First, the plasma membrane invaginates to form a cup-shaped structure that includes cell-surface proteins and soluble proteins associated with the extracellular milieu. This leads to the *de novo* formation of an early-sorting endosome (ESE) including cell-surface and soluble proteins of the extracellular space, that in some cases may directly merge with a pre-existing ESE. In this way, the trans-Golgi network and endoplasmic reticulum can also contribute to the formation and the content of the ESE [61,64,65]. ESEs can mature into late-sorting endosomes (LSEs) and generate MVBs. MVBs form by inward invagination of the endosomal limiting membrane (that is, double invagination of the plasma membrane) with the final fission and release of discrete ILVs into the MVB lumen. This process results in MVBs containing several ILVs (future exosomes). The MVB can either fuse with lysosomes to be degraded or traffick to the plasma membrane, fuse with it and release the contained ILVs as exosomes [43,61,66]. A key phase for the selection of the cargo content of the secreted EVs is the

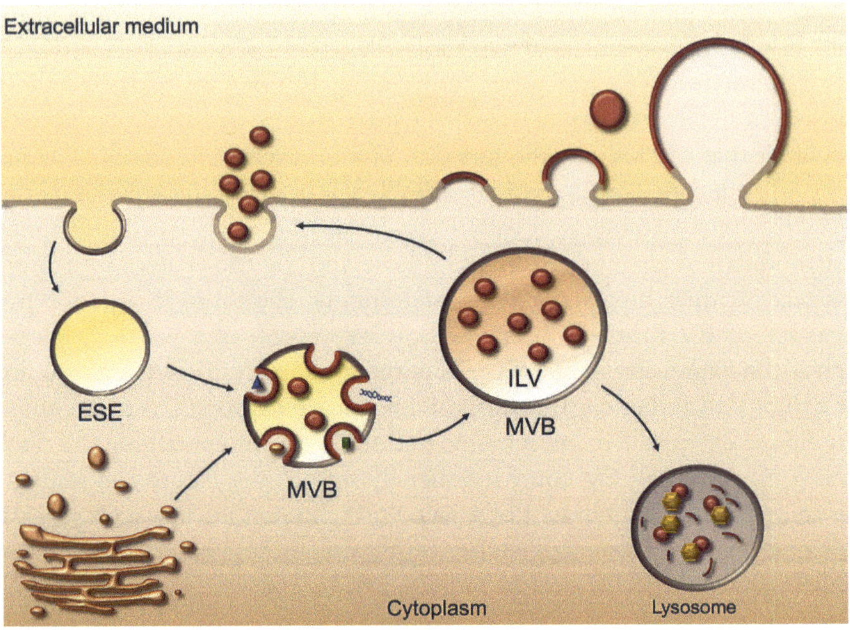


**Fig. 1** ***Principal mechanisms of biogenesis of extracellular vesicles.*** On the left is schematized the vesicle secretion pathway and its relation with the endocytic system. The cell membrane buds off to form first a cup-shaped structure and then an intracellular vesicle, called early-sorting endosome (ESE), including components of the extracellular space. The ESE, eventually fused with other ESEs from the trans-Golgi network, matures into a late-sorting endosome (LSE) and then in a multivesicular body (MVB). The MVB is generated by the inward invagination of the endosomal limiting membrane with the final fission and release of intraluminal vesicles (ILVs). This mechanism allows ILVs to encapsulate cytoplasmic molecules. The MVB can fuse with lysosomes to be degraded and its component recycled in the cell metabolism or fuse with the cell membrane, releasing outside the contained ILVs as exosomes. On the right is depicted the direct outward budding of the vesicles from the plasma membrane. First, there is the physical bending of the cell membrane, then the membrane fission with the release of the budding vesicle outside the cell. The budding vesicles are very heterogeneous in size; some of them reach 1–10 μm of diameter.

MVB and ILV formation. The Endosomal Sorting Complex Required for Transport (ESCRT) machinery is the most extensively described pathway of MVB biogenesis. It is composed of four complexes of protein subunits acting in a stepwise manner and it is generally active in local membrane remodeling also during viral budding, cytokinesis, and autophagy [67]. In the EV context, the ESCRT machinery interacts with associated proteins, such as the Alg-2-interacting protein X8 (ALIX) and Arrestin domain containing 1 (ARRDC1) protein and can be also intersected by the small

PDZ-domain protein syntenin pathway [51,61]. Alternative pathways also involve ceramide or tetraspanins such as CD63, CD81, CD82 and CD9. The generation of ceramide by neutral type II sphingomyelinase enzyme can mediate the generation of membrane subdomains which impose a spontaneous negative curvature on the membranes or could be metabolized to activate G-protein-coupled receptors that appears essential for cargo sorting into ILVs [68,69]. Tetraspanins, can regulate endosomal sorting by forming clusters and dynamic membrane platforms with different transmembrane and cytosolic proteins probably acting in the formation of the microdomains that will bud [61].

The second type of EV biogenesis consists into the direct outward budding of the EVs from the plasma membrane and requires several molecular rearrangements within the cell membrane, including changes in lipid and protein composition, and rearrangements of the underlying cytoskeleton architecture. The result is the physical bending of the membrane, which finally produces membrane budding and release of vesicles outside the cells [12,61,62] (Fig. 1). Calcium-dependent enzymatic machineries trigger rearrangements in the asymmetric distribution of membrane phospholipids between the inner and outer membrane layer and the reorganization of the cytoskeleton. In addition, activation of small GTP-binding proteins including ARF6 and ARF1 leads to the phosphorylation of myosin light chain (MLC) and actomyosin contraction, which allows the vesicles to bud off from the membranes of cancer cells [70,71]. The activity of the RHO family of small GTPases and of the RHO-associated protein kinase (ROCK), which are important regulators of actin dynamics, was also found to induce vesicle release in different populations of tumor cells [72]. Shedding of vesicles has also been shown to depend on ATP-mediated activation of purinergic P2X7 receptors, which leads to rearrangements of the cell membrane [73,74]. In tumor cells, vesicle budding and release are also linked to metabolic changes, such as the Warburg effect (preference of cancer cells for the glycolytic pathway than the much more efficient oxidative phosphorylation pathway even in aerobic conditions) [61,75].

A related mechanism of vesicle biogenesis is linked to non-apoptotic plasma membrane blebs that are frequently observed in highly aggressive cancer cells [59]. These blebs continuously expand and retract at the cell surface and contribute to cell motility [76], but can be released as shedding vesicles by actin cytoskeleton rearrangements at the vesicle budding neck, resulting in membrane scission [70–72,77]. During *in vivo* migration, tumor cells can adopt a so-called amoeboid phenotype that is associated with extensive plasma membrane blebbing and vesicle release, suggesting a role for this type of EV biogenesis during cancer invasion and metastasis [77–80]. Finally, we would like to mention the finding of small blebs of phospholipid nanotubes traveling along the nanotubes and acting as vehicles for transporting the enclosed solution [81,82]. Even though these small vesicles seem to be a distended integral part of the membrane tube and not independent vesicles entrapped within the nanotube, this mechanism points out to an extremely high flexibility grade of the cell membranes increasing the possibility of cells to communicate and exchange material.

The contribution of each mechanism of EV formation can vary depending on both the cell type and the cell physiological state. A single cell can produce different subtypes of vesicles by using different pathways [51,61]. Notably, the molecular machineries that act at the different steps of EV biogenesis are partly common, including ESCRT proteins and the generation of ceramide by sphingomyelinase. Thus, it is difficult to ascertain the origin of the produced EVs by simply impairing the function of a given mechanism involved in the biogenesis of these vesicles. Moreover, the study in detail of the pathway that regulate EV release is complicated by the intersection of the EV biogenesis pathway with other molecular pathways associated with the trafficking of intracellular vesicles. Lysosomal pathways and Golgi apparatus–derived vesicle trafficking may exert indirect effects on EV biogenesis, especially on MVBs [43].

### *2.2.2 Cargo*

The existence of different biogenesis machineries for EV production accounts for the existence of many different subtypes of EVs, characterized by different cargo content. Several mechanisms contribute to the targeting of selective membrane-associated and soluble cargoes to vesicles. Membrane cargoes, such as glycosilphophatidilinositol (GPI)-anchored proteins are present in exosomes, probably because of their affinity for lipid domains and lipid rafts or their association with tetraspanins, either by direct interaction or by entrapment in tetraspanin-enriched microdomains that could be directly involved in ILV generation [83–86]. The sequestration of cytosolic proteins into ILVs can result from co-sorting with other proteins, such as the chaperones heat shock 70 kDa protein (HSP70) and heat shock cognate 70 kDa protein (HSC70), which are found in exosomes derived from most cell types [87,88]. It has also been proposed that ubiquitylation and farnesylation of proteins can help in their segregation in the ILVs of the MVBs [89,90]. Similarly, during MV formation, cytosolic components bind to the inner leaflet of the plasma membrane through posttranslational modifications (palmitoylation, prenylation, myristoylation) and the establishment of high-order complexes, which concentrates them to the small membrane domains from which forming MVs will bud [91,92]. Lipids and other membrane-associated cargos are localized to sites of vesicle budding from the cell membrane through their affinity for lipid rafts or for oligomeric cytoplasmic proteins, by their anchoring to plasma membrane lipids. Lipids are essential molecular components of EVs. One of the first indications that lipids are important for vesicle release was provided by Trajkovic and coworkers [68], who demonstrated that exosomes were enriched in ceramide and the release of exosomes was reduced after the inhibition of neutral sphingomyelinases. The sphingolipid ceramide has often been associated to EV biogenesis and release [93]. Enzymes involved in the metabolism of glycerophospholipids or glycerolipids have also been shown to play a role in the formation and release of vesicles [94–96]. Phosphoinositides (PIPs) play important functions in the endocytic pathway, conferring a unique molecular identity to the membranes of organelles and endocytic compartments; therefore, it is possible that exosomes and MVs could be identified by specific PIPs [97,98]. EV membranes contain also a high percentage of cholesterol, that increases membrane lipid order, prevents ion leakage and affects membrane fusion [99]. A comprehensive description of sphingolipids, phospholipids, cholesterol, and other lipids detected in EVs from different sources is reported by Skotland and coauthors [100]. The lipid composition of EV membranes may play a role in conferring stability to vesicles in different extracellular environments and/or to facilitate binding to or uptake into recipient cells. Moreover, EVs contain several bioactive lipids, and also the enzymes responsible for their synthesis, that may affect the behavior of target cells both *in vitro* and *in vivo* [100–102]. In this way, EVs could act both as carriers for lipid mediators synthesized in releasing cells as well as induce changes in lipid metabolism in recipient cells. For example, the viral protein Nef released in exosomes from HIV-infected cells may be responsible for impairment of cholesterol metabolism in EV target cells [103]. This is an emerging topic and, considering the molecular complexity of EVs, it is not easy to attribute effects in the EV-receiving cells to a specific lipid [100,104].

Apart from proteins and lipids, EVs also carry RNAs, both coding and noncoding, including microRNA (miRNA) [105,106]. A number of defined RNA binding proteins (RBPs) have been suggested to mediate RNA sorting into vesicles [107–109]. Interestingly, miRNAs can be differentially sorted to vesicles through MVBs as well as directly to the plasma membrane depending on their sequence and the presence of specific motifs

[109,110], the occurrence of post-transcriptional modifications [111] or the relative abundance of the target gene in the cell, that can sequester the miRNA, preventing its incorporation in the EVs [112]. New regulators of miRNA sorting into exosomes have recently been described and include the KRAS–MEK signaling pathway acting through AGO2 [113], major vault protein [108] and Y box binding protein 1 (YBX1) [114].

Many articles have reported also the presence of fragmented genomic DNA (gDNA) [115–117], mitochondrial DNA (mtDNA) [118,119] and even parasitic DNA [120] in EVs derived from cell culture medium and plasma. The mechanisms by which DNA can be incorporated into EVs and its physiological significance remain to be elucidated. Recent studies demonstrated that upon DNA damage, fragments of gDNA can be exported to the cytoplasm, where they can trigger cytoplasmic DNA sensors leading to cellular senescence or apoptosis [121]. Remarkably, secretion of DNA fragments via EVs prevents activation of cytoplasmic DNA sensors, thereby contributing to cellular homeostasis [122]. This aspect might be particularly relevant in cancer, where elevated levels of DNA damage might require efficient removal of cytoplasmic DNA via EVs [59].

Proteomic, lipidomic and acid nucleic profiling studies identified several EV-associated molecules from different sources and in different physiological and pathological conditions. Specific databases have been developed to collect the increasing amount of data on EV cargo content as well as on the procedures related to their isolation and analysis. Among them, Vesiclepedia [123] and EVpedia [124] platforms contain info on different types of molecules found associated to EVs (proteins, nucleic acids, lipids), whereas exRNA Atlas [125] focuses on the large diversity and volume of extracellular RNA (exRNA) data.

Strikingly, also whole organelles can be transported through vesicles. Islam and coworkers [126] identified *in vivo* mitochondrial transfer through EVs from bone marrow–derived stromal cells to pulmonary alveoli after mice acute lung injury. Successively, mesenchymal stem cells (MSCs) were also found to transfer mitochondria via EVs to both macrophages [127] and renal tubular epithelial cells. Transfer of mitochondria is fundamental for the protection of the recipient cells from oxidative stress and tissue injury [128]. Before these findings, the formation of tunneling nanotubes between close cells seemed to be the unique mechanism for mitochondria transfer. The finding that mitochondria (and possibly other organelles) can be loaded and transported through vesicles increases the impact of EV-mediated cell communication.

## 2.3 Biological role

The first role assigned to EVs was the elimination of waste material from the cell of origin. Nowadays, it is widely accepted their role as important carriers of information among cells, located also at distance. In Fig. 2 the possible mechanisms of cell communication through EVs are depicted. Upon their

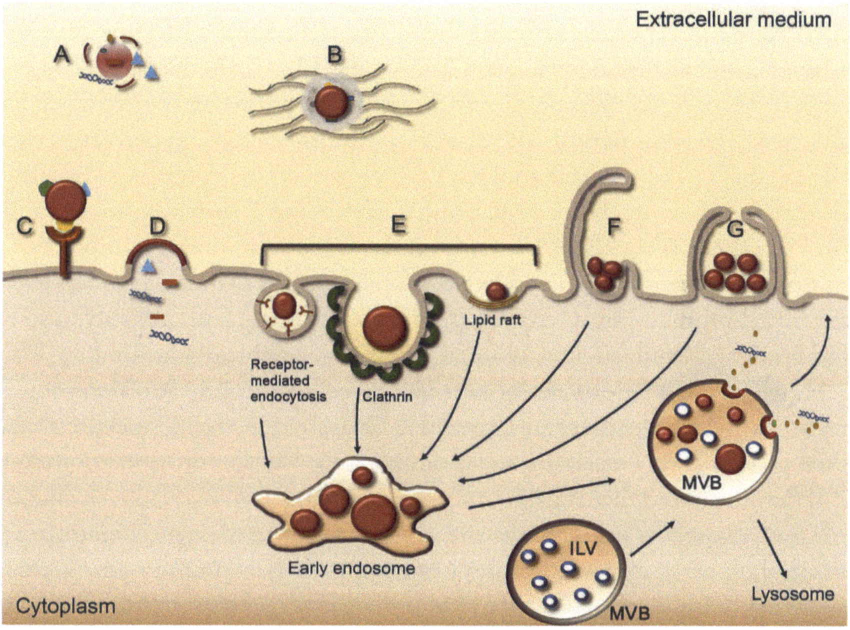


**Fig. 2** ***Biological role of extracellular vesicles.*** Extracellular vesicles (EVs) can exert their function without a direct contact with the cells (A, B), interacting with the cell surface but remaining outside (C), or being internalized after binding to the membrane, through different mechanisms (D–G). (A) Vesicles burst and release their active cargo, rendering it available to adjacent cells. (B) Enzymes on the surface of the vesicles, such as metalloproteinases and other enzymes, degrade the extracellular matrix, facilitating movements of both cells and vesicles. (C) Molecules of the membrane of vesicles can bind directly to receptors on the plasma membrane of the target cells and induce changes in their physiological state and signaling even without internalization. (D) Vesicles directly fuse with the plasma membrane. The vesicle membrane is integrated in the cell membrane and its content is released in the cytoplasm. (E) Molecules on the surface of the vesicle interact with molecules on the cell membrane and this results in the internalization of the vesicle inside an early endosome. The endocytosis can be receptor-mediated, clathrin-mediated, or occur through caveolae and lipid rafts. (F) Macropinocytosis and (G) phagocytosis of group of vesicles. The early endosome follows the endocytic pathway and can fuse to multivesicular bodies (MVBs), forming other MVBs where endocytosed vesicles and intraluminal vesicles (ILVs) coexist. Depending on MVB sorting, the internalized vesicles can be degraded into lysosomes, directly be re-secreted by the target cell or fuse to MVB membrane and release their cargo in the cytoplasm of the target cell.

release from parent cells, EVs begin their navigation in the extracellular space. Due to the different conditions they experience outside the cell of origin, (e.g. changes in pH), they can burst and release their active cargo, rendering it available for interaction with adjacent cells [129] (Fig. 2A). In this way, EVs could mediate secretion of proteins that lack signal sequence and could not be sorted to the endoplasmic reticulum. One of the secreted factors that lack a signal sequence is Fibroblast Growth Factor 2 (FGF-2), which has been reported to be included into EVs shed by tumor cells and more recently also by neurons and astrocytes [130—132]. Most EVs, however, are resistant to membrane breakdown and persist in the extracellular fluid, even for long periods of time [51]. Another cell-independent effect of EVs is due to their surface molecules, such as metalloproteinases and other enzymes involved in the degradation of extracellular matrices and invasive migration [133] (Fig. 2B). EVs tend to accumulate proximal to intercellular junctions, moving through and between cells. In this manner, EVs can leave their initial fluid and move to adjacent areas of tissues and, possibly, also to large bodies of fluid, such as blood, lymph and cerebrospinal fluid.

However, the major roles of EVs in cell-cell communication do require interaction with a recipient or target cell. Of note, the recipient cell can also be the producing cell itself, by an autocrine loop. Matsumoto and coworkers [134] demonstrated that murine melanoma B16BL6 cells are able to take up their own exosomes both *in vitro* and *in vivo* to induce their proliferation and inhibit their apoptosis, therefore promoting tumor progression. During interaction with target cells, components of the EV membranes can bind directly to receptors on the cell plasma membrane and induce changes in their physiological state without any fusion or internalization (Fig. 2C). This occurs during the immune response, when, for instance, EVs carrying on the surface MHC-peptide complexes can be captured by dendritic cells through a specific saturable receptor and presented directly to antigen-specific T cells, inducing their activation [135]. Another example is given in the nervous system by endocannabinoids found on the membrane of EVs produced by microglial cells that are able to activate their receptors on GABAergic neurons. Therefore, EV intermediation could also account for extracellular transport and signaling of hydrophobic compounds [136].

In most cases, however, the cargo of EVs needs to be internalized into the cell to exert its action. This can occur through direct fusion of the EVs with the plasma membrane (Fig. 2D) or through endocytosis/phagocytosis (Fig. 2E) [51,86,137,138]. In the first case, fusion causes the release of the EV content in the cytoplasm near the cell surface and the integration of the EV membrane in the cell plasma membrane. The endocytic mechanisms include receptor-mediated endocytosis [139], clathrin-mediated endocytosis [140], and endocytosis via caveolae and lipid rafts [141]. Moreover, macropinocytosis [142] (Fig. 2F) and phagocytosis [143] (Fig. 2G) can also occur, especially in the case of clusters of tethered EVs, such the ones produced by HeLa cells or Epstein—Barr virus (EBV)-transformed B cell lines [143]. In all these cases, intact EVs follow the endocytic pathway and fuse into the MVBs containing also the ILVs formed by the double inward of the plasma membrane (Fig. 2). Depending on MVB sorting, the internalized EVs can be degraded into lysosomes and their content recycled in the cell metabolism [144] or can directly be re-secreted by the target cell that in this way becomes the producing cell [138]. Interestingly, another mechanism has been found, still not elucidated but of primary importance, based on back fusion of EVs with the limiting membrane of the MVB, thereby releasing their cargo directly into the cytoplasm of the recipient cell, where it can freely interact with other cell components [145]. In tumor cells, fusion was enhanced by acidic pH, which is representative of what occurs in the tumor mass *in vivo* or possibly in late sorting endosomes inside target cells. In conclusion, a fusion step with either the plasma membrane or the limiting membrane of endocytic compartments must take place to permit the release of the EV cargo in the cytosol, which is necessary, for instance, for gene silencing or expression induced in the recipient cell by nucleic acids contained inside the EVs (mRNAs, ncRNAs, miRNAs) [28].

The most interesting feature of EVs is the ability to recognize and bind only specific target cells. Several studies highlighted this specificity in different contexts: in the blood platelet EVs were shown to bind monocytes and not neutrophils [146], in the nervous system EVs from neurons or glia are also very selective [147,148], and EVs from mesenchymal cells recognize mesenchymal cells, but not MSCs, neither endothelial cells or granulocytes [149]. The specificity of EV targeting to particular recipient cells is conferred by the interaction of the surface molecules present on both EVs and target cells. In EVs, such a role has been proposed for tetraspanins [150], glycoproteins [151], proteoglycans [152], lectins, and especially integrins which are present at variable density on the membrane of different EV subtypes [51]. Integrin composition of EVs released by cancer cells *in vivo* mediates their targeting to specific organs such as lung and liver and the consequent formation of organotropic metastasis [153]. In target cells the binding to EVs could involve surface receptors, adhesion molecules, such as integrins again, proteoglycans, as well as extracellular matrix proteins (e.g. fibronectin and laminin) [138,154]. The extracellular matrix might act as a 'zipper' between specific integrins present on the membranes of EVs and target cells. The EV lipid composition can also have an impact on recipient-cell targeting [61].

## 3. Role of EVs in tumor growth and metastasis

### 3.1 Impact of oncogenic transformation on vesiculation

Mutations or epigenetic alterations affecting oncogenes, tumor suppressors and their corresponding regulatory pathways drive both cell-intrinsic and non-cell-autonomous aspects of cancer pathogenesis [49]. Indeed, oncogenic mutations impact vesicle production and release, as well as their uptake. Activation of oncogenic signaling pathways, for instance by *H-RAS* and *EGFRvIII* have been shown to increase EV production and to affect cargo content in cancer cells [62,155−158]. In addition, oncogenic transformation due to stimulation of EGFR activity or *K-RAS* mutations were found to enhance macropinocytosis-mediated EV uptake by cancer cells [142,159]. Other studies demonstrated a strongly increase in EV uptake by cancer cells expressing *H-RAS* or *v-SRC* oncogenes, in comparison to their isogenic, non-transformed epithelial cell counterparts (IEC18) [160]. The ability of oncogenes to impact EV release and uptake suggests that effector molecules involved in these processes might be regulated by specific transforming signaling pathways. Intriguingly, RAS is known to upregulate the expression of syndecan-1 in IEC18 cells [161], which may impact the syndecan-1/syntenin/ALIX EV biogenesis pathway [162]. Similarly, oncogenic v-SRC, known to drive vesiculation of IEC18 -cells, is able to activate ADP-ribosylation factor 6 (ARF6), a small RAS-like protein GTPase implicated in vesicle production [94,163]. *K-RAS*, *H-RAS* and *BRAF* oncogenes are also known to regulate the RHO pathway which has a documented role in EV biogenesis [164,165]. Moreover, analysis of large databases containing gene expression data on a variety of tumors, such as The Cancer Genome Atlas (TCGA), has also shown the segregation of genes involved in vesiculation with specific cancer subtypes, each defined by a unique repertoire of oncogenic mutations [158]. Genes implicated in vesiculation might also possess a direct oncogenic activity, as found for the small GTPase Rab35 [166,167].

A large body of evidence points to increased and modified EV release in cancer cells compared to non-malignant counterparts [158]. In agreement, increased amount of circulating vesicles have been detected in plasma of cancer patients, with considerable contribution of EVs released from cancer cells themselves [168–170]. In principle, cancer EVs should be different by the EVs released from the normal counterparts, being enriched in oncoproteins and oncogenic DNA or RNA. Therefore, studies on differential segregation of molecules between cancer and normal EVs might lead to the discovery of several new biomarkers for the diagnosis and prognosis of many cancer types. The transfer of this oncogenic materials through EVs to non-malignant cells causes the acquisition of a tumorigenic phenotype in the recipient cells. This horizontal malignant transformation is a very intriguing concept per se. Even though it is expected to be transient in the expanding cell population, due to molecular degradation and dilution with cell division, it seems to have a powerful and perhaps still underestimated impact on tumor biology. In the following subparagraphs we will discuss how EV-mediated horizontal transfer can affect cancer growth and progression.

## 3.2 EVs in tumor microenvironment

Primary tumors develop from successive cellular and genetic modifications that result in the loss of the cellular identity and enhanced ability of tumor cells to survive, proliferate, and disseminate in the body. Under physiological conditions, tissue homeostasis is finely tuned and cellular proliferation is inhibited by cell-cell contacts. During tumorigenesis, cells lose the normal tissue architecture and contact inhibition is abolished, leading to unlimited proliferation that enable tumor growth and invasion of neighbor tissues. At late stages, tumor cells can also acquire mesenchymal-like phenotype with high motility capacity and leave the primary tumor to disseminate in the body, establishing metastasis in other tissues [171]. Tumors comprise a heterogeneous population of cells that can differ for phenotypic characteristics, genetic mutations, proliferation rate and drug resistance. Moreover, several noncancerous cellular types contribute to create the surrounding TME that can communicate with cancer cells and act in supporting or antagonizing tumor development (Fig. 3). Cell communication is fundamental for tumor development and the maintenance of TME, therefore EVs can play a major role in regulating interactions between the different cell types.

### *3.2.1 Cancer stem cells*

Several studies showed that only a small cell population within the tumor mass has tumorigenic ability, and can regenerate the original tumor both

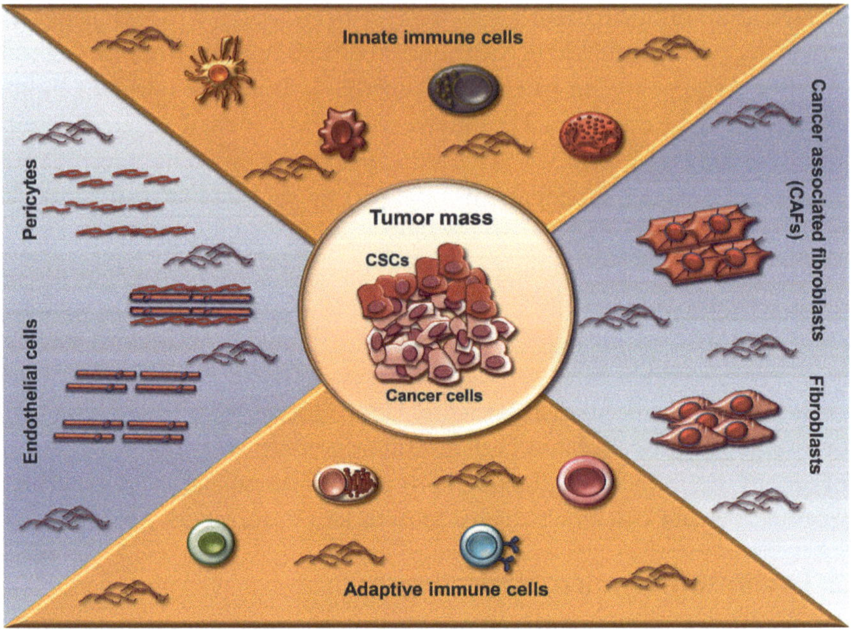


**Fig. 3** ***Composition of tumor microenvironment.*** The Figure shows a graphic representation of the environment around solid tumors that is denoted tumor microenvironment or TME. At the center is the tumor mass formed by a heterogeneous cell population, which include cells with stem-like features known as cancer stem cells or CSCs (top), and the non-stem cancer cell population (bottom). Several non-cancerous cellular types contribute to create the surrounding TME, including endothelial cells, pericytes, fibroblasts, cancer-associated fibroblasts (CAFs), innate and adaptive immune cells. Bidirectional cell communication among cancer cells in the tumor mass, as well as between cancer cells and the surrounding cell types, leads to profound morphological and functional changes, influencing tumor malignancy. In advanced tumors, endothelial cells actively form new vessels and normal fibroblasts turn into CAFs, highly supportive of neoplastic processes. Innate and adaptive immune cells can either act to protect the host by attempting rejection of the tumor or stimulate proliferation and dissemination of cancer cells, thereby accelerating the progression of the disease.

*in vitro* and *in vivo* [172]. These cells are named cancer stem cells (CSCs) and, similar to normal stem cells, have the ability to self-renew and through asymmetric division can maintain stem cell pool or differentiate in a variety of cell types to generate a cellular bulk that promotes tumor growth. Therefore, CSCs are the main responsible for cancer recurrence and, interestingly, also for resistance to chemo/radiotherapy as well as tumor invasion and metastasis [173]. CSCs reside in a distinct microenvironment called "CSC niche", where they receive the signals required for the maintenance of stem-like phenotype and survival [174] (Fig. 3). An important role in cancer niche homeostasis is played by hypoxia. In the primary tumor, uncontrolled proliferation leads to a hypoxic state within the tumor mass due to an inadequate vascularization. Hypoxia induces in CSCs the activation of several signaling pathways (e.g. TGF-β, HIF1-α and Wnt/β-catenin) that can trigger pleiotropic pro-tumorigenic effects such as cancer cells dedifferentiation and angiogenesis induction. In response to these pathways, CSCs produce pro-angiogenic and immunosuppressive factors that promote vessel formation and inhibit immunosurveillance, allowing tumor growth [175,176]. Mao and coworkers [177] showed that Wnt/β-catenin pathway is involved in cell proliferation and maintenance of stemness and that its activation induces dedifferentiation of non-stem cancer cells toward a stem-like phenotype, such as expression of CSC markers and acquisition of a self-renewal capacity. Moreover, CSCs through Wnt pathway activation play also a fundamental role in triggering EMT, considered an early step in the invasion-metastasis process as it gives at cancer cells increased motility and invasiveness. CSCs seem also to be involved in the formation of a tumor-supportive microenvironment known as pre-metastatic niche in target tissues, by secreting soluble factors that can reproduce a hospitable TME in order to help the seeding of tumor cells in secondary sites [178].

Emerging evidence supports the idea that CSC-derived EVs (CSC-EVs) can function as powerful regulators of TME, modulating each step involved in tumor progression, including tumor growth, angiogenesis, immunosuppression, drug resistance and even metastasis dissemination. First of all, CSC-EVs are able to induce reprogramming of non-stem cancer cells into a CSC-like phenotype in a variety of cancer types [179]. Sun and colleagues have shown that EVs released by glioblastoma stem cells mediate the reprogramming or dedifferentiation of non-stem glioma cells by transferring Notch1 protein [180]. Moreover, the ability of Claudin7-loaded EVs released by CSCs to promote reprogramming of non-metastatic tumor cells by inducing EMT has been reported for colorectal cancer-initiating cells as well as for pancreatic CSCs [181,182] (Fig. 4). Growing evidence suggests that CSCs can prepare target tissues via vesicles, promoting the establishment of the pre-metastatic niche. In renal cancer, vesicles released from CSCs contain pro-angiogenic mRNAs (and also miRNAs), such as the one coding for VEGF, FGFs, angiopoietin1, ephrin A3, MMP2 and MMP9 matrix metalloproteinases, that are absent in vesicles released from non-stem tumor cells and that may be involved in tumor progression. These vesicles are able to induce in endothelial cells a variety of effects, including cell proliferation,

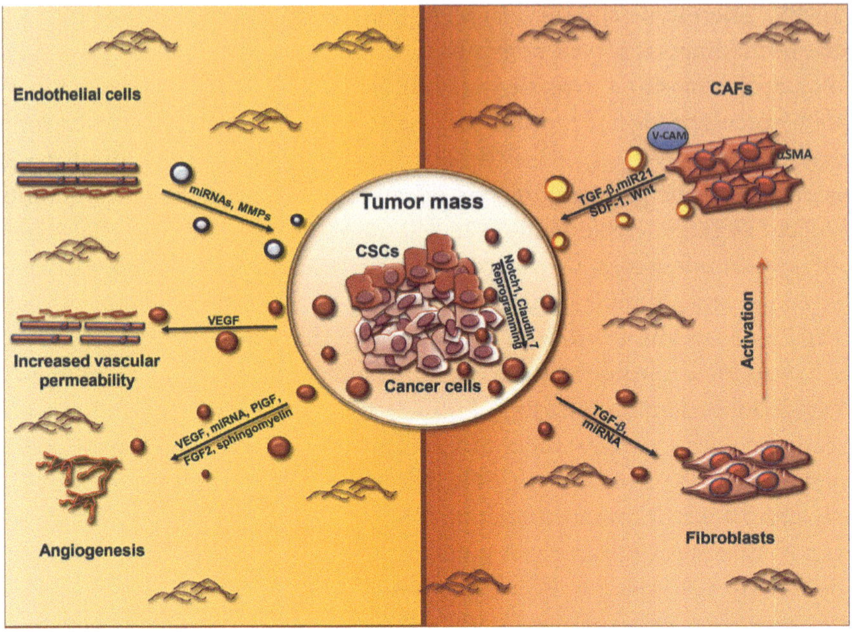


**Fig. 4** ***EV-mediated cell communication in the tumor microenvironment.*** The Figure reports some of the main interactions mediated by extracellular vesicles (EVs), as examples of the active EV-mediated cross-talk between tumor cells and the surrounding microenvironment. At the center, cell communication between cancer stem cells (CSCs) and non-stem cancer cells is highlighted. CSC-EVs are able to induce reprogramming of non-stem cancer cells into a CSC-like or mesenchymal phenotype. Tumor cells send also EV-associated paracrine signals to the surrounding endothelial cells (left) and fibroblasts (right), strongly influencing their morphology and function. EVs released by tumor cells can target endothelial cells, inducing an increase in vascular permeability and angiogenesis. On their side, endothelial cells release vesicles that may systemically direct the function of heterotypic cell types either in the circulation or in different organ microenvironments. Furthermore, cancer EVs may target normal fibroblasts recruited by the tumor and induce their activation in cancer-associated fibroblasts (CAFs), thereby increasing α-SMA expression. Pro-tumorigenic EVs released by CAFs are delivered to cancer cells, thus sustaining tumor growth, suppressing apoptosis, conferring chemoresistance and promoting cell migration and invasion. The specific molecules delivered by EVs involved in each process are indicated by arrows.

organization, invasion of extracellular matrix, and resistance to apoptosis, contributing to the establishment of a pre-metastatic niche in lung [183]. Finally, it is widely accepted that CSCs are resistant to most therapeutic strategies: radiation and chemotherapy treatments eliminate the bulk of the population of non-stem cancer cells but not CSCs [184]. Several signaling pathways have been involved in CSC drug resistance, which include TGF-β, PI3K/AKT, NF-kB, Signal Transducer and Activator of Transcription 3 (STAT3), Wnt/β-catenin, Hedgehoh and Notch [184]. Cancer EVs can transfer drug resistance from one cell to another by delivery of proteins, miRNAs and lncRNAs that cause downregulation of promoters of apoptosis and cell differentiation [185]. For instance, Santos and coauthors show that breast CSCs can modulate chemosensitivity of other tumor cells by delivering of miR-155 through EVs. The authors do not explore the molecular pathway triggering the chemoresistance, however they suggest that TGF-β could be involved, as it is a target of miR-155 [186]. CSC-EVs can also induce chemoresistance through a competition mechanism based on binding with circulating therapeutic antibodies. Breast carcinoma cell lines overexpressing *HER2* oncogene, coding for the human epidermal growth factor receptor 2, can release EVs carrying HER2, which can bind Trastuzumab (the therapeutic antibody that blocks HER2), sequestering it from circulation and interfering with the ability of this antibody to inhibit proliferation in breast cancer cells [187]. Several *in vitro* studies on cancer cell lines show that EVs can modulate drug resistance by additional mechanisms, including drug seizure and increased outflow from the cancer cells [188,189]. However, above studies do not investigate the specific tumor populations responsible for EV release, thus the contribution of CSCs in these mechanisms remains to be clarified.

### 3.2.2 Interaction of cancer cells with cancer-associated fibroblasts

Cancer-associated myofibroblasts or CAFs are the major cellular constituents of the tumor stroma and are characterized by the expression of α-smooth muscle actin (α-SMA), associated to a contractile phenotype (Fig. 3). CAFs are responsible for the deposition, remodeling and stiffening of extracellular matrix (ECM), very relevant to tumor growth and dissemination. Furthermore, CAFs are involved in reciprocal signaling interactions with cancer cells and infiltrating leukocytes through soluble factors, direct cell-cell adhesion and vesicles [190]. Although the origin of CAFs is still controversial, they may arise from normal tissue fibroblasts, recruited by the developing tumor and subsequently activated. However, they may also derive from hepatic and pancreatic stellate cells as well as from bone marrow stem cells [191]. Interestingly, MSCs, derived from bone marrow and transplanted into tumour-bearing mice may exhibit a CAF phenotype, indicating that bone marrow-derived cells may acquire a CAF phenotype, in the presence of neoplastic cells [192]. In normal development and physiology, fibroblasts produce connective ECM and actively participate to tissue repair following damage, also cross-talking with local epithelial cells. During wound healing, fibroblasts produce TGF-β and express α-SMA, therefore converting themselves into myofibroblasts. In the initial stages of tumorigenesis, there is evidence that fibroblasts circumventing early tumor lesions exhibit a clearcut tumor-suppressive activity [193]. Emerging evidence shows that, during tumor development and establishment, fibroblasts start exhibiting myofibroblastic characteristics, also in response to signals from cancer cells, like TGF-β, IGF-1 and others [194]. However, the multiple mechanisms by which developing tumor cells promote the conversion of healthy fibroblasts into CAFs is still not completely understood.

Tumor EVs contain a variety of molecules and strongly affect the physiology of recipient cells in TME (Fig. 4). One known case is that of tumor-derived vesicles containing TGF-β associated to the transmembrane glycan betaglycan and triggering a differentiation program of fibroblasts toward a myofibroblastic phenotype, with a concomitant upregulation of α-SMA expression. This complex elicits SMAD-dependent signaling in fibroblasts with a kinetic and extent of the response higher than that elicited by soluble TGF-β [195]. Active TGF-β is found also in bladder cancer-derived EVs: these vesicles are internalized by fibroblasts, promoting their proliferation and CAF markers expression, and activating SMAD-dependent signaling. Strikingly, EV-associated TGF-β represents the majority of the total TGF-β in the cancer cell medium (between 53.4% and 86.3%) [196].

Exposure of MSCs, derived from adipose tissue to breast cancer-derived vesicles, results in a myofibroblastic phenotype, associated to an increased expression of the SMAD-mediated tumor-promoting factors TGF-β, VEGF, Serum-Derived Factor-1 (SDF-1) and C-C Motif Chemokine Ligand 5 (CCL5) [197]. MDA-MB-231 breast carcinoma and U87 glioblastoma cell lines are found to secret EVs that are able to confer anchorage-independent growth and enhanced survival capability to normal fibroblasts and epithelial cells. Although the mechanism underlying these effects is unclear, the data indicate that it requires the cross-linking activity of tissue transglutaminase and the cross-linking substrate fibronectin [165]. An important breakthrough is the discovery of nucleic acids in EVs. A large body of evidence shows the occurrence of EV-associated miRNAs in cancer, with the ability of targeting peritumoral fibroblasts and driving them to a CAF phenotype. In prostate cancer, the CSC population and the bulk tumor cells secrete several EV populations containing different subsets of miRNAs. The differential miRNA content of these EVs may contribute to local invasion and pre-metastatic niche formation, by modulating fibroblast migration [198,199]. In breast cancer, EVs may contain miR-105, which is upregulated by the oncoprotein MYC in cancer cells and activates MYC signaling in recipient CAFs. As a result, CAFs metabolism is affected in a peculiar way: when nutrients are sufficient, miR-105 enhances glucose and glutamine metabolism, thus supporting adjacent cancer cells. When nutrient levels are low, CAFs convert metabolic wastes, including lactic acid and ammonium, into energy-rich metabolites [200]. Fong and coworkers showed that the EV-mediated transfer of miR-122 to stromal cells suppresses glucose uptake by downregulating the glycolytic enzyme pyruvate kinase, resulting into an increasing nutrient availability in the pre-metastatic niche for breast cancer cells [198].

In solid cancers, there are many examples of EV-mediated bidirectional communication between the ever-evolving TME and tumor cells (Fig. 4). As potent mediators of cell-cell communication, EVs may be produced also by CAFs which undergo an enhanced secretion and delivering of EVs to cancer cells, compared to normal fibroblasts. However, the specific mechanisms through which healthy stromal cells are triggered to release vesicles that promote the malignant behavior of cancer cells is not completely understood. In agreement with the cancer supportive role exerted by CAFs, EVs of myofibroblastic origin stimulate tumor progression. In breast tumors, CAFs secrete vesicles promoting protrusive activity, motility, and dissemination of breast cancer cells through the Wnt-Planar Cell Polarity (PCP) signaling pathway. In particular, Wnt ligands produced by breast cancer cells tether to fibroblast vesicles triggering Wnt-PCP signaling in cancer metastasis [201]. Moreover, EVs derived from fibroblasts can induce the dedifferentiation of cancer cells to promote chemoresistance in colorectal cancer via activation of Wnt signaling pathway, increasing the percentage of CSCs [202]. Tumor-promoting CAFs could not only exert paracrine signaling to cancer or other stromal cells, but can undergo autocrine signaling. A self-sustaining TGF-β and SDF-1 autocrine signaling loop initiates and maintains the differentiation of fibroblasts into myofibroblasts and the concurrent pro-tumoral phenotype. This signaling mechanism may represent an attractive therapeutic target to interfere with the progressive acquisition of the CAF phenotype [203]. In ovarian cancer, tumor cells usually invade the peritoneum, enriched in stromal cells. EVs and cell lysates isolated from cancer-associated adipocytes (CAAs) and CAFs contain elevated levels of miR-21. Interestingly, miR21-associated EVs are delivered to the cancer cells, bind to Apoptotic Protease Activating Factor-1 (APAF1), suppress apoptosis and confer chemoresistance. This suggests that the inhibition of stromal-derived miR-21 may be a novel therapeutic strategy for metastatic ovarian cancer [205]. CAFs have also the ability to transfer a remarkable amount of lipids to neighboring cells, in a vesicle-dependent fashion. As a consequence of this phenomenon, tumor cells increase their proliferation rate, indicating that EV-mediated trafficking indeed supports cancer cells [206]. To study EVs *in vivo*, vesicles may be tracked in cells, appropriately engineered for secreting fluorescent EVs. These fluorescent EVs, produced by mouse triple-negative breast cancer 4T1 and 4TO7, are incorporated by CAFs. One of the most abundant microRNAs associated to these vesicles is miR-125b, that is transferred to normal fibroblasts in TME and contributes to their development into CAFs [204].

In general, the relevance of CAFs to the tumor progression is also shown by the effects of the experimental drug minnelide, a water-soluble promising pro-drug isolated form the herb *Tripterygium wilfordii*. Transcriptomic profiling conducted after treatment with minnelide revealed the deregulation of the TGF-β signaling pathway in CAFs, associated to an apparent reversal of their activated state. In an established mouse model of pancreatic cancer, administration of minnelide results in the depletion of active stromal fibroblasts and tumor regression, with implications for stromal-based anti-cancer strategies [207]. These issues, taken together, have many implications motivating an intense future research in the fields of tumor-stroma communication, EV function and anti-cancer therapies.

### 3.2.3 Interaction of cancer and endothelial cells

In the TME, endothelial cells are devoted to the formation of new vessels to provide oxygen and fresh nutrients for the growing tumor mass. Tumor angiogenesis, that refers to the formation of new blood vessels within tumors is considered a key feature of tumor progression. In the absence of neoangiogenesis, tumor mass cannot grow beyond a few millimeters, a size that would not be life-threatening. Two cell types participate to new vessels formation, the endothelial cells, lining inside the vessels and the perivascular pericytes, playing a role in blood vessel contraction [208] (Fig. 3). Regardless of cancer type, tumor-associated blood vessels exhibit a morphology different from normal blood vessels. In normal blood vessels, endothelial cells form regular monolayers with tight junctions, whereas in tumors, they are disorganized and loosely connected, often extending across the lumen. Tumor-associated endothelial cells are characterized by a peculiar gene expression profile. In endothelial cells from invasive epithelial ovarian cancers, specific sets of genes are found overexpressed which, if subjected to silencing, block endothelial cell migration and tube formation [209]. Several efforts have been made to develop anti-cancer therapeutics targeting these abnormalities.

In physiological conditions, the stimulatory and inhibitory signals are balanced, so that blood vessels form only when and where are needed, i.e. during growth and healing. Endothelial cells in adults are mostly quiescent, but their proliferation can be induced by angiogenic factors like Vascular Endothelial Growth Factor (VEGF) A or by hypoxic conditions, occurring in pathologies like cancer. VEGFs are a well-known family of secreted polipeptides, with a highly conserved receptor-binding cystine-knot structure similar to that of the platelet-derived growth factors. There are five VEGF family members in mammals (VEGF A, B, C, D) and the Placental Growth Factor (PlGF), binding to three tyrosine kinase receptors (VEGFR-1, -2 and -3) and eliciting endothelial cell proliferation, migration and permeability. Factors belonging to this family are secreted by cancer cells, eliciting new vessels formation and enhancing vascular permeability by a paracrine mechanism [208]. The role of angiogenic factors in the promotion of tumor-associated angiogenesis has been the subject of intense investigation.

Recent evidence highlights the important role of EVs carrying angiogenic factors to promote endothelial cell migration, invasion, and tube formation, as well as to induce *in vivo* neovascularization (Fig. 4). Ko and coauthors described the responses of endothelial cells to vesicles carrying on their surface the 189 amino acids isoform of VEGF. Interestingly, high levels of EV-VEGF are associated with disease progression in bevacizumab-treated ovarian cancer patients, raising the possibility that resistance to bevacizumab might stem in part from elevated levels of EV-VEGF [210]. Relevant amounts of VEGF in vesicles, together with MMP2 and MMP9 matrix metalloproteinases, are reported to be secreted also by CABA I and A2780 human ovarian carcinoma cell lines. Interestingly, active VEGF is released following vesicle burst induced by acidic pH (a characteristic of the TME), leading to the stimulation of endothelial cell motility and invasion [129]. In human renal cancer and in glioblastoma, EVs released by CSCs contain several proangiogenic mRNAs such as VEGF, FGFs, angiopoietin1, ephrin A3, MMP2, and MMP9. Once internalized by normal endothelial cells, these mRNA stimulate vessel formation, sustaining tumor growth [183,211] (Fig. 4). A different proangiogenic mechanism is reported for epithelial ovarian cancer cells, involving the vesicle-associated CD147, a membrane-spanning molecule, known to regulate the expression of MMP in peritumoral stromal cells. In this study, the authors show that CD147 accumulates at high levels in tumor cells as well as in tumor vesicles. The CD147-enriched vesicles are capable of promoting tube formation in human umbilical vein endothelial cells (HUVECs) in culture [212]. Another study shows that, following RAS-induced EMT, epithelial cells release vesicles containing VEGF-associated proteins, the small GTPase Rac1, and the serine threonine kinase PAK2, all proangiogenic factors [213].

Emerging evidence shows that miRNAs play a key regulatory role in tumor angiogenesis. Therefore, miRNAs might have a great impact in clinic and could be used as possible cancer biomarkers as well as powerful targets in therapeutic interventions against *in vivo* tumor angiogenesis [214]. Among miRNAs found to be involved in angiogenesis is miR-92a-3p, that is secreted in vesicles by colon cancer cells and is delivered into HUVEC endothelial cells. This transfer leads to increased proliferation, motility and tube formation, possibly, through the downregulation of *Dkk-3* target gene, a direct target of miR-92a-3p. Although the complete molecular mechanism by which miR-92a-3p exerts these proangiogenic effects is still unclear, the data suggest the involvement of a partial endothelial-to-mesenchymal transition [215]. Another example of vesicle-associated miRNA is miR-142-3p, known as tumor suppressor. This miRNA is enriched within EVs produced by lung adenocarcinoma cells and may be transferred to both endothelial cells and fibroblasts to promote tumor-associated changes. miR-142-3p delivery in endothelial cells promotes angiogenesis through inhibition of TGF-βR1 receptor, whereas in lung fibroblast cells promotes the CAF phenotype in a TGF-β independent manner [216]. Similarly, in lung carcinoma, hypoxic cancer cells release EVs enriched with miR-23a, that is internalized by endothelial cells and lead to the accumulation of Hypoxia-Inducible Factor 1α (HIF-1α) and angiogenesis induction [217]. EV-associated miRNAs may affect also endothelial integrity: for instance, transfer of metastatic breast cancer-secreted miR-105 to endothelial monolayers destroys the integrity of the tight junctions that are natural anti-metastatic barriers. Mechanistically, miR-105 targets and downmodulates the tight junction protein Zonula Occludens-1 (ZO-1). As a proof of concept, overexpression of miR-105 in non-metastatic cancer cells induces vascular permeability and metastases formation. In contrast, inhibition of miR-105 in highly metastatic tumors reduces their metastatic potential. Interestingly, miR-105 can be detected in the circulation at the pre-metastatic stage, and its levels in the blood and tumor are associated with ZO-1 expression and metastatic progression [218]. In another report, tumor-derived EVs (TDEVs) were found to contain tetraspanins

(Tspan8-alpha4/beta1) and activate angiogenesis. Endothelial cells or lung fibroblasts exposed to these EVs exhibit dramatic changes of their mRNA, miRNA and lncRNA profiles [219]. Glioblastoma cells also release vesicles containing angiogenic mRNA, miRNA and proteins. These EVs may be taken up by brain microvascular endothelial cells and the associated RNAs may be translated into recipient cells, as shown by experiments with an mRNA encoding a reporter protein. Glioblastoma-derived EVs are proangiogenic, but they also stimulate glioma cell proliferation, indicating a self-promoting aspect. Also, the finding that specific glioma mRNA variants and miRNAs can be detected in serum EVs of glioblastoma patients, opens new diagnostic possibilities. This is the case of the tumor-specific EGFRvIII that was detected in serum vesicles and could provide diagnostic information, helping in therapeutic decisions for cancer patients with a non-invasive procedure [220]. Finally, EVs carry not only proteins and nucleic acids, but also bioactive lipids that can function as angiogenesis promoters [221]. One of the major lipids in mammalian plasma membrane is sphingomyelin that has been identified as the active component of pro-angiogenic and pro-migratory EVs. These biological activities disappear following heat-treatment or lipids extraction of the vesicles [222].

Endothelial cells are also able to release vesicles in the TME affecting the phenotype of surrounding cells (Fig. 4). After VEGF and FGF-2 stimulation, endothelial cells release EVs carrying on the surface MMPs that initiate the proteolysis necessary for tumor invasion and angiogenesis [223]. Circulating EVs from endothelial cells have been isolated from tumor-bearing mice and found to carry elevated levels of specific miRNAs (miR-30c, miR-126, miR-146a, and miR-125b), compared to control animals. These purified EVs expressed expected EV markers and were transferable to multiple recipient cells. These results suggest that endothelial cells may systemically direct the function of heterotypic cells either in the circulation or in different organ microenvironments via specific EVs [224]. A set of differentially expressed miRNAs was identified in microvascular endothelial cells co-cultured with tumor cells. Unexpectedly, most miRNAs were derived from tumor cells, packaged into vesicles, and then directly delivered to endothelial cells. Among these, miR-9 elicits modifications of endothelial cells morphology in culture, activating the JAK-STAT pathway and promoting cell migration and tumor angiogenesis [225]. Although anti-angiogenic therapies are effective against different tumors, they are limited by subsequent tumor vasculogenesis and progression. To mimic carcinoma-associated endothelial cells, Zeng and coauthors took advantage of their previous findings concerning miR-9 that is significantly elevated in endothelial cells associated to hepatocellular carcinoma (HCC). Therefore, HUVECs engineered to express miR-9 are taken as a model for studying the effects of anti-angiogenic drugs. Administration of vandetanib results in the autophagy of HUVECs, but also in the release of VEGF-enriched vesicles. This EV-mediated mechanism may be responsible for the failure of this anti-angiogenic drug to cause lasting remission of angiogenesis [226]. In general, a deep knowledge of endothelial cell function and abnormalities in tumor-associated vasculature will provide crucial information to design new and powerful anti-angiogenic strategies.

## 3.3 EVs in epithelial-mesenchymal transition and metastasis progression

Epithelial and mesenchymal cells derive from different embryonic layers and are very distant from each other in both morphological and functional aspects. Epithelial cells derive from either the ectoderm or endoderm, the first two embryonic layers to form during development. They are characterized by the establishment of close contacts with their neighbors, through the sequential arrangement of adherens junctions, desmosomes, and tight junctions, adopting an apico-basal polarization. Epithelial cells have a flat or polygonal shape and are cohesively arranged in single cell layers or sheets, acquiring the ability to function as barrier or in absorption. Conversely, mesenchymal or stromal cells derive from the third embryonic layer, the mesoderm and present characteristics that, in contrast to those of epithelial cells, enable their migration and invasion of tissues. They have neither uniform composition nor cellular adhesion and are morphologically irregular, displaying an elongated, spindle-like shape that allows for less rigid topological functionality. Epithelial cells assure the tissue integrity, which is essential to maintaining the body composition, whereas mesenchymal or stromal cells are organized in a three-dimensional extracellular matrix, are associated to great migratory freedom, and comprise connective tissues adjacent to epithelia and supporting them [227].

Despite these great differences, transitions of cells between epithelial and mesenchymal states take place in a highly plastic and dynamic way [228]. A shift toward the mesenchymal state, in a process referred to as EMT, was first apparent to developmental biologists in the late nineteenth century, by clear histological evidence that, during gastrulation, mesenchymal cells originated from the adjacent embryonic ectoderm [228,229]. It is now well known that the EMT and its reverse process, named mesenchymal-epithelial transition (MET), occur at several rounds during embryogenesis and organogenesis to guarantee the final differentiation of specialized cell types and the acquisition of the complex three-dimensional structure of internal organs [227,228,230,231]. During normal adult life, various mechanisms guarantee epithelial homeostasis and protect the epithelial phenotype in order to maintain tissue integrity and organ function, with few exceptions such as in the mammary gland during puberty and pregnancy, where remarkable epithelial plasticity is observed [228]. Notably, EMT/MET processes are reactivated during wound healing and in pathological conditions, such as fibrosis and cancer metastasis. Indeed, the significant parallels between cell plasticity in embryonic development and carcinoma progression led to the proposition that EMT/MET is a central driver of epithelial-derived tumor malignancies [232]. EMT has been shown to trigger the dissociation of carcinoma cells from primary carcinomas, which subsequently migrate and disseminate to distant sites. Significantly, it is the MET that is then believed to trigger the cessation of migration, inducing the same cells to proliferate and seed the new tumor [228].

EMT/MET processes are tightly controlled and coordinated by multiple pleotropic factors that target and influence downstream effector molecules and signal transduction pathways [233]. In particular, EMT implies a profound reprogramming of gene expression mainly orchestrated by specific "master" transcription factors, known as EMT-inducing transcription factors (EMT-TFs; i.e., ZEB1, SIP1/ZEB2, Twist1, Twist2, E12/E47, Tbx3, the Snail family members Snail2 or Slug, Snail3 and, in particular, Snail1) [233–236]. The EMT-TFs primarily act as repressors of the epithelial genes and may guide the recruitment of the epigenetic machinery to the chromatin context, thus allowing the proper regulation of gene expression [237–239]. The orchestrated interplay of EMT-TFs, inducers (e.g. HGF [240], PDGF [241]) and signaling pathways (e.g. TGF-β [242,243] and Notch [244]) facilitate the shift in plasticity [245]. The current view is that the EMT/MET is not a binary process, a simple shift between two alternative states, but rather occurs in a gradual manner characterized by several cellular states exhibiting intermediate morphological, transcriptional, and epigenetic features, between epithelial and mesenchymal cells. The intermediate states between epithelial and fully mesenchymal states have been referred to as partial, incomplete, or hybrid EMT states [246]. The complexity of the EMT/MET phenotypes reflects the complexity of the regulatory circuits that, beyond the transcriptional control, also involve several ncRNAs, including miRNAs (e.g., miR-200 family or miR-34)

[247–249] or lncRNAs (e.g., HOTAIR) [237]. Communication among neighboring cells and also at distance is critical for the onset and the progression of the EMT/MET programs which are fundamental during the processes of cancer invasion and formation of metastasis.

EVs have been involved in the horizontal transfer of an assortment of constituents, such as proteins (inducing factors, membrane receptors, signaling transducers, TFs) and RNA (mRNA, lncRNA, miRNA) acting as modulator of gene expression and phenotypic features of the cells. Therefore, a lot of efforts have been made to isolate and characterize EVs produced by cells before and after EMT/MET processes. First of all, both quantitative and qualitative changes have been found in EVs obtained from various EMT models [245]. Following induction of EGFR coupled with the blockade of E-cadherin, both human squamous carcinoma A431 and colon epithelial DLD-1 cells acquire mesenchymal traits and notably release a higher number of EVs [250]. Proteomic analysis of EVs obtained from EMT-induced A431 cancer cells resulted in the identification of profound qualitative differences in comparison with parental A431 epithelial cells, mainly in pathways associated with cell growth, signaling and motility [251]. Similarly, proteomic analysis of EVs (both membrane and luminal content) derived by *in vivo* metastasis model of human bladder carcinoma cell line T24 and its two isogenic derivative cell lines SLT4 and FL3, showed the upregulation of EMT-associated proteins in both membrane fraction (e.g. vimentin and hepatoma-derived growth factor) and luminal fraction (casein kinase II $\alpha$ and annexin A2) of metastatic vesicles [252]. Global proteomic profiling of the EMT model of MDCK (epithelial) cells and oncogenic *H-RAS*-transformed MDCK (21D1) cells and derived EVs led to the identification of prototypic expression of EMT hallmark proteins (E-cadherin, EpCAM and vimentin) in both cell and vesicles proteomes [253]. In addition, 21D1 EVs were enriched with several metastatic niche-associated proteins, such as proteases and integrins, and expressed specific proteins absent in the epithelial counterparts, such as splicing complex components (e.g. SF3B1, SF3B3 and SFRS1) and the key transcriptional regulator YBX1, whose overexpression in MDCK cells has been demonstrated to induce a partial EMT phenotype [254].

Notably, EVs released by cancer cells are not only molecularly different from ones derived from the normal counterpart, but this difference has a functional significance. EVs released by cancer cells are able to direct a shift in plasticity of normal epithelial recipient cells toward a mesenchymal phenotype. EVs isolated from MDA-MB-231 cells stimulated with linoleic acid (that induces the proliferation, motility and invasive potential of breast cancer cells) caused decrease in the expression of epithelial markers and increased expression of mesenchymal markers in the MCF-10A non-tumorigenic mammary epithelial cell line [255]. EVs from highly metastatic lung cancer cells induced an EMT associated with migration, invasion, and proliferation in recipient human bronchial epithelial cells. Similarly, urothelial cells conditioned with bladder cancer EVs showed elevated expression of mesenchymal markers (α-SMA, S100 Calcium Binding Protein A4 or S100A4), diminished expression of epithelial markers (E-cadherin and β-catenin) and enhanced invasion and migration abilities compared to normal urothelial cells. EVs isolated from patient bladder cancer samples also induced the expression of mesenchymal markers in their normal counterparts [256].

Several studies focused on the identification of the molecules and/or the signaling pathways involved in the mesenchymal shift of recipient cells. Higginbotham and coauthors [257] identified in vesicles released by human breast and colorectal cancer cells full-length, signalling-competent EGFR ligands, such as amphiregulin, able to increase the invasiveness of recipient cancer cells. Chen and coworkers [258] showed that EVs derived by highly metastatic hepatocellular carcinoma MHCC97H cells induced EMT and consequent migration and invasiveness properties to low metastatic HCC cells via MAPK/ERK signaling pathway activation [258]. Le and coworkers [259] identified in EVs obtained from both metastatic mouse and human breast cancer cells, the presence of miR-200 that regulates the expression of the EMT-inducing transcription factors Zeb1 and Zeb2 [260]. Moreover, miR-200 was found increased in the sera of mice with metastatic tumors and was also shown to be transferred via metastatic breast cancer EVs to non-metastatic cells, accompanied by changes in gene expression and EMT. Strikingly, EVs obtained from miR-200-expressing tumors facilitated metastasis of cells with lower metastatic potential, thereby endowing them with the capacity to colonize distant tissues [259].

## 4. EVs in cancer immune response

The immune system is classically compared to the police compartment of body, as its mainly established role consists in patrolling tissues and provide instant defense against potential pathogen invasions. However, its role in the regulation of tissue homeostasis and wound healing is also crucial for maintaining body integrity and safety [261]. Thus, to perform these opposite functions, inflammatory processes triggered by infections need to be compensated by suppressive mechanisms that turn off inflammation and begin repairing tissue damage [261]. The immune system comprises a variety of cell subsets that can be mainly divided in two branches: effector cells and suppressive cells. The role of effector cells is to protect body from any type of danger, such as bacterial and virus infections or tumorigenesis. This branch includes macrophages, natural killer (NK), T and B lymphocytes, dendritic cells (DCs), and polimorphonuclear leukocytes (neutrophils, basophils, eosinophils) [262]. Suppressive cells have the role to restrain the immune response to maintain immune tolerance and attenuate inflammation in order to suppress pathological autoimmune and inflammatory immune responses to prevent damage to body tissues. This branch mainly includes T and B regulatory (Treg and Breg) cells [263,264]. Therefore, suppressive cells also contribute to wound healing and tissue repair [265]. For example, macrophages can be conventionally divided into two subgroups, M1 and M2 polarization states. M1-type macrophages are considered inflammatory cells and play important roles in the innate response to invading pathogens. By contrast, M2-type macrophages are considered anti-inflammatory cells and can be activated by a variety of interleukines (ILs), such as IL-4, IL-10, IL-13, and glucocorticoids, all of which play important roles in tissue repair [266]. Similarly to macrophages, neutrophils and DCs can exist in pro-inflammatory as well as in anti-inflammatory polarization [267,268].

Ideally, immune system is able to recognize and eliminate tumor cells. Indeed, cancer cells express several tumor-specific antigens on their surface that have the potential to elicit tumor-specific immune responses [269]. These tumor-associated antigens (TAAs) are phagocytosed by professional antigen-presenting cells (APCs), such as macrophages and DCs. APCs release pro-inflammatory cytokines, such as IL-6, IL-12 and IL-23 as well as tumor necrosis factor-$\alpha$ (TNF-$\alpha$), with strong tumoricidal functions and recruit adaptive immune cells [270]. Moreover, APCs present antigens and provide costimulatory signals for activation of T, NK and B limphocytes, promoting the anti-tumor immunity. Killing of tumor cells is mainly accomplished by cytotoxic immune cells, such as NK cells and $CD8^+$ T cells or CTLs, which can induce apoptosis of target cells through a large range of mechanisms, including release of perforins and granzymes cytotoxix proteins, Fas ligand (FasL)-mediated apoptosis, and secretion of the pro-inflammatory signals interferon-$\gamma$ (IFN-$\gamma$) and TNF-$\alpha$ [271,272]. The role of B cells in TME is less understood, as there is experimental evidence supporting both the positive and negative effects of this cell type on cancer progression [273]. B lymphocytes contrast tumor progression by producing TAA-specific antibodies, that bind tumor cells and directly promote their eradication through antibody-dependent cell-mediated cytotoxicity (ADCC) involving also NK and T cells. Moreover, B lymphocytes can activate the innate immunity against cancer through antibody-mediated activation of the complement system [274]. B cells in the TME may also serve as APCs, promoting the activation of T cells against tumor cells, or secrete effector cytokines (e.g. IFN-γ), that contribute to the activation, survival and proliferation of tumor-infiltrating T and NK cells [275]. On the other side, pro-tumorigenic B cell effect is performed by immonoregulatory B or Breg cells that can suppress T and/or NK cell responses through the release of anti-inflammatory mediators, such as IL-10 and TGF-β, or through the expression of inhibitory molecules, including programmed cell death ligand 1 (PD-L1) and proteases, which directly impair T cell function [276]. In addition, $CD19^+$ $CD25^+$ Breg cells can promote tumor growth and metastasis by TGF-β—dependent conversion of non-Treg to Treg cells [277].

It is known that cancer cells can adopt a variety of strategies to escape anti-tumor immunity. To avoid immune-mediated elimination, cancer cells may lose their antigenicity through the immune selection of cancer cells that lack or mutate TAAs as well as through the acquisition of defects in antigen presentation [278]. Once escaped from initial elimination, tumor cells can modulate the function of immune system to promote tumor growth [279]. The role of different immune subsets in tumor biology is still under investigation, as some immune cells can exert both tumor-suppression and tumor-promotion functions, depending on the stage of the tumor. However, it is widely recognized that DCs, NK, T and B lymphocytes as well as activated macrophages perform an immuno-antagonizing function. On the contrary, myeloid-derived suppressor cells, tumor-associated macrophages, neutrophils, $CD4^+$ T helper limphocytes, regulatory DC (DCreg), Breg and Treg cells are key immunosuppressive cells that can help in promoting tumor progression [266,267,280]. Therefore, although immune effector cells initially act to inhibit tumor development, increasing evidence suggests that infiltrating immune cells may help to create an "immune privileged" site. Indeed, once tumor cells escape immunosurveillance, they can recruit the regulatory branch of immune cells in TME, which contributes to create a local suppression of tumor-specific immune responses [281].

As illustrated in Fig. 5, TDEVs have been reported to act both stimulating and suppressing immune response, thus exerting an anti-tumor effect

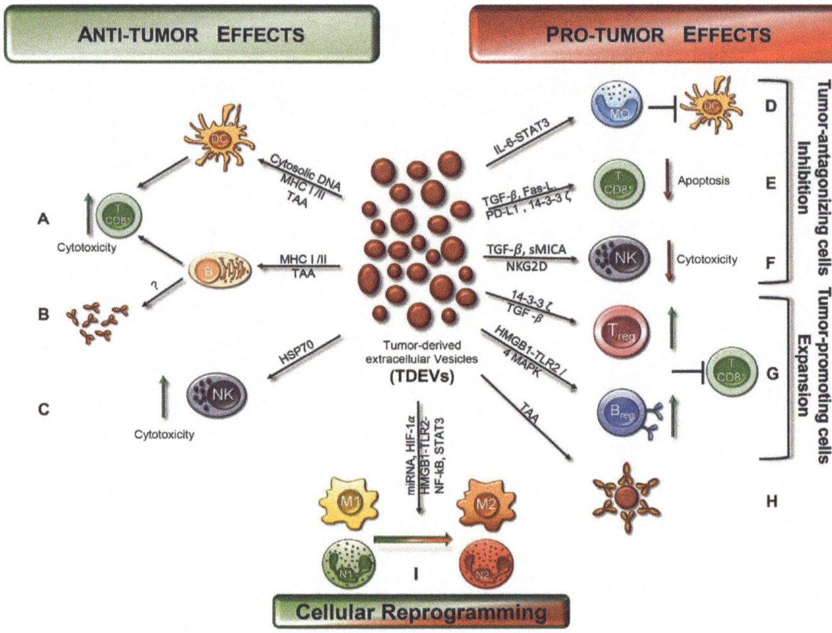


**Fig. 5** ***Effects of tumor-derived extracellular vesicles on the immune system.*** Tumor-derived extracellular vesicles (TDEVs) can exert opposite functions on the immune system, promoting both anti-tumor and pro-tumor responses. On the left is the anti-tumor response (A—C), that is achieved through the activation of dendritic cells (DCs), B cells and Natural Killer (NK) cells. (A) Activation of dendritic and B cells results in an enhanced stimulation of CD8$^+$ T lymphocytes, increasing their cytotoxicity against tumor cells. (B) Activation of B cells can also induce the production of anti-tumor antibodies. (C) Activation of NK cells induces a cytotoxic response. On the right is the pro-tumor response (D—I). It can be achieved through different mechanisms based on inhibition of tumor-antagonizing cells (D—F), expansion of T regulatory (Treg) and B regulatory (Breg) tumor-promoting cells, leading to inhibition of CD8$^+$ T cells (G) or by capture of antibodies direct against tumor cells (H). Inhibition of tumor-antagonizing cells is achieved by blocking of DC differentiation from monocytes (MO) (D) or by impairment of cytotoxic function of CD8$^+$ T lymphocytes (E) and NK (F) cells. (I) TDEVs can induce the reprogramming of macrophages and neutrophils, converting them from a pro-inflammatory polarization (M1/N1) to an anti-inflammatory polarization (M2/N2). The specific molecules involved in each process are indicated by arrows.

(Fig. 5A—C) or a pro-tumor effect (Fig. 5D—I). Tumor EVs can mediate their function being internalized by the cells, by direct contact between surface components of the vesicle and the target cell or by antibody decoy mechanisms. The anti-tumor activity of TDEVs is mainly promoted by vesicle internalization by APCs, such as DCs and B cells (Fig. 5A). TDEVs are enriched at the surface with tumor antigens that are taken up by DCs, increasing the antigen presentation to CTLs. Activated CTLs can recognize and induce apoptosis of cancer cells, enhancing the anti-tumor immune response [282]. An important role in stimulating anti-tumor response is played by DNA encapsulated in EVs, which can be uptaken by APCs. Once in the cytoplasm, cytosolic DNA activates two of the main components of the DNA sensor machinery that are the Stimulator of Interferon Genes (STING) transmembrane protein and cyclic GMP–AMP synthase (cGAS). Activation of STING pathway leads to the production of type I interferons and other cytokines, which contributes to effector T cell recruitment into the inflamed TME, resulting in induction of antitumor T cells response [283]. TAA-bearing TDEVs can also be uptaken by B cells [284], which expose tumor antigen on plasma membrane associated with Major Histocompatibility Complex Class I and II (MHC-I/MHC-II) molecules, making them accessible for the binding with T cells. Thus, B cells can function as APCs to trigger the activation of T lymphocytes, promoting anti-tumor cytotoxicity [285]. TDEVs are enriched with tumor-derived antigens, thus having the potential to directly bind the B cell receptors on B cell surface, triggering the production of anti-tumor antibodies (Fig. 5B). However, although the occurrence of autoantibodies have been reported for many types of cancer, it remain still unclear if EVs can contribute to humoral response [275]. In addition, TDEVs enriched in HSP70 heat shock protein or other specific tumor antigens can promote the migratory and cytolytic activity of NK cells (Fig. 5C) as well as TNF-α production by macrophages. Interestingly, the EV-mediated lytic activity of NK cells was lockable by HSP70-specific antibody, indicating that the activation mechanism lies on contact between surface components of the vesicle and the target cell [286].

However, the main effect exerted by TDEVs is the interference with the anti-tumor immunity, obtained through the inhibition of the activity of tumor-antagonizing immune cells (Fig. 5D–F) as well as through the expansion of tumor-promoting immune cells (Fig. 5G). The inhibition of immune response can be achieved through direct uptake by immune cells of TDEVs containing proteins or nucleic acids that interfere with cellular differentiation and/or activation of the recipient cells [287–289]. For instance, TDEVs inhibit the differentiation of human and murine myeloid precursors into DCs *in vitro* (Fig. 5D). DC precursors capture tumor EVs and subsequently induce the production of IL-6 and other cytokines, which leads to the activation of STAT3, inhibiting functional maturation of tumor-associated DCs [287]. Wang and coworkers demonstrated for the first time that the protein 14-3-3ζ, associated in HCC with increased proliferation, EMT and metastasis, can be also delivered to T lymphocytes through EVs, inducing the impairment of their anti-tumor activity (Fig. 5E). Moreover, the uptake of 14-3-3ζ deviated the differentiation of T cells from effective ones to regulatory [289]. TGF-β is an immunosuppressive cytokine found to play an important role in regulation of immune cells by TDEVs [290–292]. TGF-β can be expressed in a secreted or membrane-bound form and it is involved in the differentiation of Treg lymphocytes in TME [293]. In addition, Treg cells can also release TGF-β to induce immunosuppression, as this cytokine exerts anti-proliferative functions on multiple types of immune cells, including T and B lymphocytes and NK cells. Indeed, TGF-β can suppress T cell proliferation by inhibiting the production of IL-2, a lymphokine known to potently activate T and NK cells, together to other cells of the immune system [291]. TGF-β has been also described as a potent downregulator of the C-type lectin-like receptor that belongs to the NK group 2 (NKG2) of receptors as member D (NKG2D), an activating receptor expressed on virtually all NK cells [292]. TGF-β can be also encapsulated in tumor EVs. In hypoxic condition, NK lymphocytes can be inhibited by delivery in TDEVs of TGF-β that decreases the cell surface expression of the activated receptor NKG2D, thereby inhibiting NK cell functions [288] (Fig. 5F).

Moreover, TDVEs can induce immune suppression even without cell uptake, only by direct contact among surface molecules present on EVs and target cells. Clayton and coworkers showed that the membrane-bound form of TGF-β expressed on tumor EVs plays a double role in converting anti-tumor immune response toward a pro-tumor activity. On one hand, EVs bearing TGF-β impair proliferation in response to IL-2 of $CD8^+$ T cells and NK cells, exerting an antiproliferative effect on these cell subsets. On the other hand, they promote proliferation of $CD4^+$ $CD25^+$ Treg cells, enhancing their inhibitory functions on other immune cells (Fig. 5G) [290]. In HCC, tumor EVs express the high mobility group box 1 protein (HMGB1) on their surface and can promote the expansion of a particular Breg lymphocyte subset (Fig. 5G). HMGB1 is a DNA-binding nuclear protein, which can bind with high affinity to several receptors, such as Toll like receptor-2 and -4 (TLR-2 and TLR-4) on immune cells, where it can activate the MAPK pathway and proliferation. EV-mediated expansion of Breg cells results in an immunosuppressive microenvironment by secreting IL-10 and impairing $CD8^+$ T cell functions, thus providing favourable conditions for HCC progression [294]. Several studies reported that molecules expressed on the membrane of TDEVs are also able to directly bind receptors on immune cells and induce cellular death or impair their functions (Fig. 5E and F). Indeed, TDEVs are enriched in death ligands, such as Fas ligands and TNF-α, which can activate T cell membrane Fas receptor (FasR) and TNF receptor 1 (TNF-R1) respectively, inducing apoptosis of the cytotoxic T lymphocytes [295]. Another cell death ligand found on the surface of vesicles from plasma samples of patients with a variety of cancers is PD-L1, which can bind programmed cell death protein 1 (PD-1) on T cells, inducing dysfunction, exhaustion and neutralization of T lymphocytes in order to protect tumor cells from T cell-mediated cytotoxicity [296]. In melanoma, IFN-γ induces PD-L1 release in vesicles, which suppresses the function of $CD8^+$ T cells and facilitates tumor growth [297]. TDEVs can also express the MHC class I-related chain (MIC) A and B ligands able to bind NKG2D on NK cells. MICA ligands are usually upregulated in cancer cells, inducing NK activation that results in NK anti-tumor cytotoxic activity. Acting as a decoy, MICA-harbouring EVs secreted by several cancer types downregulate the *in vitro* NKG2D receptor-mediated cytotoxicity and thus impair NK cell function [298]. Finally, TDEVs can exert immunosuppression in a cell-independent manner, by acting as decoy for anti-tumor antibodies (Fig. 5H). Battke and coauthors demonstrated that TDEVs expressing TAAs can efficiently bind tumor-reactive antibodies, and dramatically reduce their binding to tumor cells. As a consequence, antibody sequestration reduces the antibody-dependent cytotoxicity by immune effector cells, such as NK and CTLs [299]. This new mechanism of tumor escape can have deep implications in immunotherapy.

Beside immune-suppression mechanisms, TDEVs can also reprogram immune cell phenotype, converting them from tumor-antagonizing immune cells into tumor-promoting immune cells (Fig. 5I) [300,301]. Several studies on a variety of cancer types demonstrated that TDEVs contain miRNAs, such as miR-2-3p miR-125b-5p, miR-1246, miR-103 and miR-146, that are able to reprogram macrophages from an anti-tumor activated M1 phenotype toward a pro-tumor M2 phenotype [301,302]. These miRNAs can induce a range of effects from macrophage activity inhibition to cell death induction [301]. For instance, in HCC, miR-23a-3p upregulates PD-L1 expression in macrophages via STAT3 signaling. PD-L1 binds PD-1 on T cells, inhibiting their anti-tumor cytotoxic effects [302]. The same molecular pathway is involved in the reprogramming of monocytes from glioblastoma-derived EVs, which contain phosphorylated STAT3. Once internalized in the cytoplasm, EVs change the morphology and actin cytoskeleton of the monocytes, triggering the initial steps of diapedesis, which is necessary for migration of these cells from the circulation into the TME. Thus, monocytes become polarized to M2 macrophage phenotype, upregulate PD-L1 and secrete cytokines that promote tumor growth [303]. In a study on epithelial ovarian carcinomas (EOC), the authors report that hypoxia increases the miR-21−3p, miR-125 b-5p and miR-181 d-5p levels in EOC-derived EVs, which are endocyted by non-polarized macrophages and promote the acquisition of a M2 phenotype via the induction of HIF-1α and HIF-2α. Finally, these macrophages promote EOC cell proliferation and migration [304]. A similar mechanism has been shown in a study on lung cancer cells, where authors show that miR-103 can be transferred from hypoxic cancer cells to macrophages, resulting in the enhancement of M2 polarization by the downregulation of PTEN [305]. Similarly, TDEVs can promote pro-tumorigenic N2 polarization of neutrophils (Fig. 5I). Indeed, Zhang and coauthors demonstrated that gastric cancer-derived EVs induce autophagy and pro-tumor activation in neutrophils through the HMGB1/TLR4/NF-kB signaling pathway, which finally promotes the proliferation and migration of gastric cancer cells [300].

The effect of TDEVs on immune system plays an important role also in invasion and metastasis dissemination. Fabbri and coworkers found that miR-21 and -29a released by tumor cells in EVs can bind as ligands to TLR-7 and TLR-8 on macrophages, leading to TLR-mediated NF-κB activation and secretion of pro-metastatic inflammatory cytokines such as TNF-α and IL-6 [306]. TDEVs may also prime the metastatic niche by modulating tumor-infiltrating lymphocytes to foster immune evasion and suppression within the metastatic microenvironment. Indeed, Liu and coauthors demonstrated that TDEVs enriched in small nuclear RNAs can be taken up by lung epithelial cells and induce chemokine secretion via activating TLR-3 receptor, promoting neutrophil recruitment in the lung pre-metastatic niche [307]. Once recruited in the niche, neutrophils can contribute to metastasis establishment exerting pro-tumor properties, such as induction of tumor angiogenesis, supplying of pro-tumor cytokines and promotion of immune evasion by immunosuppression of T cells [308]. Moreover, macrophages can be instigated to release cytokines that promote the establishment of a pro-inflammatory microenvironment that supports metastasis. Specifically, pancreatic tumor-derived vesicles loaded with macrophage inhibitory factor (MIF) are uptaken by Kupffer cells, the resident macrophages in the liver, and promote TGF-β secretion, which stimulates neighboring hepatic stellate cells to secrete fibronectin. Fibronectin subsequently promotes the recruitment of bone marrow-derived macrophages and neutrophils in the liver, completing the formation of the pre-metastatic niche [309]. Finally, it has been shown that TDEVs can also act modulating the dissemination of immune cells in body fluids. Pucci and colleagues showed that melanoma-derived EVs can mediate the impairment of subcapsular sinus $CD169^{+}$ macrophages, which form a physical layer that block TDEV dissemination in spleen and lymph nodes. The disruption of this barrier allows TDEV entrance in the lymph node cortex, where they interact with B cells and foster tumor progression by inducing autoantibodies production [310].

The effects of immune cell-derived EVs on cancer growth and progression are less investigated. As shown in Fig. 6, EVs from immune cells can

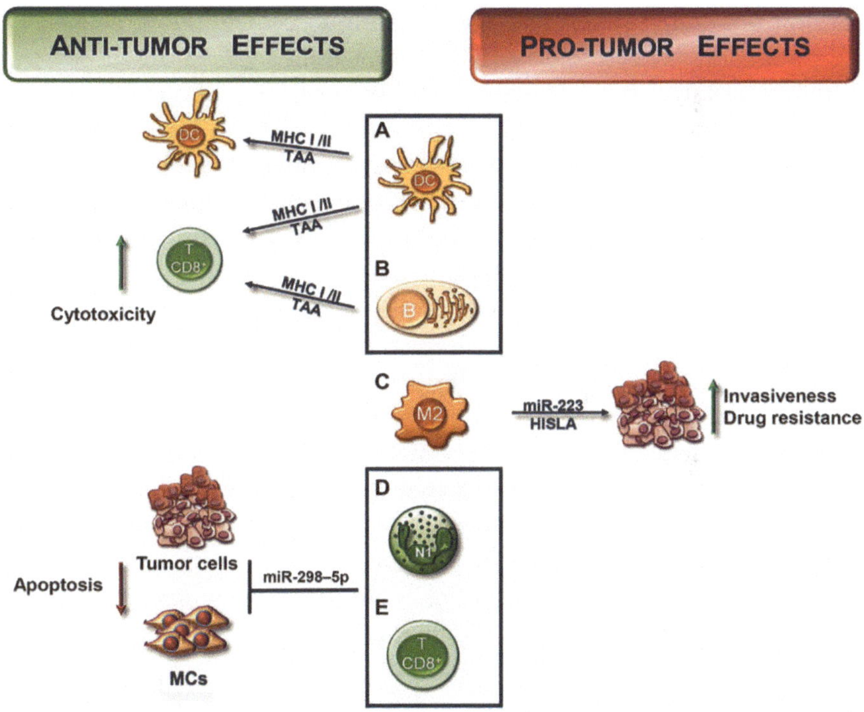


**Fig. 6** ***Effects of immune cell-derived extracellular vesicles on cancer growth.*** Extracellular vesicles (EVs) from immune cells can promote anti-tumoral (A, B, D, E) as well as pro-tumoral effects (C). Activated DCs (A) release EVs carrying the Major Histocompatibility Complex (MHC) class I/peptide complexes that target naive DCs for efficient $CD8^{+}$ T cell priming. Moreover, both DCs (A) and B cells (B) secrete EVs able to stimulate naive $CD8^{+}$ T cell proliferation and potentiate the cytotoxic response. EVs secreted by N1 neuthophils (D) and $CD8^{+}$ T cells (E) directly target both tumor and mesenchymal cells (MCs) in the tumor microenvironment, inducing their apoptosis. On the contrary, M2 macrophages (C) secrete EVs able to increase invasiveness and drug resistance in tumor cells. The specific molecules carried by EVs involved in each process are indicated by arrows.

promote both an anti-tumor (Fig. 6A and B, D, E) or pro-tumor response (Fig. 6C). The anti-tumor effects are achieved mainly by exchange of EVs between immune cells to boost cytotoxic response against tumoral cells (Fig. 6A and B). Activated DCs and B cells can secrete EVs able to *in vivo* stimulate naïve $CD8^+$ T cell proliferation and differentiation into cytotoxic T lymphocytes, inducing an efficient antitumor immunity [311]. Moreover, EVs could be mediators of antigenic exchange between DCs or APCs. Indeed, Andrè and coauthors showed that DCs release abundant MHC class I/peptide complexes transferred within EVs to other naive DC for efficient $CD8^+$ T cell priming [312]. EVs from B cells also contain the MHC I and II molecules, as well as costimulatory signals that activate T cells [313].

In addition, immune cell-derived EVs can directly target both tumoral and non tumoral cells in TME [314,315]. Indeed, neutrophils can release vesicles that display cytotoxic activity *in vitro* against several tumor cell lines as well as activated, but not resting, immune cells (Fig. 6D) [314]. Seo and colleagues demonstrated that EVs from activated $CD8^+$ T cells are able to induce depletion of tumoral mesenchymal cells, interrupting tumor invasion and metastasis. Apoptosis of tumoral mesenchymal cell populations is achieved through the delivery of miR-298—5p (Fig. 6E), however, the exact mechanism of action needs to be elucidated [315].

Finally, a tumor promoting effect of EVs through the direct uptake of vesicles by tumor cells has been described only for EVs released by macrophages (Fig. 6C). A study in 2011 first showed that IL-4—activated macrophages can promote the invasiveness of breast cancer cells through EV-mediated delivery of oncogenic miR-223 [316]. In addition, recently, Chen and colleagues have demonstrated that lncRNAs encapsulated in EVs can shuttle from tumor associated macrophages (TAMs) to reprogram breast cancer cell metabolism. In particular, a myeloid-specific lncRNA, HIF-1$\alpha$—stabilizing long non-coding RNA (HISLA) is released via EVs from TAMs to tumor cells to stabilize HIF-1$\alpha$ and to enhance aerobic glycolysis and apoptotic resistance [317].

It is now clear that interactions between cancer cells and immune cells change during different stages of tumorigenesis and rely on a fine equilibrium between mechanisms triggering immune-activation and immune-suppression [262]. Overall, the above data confirm that EVs contribute to the complexity of this crosstalk and highlight new mechanisms by which TME can modulate the immune response to support tumor growth and progression. Furthermore, the finding that EVs derived from immune cells may have a direct effect on tumor cells pave the way for the development of new strategies in cancer therapy.

# 5. EV-based clinical approaches against cancer

The emerging role played by EVs in modulating tumor biology raised the idea of their use in cancer clinical applications. The different strategies adopted have been summarized in Fig. 7.

## 5.1 EVs as biomarkers

EVs released by cells can actively travel in intercellular matrices and eventually reach the circulation, or can be released directly in biological fluids. EVs have been isolated from several biofluids, including plasma [318–321], urine [322,323], saliva [320,324,325], nasal secretions [326], bile [327], breast milk [320,328], aqueous humor [329], semen [330], amniotic fluid [331], cerebrospinal fluid [332], synovial fluid [333], pleural effusions [334] and bronchoalveolar lavage [335,336]. Biofluids contain highly heterogeneous EV populations, of variable size and originating by diverse mechanisms [13]. Notably, different EV types exhibit distinct molecular cargo and functional characteristics [12]. As reported above, the EV cargo comprises a repertoire of macromolecules (DNA, RNA, proteins and lipids) that is reflective of

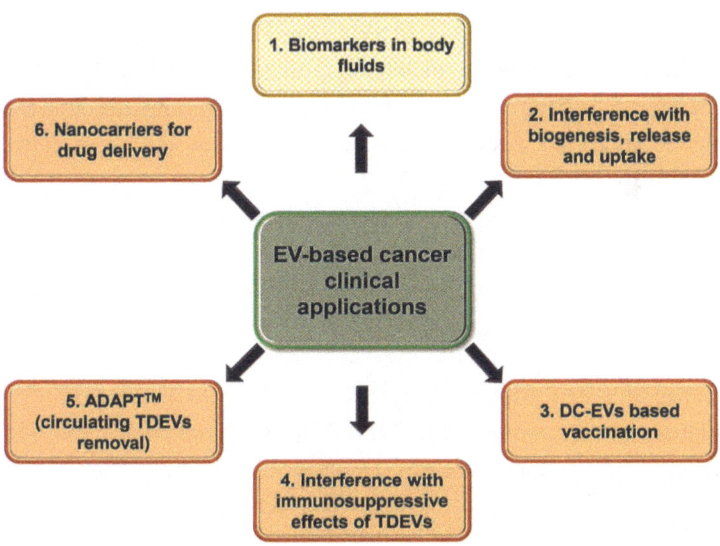


**Fig. 7** ***Extracellular vesicle-based approaches in cancer clinics.*** Summary of the main extracellular vesicle (EV)-based applications in cancer clinics in both diagnostics (top box) and therapeutics (all the other boxes). The specific applications are described in the text. Abbreviations: *DC,* dendritic cell; *TDEVs,* tumor-derived extracellular vesicles; *ADAPT,*™ adaptive dialysis-like affinity platform technology.

both the cell of origin and the type of EVs and is protected from enzymatic degradation by a lipid membrane envelop. The heterogeneity and stability of cargo makes possible the simultaneous profiling of different components and renders analysis of EVs in body fluids extremely informative for clinical studies [337]. The discovery of suitable biomarkers, tumor specific and easily detectable at earlier stages and through non-invasive methods, represents the gold standard in diagnostic cancer research. Besides studies on circulating tumor cells, cell-free oncogenic DNA and RNA, investigation on EVs is rapidly gaining recognition in cancer diagnosis. A large body of evidence indicates that cancer cells produce higher amount of EVs, involved in the creation of a microenvironment that favors tumor growth, EMT and metastasis formation [338]. Direct counting and characterization of EVs in liquid biopsies and/or profiling of their molecular cargo might have a great diagnostic and prognostic value for several types of cancers [339]. However, an important issue to deal with is the production of various types of EVs also by benign cells that makes the biomarker analysis complicated. Cancer-derived EVs float in a sea of physiologically normal EVs [340], therefore, discriminating TDEVs from the normal ones is a fundamental challenge for such type of studies. Moreover, body fluids contain also cells and non-EV particles such as lipoproteins, that is necessary to remove during EV purification to avoid misleading results. Due to the high concentration of soluble proteins and lipoproteins, blood, plasma and serum have a high viscosity and density, which hampers the concentration, isolation and detection of EVs, rendering necessary to develop new protocols that reduce contamination and favor following EV analysis [318].

Even with these limitations, in the last decade, circulating vesicles have been found to be a very promising source of cancer-associated molecules as candidate biomarkers for many type of tumors, such as glioblastoma, melanoma, prostate, pancreatic, colorectal, breast and lung cancers [53,341,342]. Among best potential cancer EV biomarkers there are: (i) the PD-L1 ligand, able to mediate T cell immunosuppression, for gliobastoma [343], head and neck squamous cell carcinoma (HNSCC) [344], melanoma [345,346] and non-small cell lung carcinoma (NSCLC) [346]; (ii) the developmental endothelial locus-1 (DEL-1) protein, promoting adhesion of endothelial cells and angiogenesis in breast cancer [347]; (iii) the macrophage migration inhibitor factor MIF as prognostic factor for pancreatic ductal adenocarcinoma (PDAC) metastasis [348] and (iv) mutated forms of the EGFR tyrosin kinase in glioblastoma [62,105] and lung carcinoma [349]. An interesting history regards the membrane-anchored proteoglycan Glypican 1 (GPC1), that was first reported in a highly cited paper as a very promising EV-associated biomarker for early detection of PDAC [152]. Subsequent studies using different methodological approaches to detect GPC1, such as liquid chromatography-tandem mass spectrometry [350] and ELISA [351] instead of flow cytometry, did not confirm previous results. However, two more recent studies closing resembling the original study methodology, produced also findings in agreement with that study [352,353]. All these analyses focused on GPC1 in pancreatic cancer, but with different study designs, sample sizes, and methodologies. The lack of a standard analytical approach makes it extremely challenging to directly compare the results of different studies, and illustrates the fundamental issue in the development of EV biomarkers that is the need of standardization and replication studies [354]. Currently, clinical trials (NCT03032913 and (NCT03410030) are ongoing to evaluate GPC1 performance as a clinical biomarker for PDAC. Other clinical trials focus on the feasibility of detecting *EGFR* mutation (NCT03236675) or *PD-L1* mRNA (NCT02890849) in plasma EVs isolated from patients with NSCLC [354].

Several studies identified in circulating EVs from cancer patients the presence of oncogenic mRNAs and more recently double-stranded DNA fragments, carrying cancer driver mutations (e.g. TP53, K-RAS, EGFRvIII) [105,116,152,355–358] as well as neoantigens due to genomic rearrangements insertions, deletions, inversions and other mutational signatures [359]. Also lncRNAs, such as HOX transcript antisense RNA 1 (HOTAIR1) and metastasis associated lung adenocarcinoma transcript 1 (MALAT1) that are specifically associated to cervical cancer, have been found significantly enriched in EVs from cervicovaginal lavages of cervical cancer patients [360]. Oncogenic EV-associated RNAs and DNA offer an extraordinary level of cancer specificity and their detection could be very helpful in discriminating TDEVs from other vesicles and in subsequent quantitative analysis [53].

The protein and nucleic acid content of EVs has been thoroughly studied in an attempt to identify more and more diagnostic and prognostic biomarkers [361]. Over the recent years, the extensive miRNAs profiling studies demonstrated significant differences in miRNA levels in cancer cells respect to the normal ones. Elevated levels of oncogenic miRNAs, such as miR21, miR155 and the miR17-92 cluster were found in circulating EVs from cancer patients. As an example, miR21 was found enriched in EVs from urine samples of bladder cancer patients, in serum of patients affected by colorectal or ovarian cancer and in the cerebrospinal fluid of glioma patients. Other EV-miRNAs associated with multiple cancer types include miR155 and miR1246 [362]. Unlike proteins, microRNAs ca be readily amplified and coupled with high-throughput multiplexed RNA profiling. The identification of an EV-miRNA signature might be really promising in cancer diagnosis and prognosis and also be associated to other biomarker detection [43,53,350,363]. miRNA expression is also evaluated as prognostic and predictive marker in one study (NCT02514681) on pertuzumab retreatment in previously treated HER2-positive metastatic advanced breast cancer, even though the study design does not specify which miRNAs are analyzed [354]. The diagnostic potential of phosphoproteins in circulating EVs from breast cancer patients has also been reported [364], as well as exosome surface protein analyses in PDAC [365].

There are also a few studies that have analyzed the EV lipid profile to get insights in the lipid signature associated with specific diseases. EVs contribute with their lipid molecules or their lipid-related enzymes to several pathophysiologies, including inflammation, atherogenesis and tumor development [366]. Tumorigenesis induces dramatic changes in EV lipid composition, suggesting that increasing our knowledge of the EV lipidome may lead to improved diagnostic and prognostic cancer biomarkers [100,366]. High-throughput lipidomic studies on urinary samples from prostate cancer patients and healthy volunteers revealed significant differences in the lipid composition of isolated EVs [367,368]. In one study on 15 patients, the largest differences were observed for lactosylceramide, which was highest in the patient group; however, no significant differences in lipid classes between cancer patients and healthy controls were found [367]. Another study, using different methodologies and analyzing less patients, revealed an increase in phopspatidylglycerol and triacylglycerol in prostate cancer patients [368]. More recently, other lipidomic studies on EVs obtained from other biologic fluids, such as plasma and serum, also reported a differential lipid distribution in cancer patients respect to healthy controls, and in some cases a significant correlation with tumor stage and overall patient survival [100,369]. More mass spectrometry analyses will allow to further explore the potential use of EV lipids as biomarkers. Moreover, it is important to have guidelines to minimize the impact of pre-analytical variables that can influence lipid determinations as well as standardized and well described methodological approaches to assure reliability and reproducibility of the data [100,370].

In summary, to exploit EVs in body fluids as novel biomarkers in cancer, large scale, rapid, high-throughput, and reproducible methods for their isolation, detection, and characterization are required [338]. Moreover, a lot of effort is being made in improving the methodologies to selectively purify single EV populations based on their subcellular origin, size, cargo, and donor cell type and status [337,371]. Because most if not all studies on EVs are single-centre studies, their clinical relevance remains limited. The need to improve standardization and reproducibility of EV research is urgent [372]. ISEV published position papers detailing "minimal experimental requirements for definition of extracellular vesicles and their functions" [55,57]. The ISEV board highlighted the need to consider all the issues related to purification and characterization of EVs when making strong conclusions on their involvement in any physiological or pathological situation, or when proposing EVs or their molecular cargo as biological markers [55]. Accordingly, ISEV organized a biomarker workshop in Birmingham (UK) in November 2017, and during that workshop several working groups were created to focus on a particular body fluid. The first report of the blood EV work group was published in 2018 and is based on responses by work group members to a questionnaire in order to discover the contours of a roadmap. Since blood is such a complex body fluid, the report might also be valuable for exploring body fluids other than blood [372].

In conclusion, the identification of EV biomarkers is a very promising as well as challenging objective. The first challenge consists in the identification of common standard procedures for the isolation and characterization of EVs from liquid biopsies as well as from preclinical studies, so to guarantee the reliability and reproducibility of the data. The second one is the development of a multicomponent, combinatorial approach, that considers all the EV molecules specifically associated to cancer, so to enhance the specificity and the sensitivity of an EV-based diagnostics.

## 5.2 EVs in cancer therapy

### *5.2.1 Interference with TDEV biogenesis, release or uptake*

As vesicle production and uptake is usually increased in cancer cells, one of the first lines of research to be explored was the use of drugs able to interfere with biogenesis, release or uptake of TDEVs, so to inhibit the EV-mediated communication inside the tumor and its microenvironment as well as at distant sites [63]. Different drugs have been found to reduce EV release, by interfering with their biogenesis mechanisms, through specific or more general mechanisms. Inhibitors of endocytosis pathway at different levels can affect secretion of EVs and desirably contrast their pro-tumorigenic effect on target cells. It has been shown that both breast cancer cells and HeLa cervical carcinoma cells treated with Y-27632, an inhibitor of RhoA/ROCK-dependent signaling pathway, produce a lower number of MVBs and late endosomes [72]. Vesicle secretion through the endocytic pathway can be also induced by increased concentration of intracellular calcium. Dimethyl amiloride, an inhibitor of the $H^{+}/Na^{+}$ and $Na^{+}/Ca^{2+}$ exchangers, that counteracts calcium increase, was indeed found to inhibit vesicle secretion [373]. Moreover, transfer of EV-associated domains into the lumen of the endosome requires the sphingolipid ceramide [374]. GW4869 sphingomyelinase-2 inhibitor, interfering with ceramide production, is able to reduce EV secretion in tumor cells and, interestingly, this inhibition has a great effect *in vivo*, by affecting tumor growth, lowering also the number of lung metastases in tumor-bearing mice [306]. Other studies attempted to inhibit vesiculation, by interfering with different steps of vesicular body processing. In particular, Rab family was extensively investigated: RAB27A and RAB27B small GTPases regulate vesicle exocytosis. *RAB27A* knockdown via RNAi results in reduced tumor growth and decreased lung dissemination, affecting EV-mediated neutrophil mobilization needed to metastasis establishment [375].

Development of therapeutic approaches that interfere with TDEV uptake by recipient cells can be even more challenging. EV internalization can occur through multiple routes, moreover molecules or pathways involved in the regulation of these mechanisms are only partially identified [51]. Adhesion molecules seem to play a pivotal role in regulating recipient cell interactions with EVs. Annexin V, which blocks phosphatidylserine exposed on the EV surface can abrogate the uptake of EVs by U373 malignant glioma cell line [62]. In addition, treatment with heparin, a competitive inhibitor of cell surface receptors dependent on heparan sulfate proteoglycans (HSPG) coreceptors, decreases EV uptake in bladder cancer cells [376]. However, interfering with molecular pathways related to both intracellular and extracellular vesicular trafficking, underlying several responses of the cells, could produce unknown and uncontrolled sides effects. Therefore, converting these findings in therapeutic clinical approaches is extremely complex and challenging.

### 5.2.2 Immunotherapy

The discovery that EVs can effectively modulate the immune response has paved the way for their use in cancer immunotherapy. Over the past 20 years, much interest has also been placed on the development of vesicle-based cell-free vaccines as alternative approach to cell therapy for suppressing tumor growth. The major therapeutic approach investigated relies on the use of EVs derived from activated DCs, considered the most immunostimulatory vesicles released by the immune system. DC-derived EVs express MHC-I, MHC-II and costimulatory molecules able to induce the activation of a variety of immune cells, including T and B lymphocytes and NK cells both *in vitro* and *in vivo* [377]. To specifically direct the immune response toward cancer cells, DCs can be cultured in presence of TAAs or TDEVs. Vaccination studies performed with DC-derived EVs, in which DCs were pulsed with tumor peptides or with EVs derived from cancer cells, have shown that EVs can trigger a potent anti-tumor response in a model of murine breast cancer as well as leukemic cells [378,379]. Based on this idea, immature DC-derived exosomes have been applied in clinical trials for melanoma [380] and NSCLC [381]. Both studies used EVs purified from patient monocyte-derived DCs that had been pulsed with antigenic peptides expressed by the patient cancer. Despite several studies have confirmed the safety of DC-derived EV administration, the results have shown a poor clinical efficacy and a weak specific immune response elicited by this technique. To advance T cell stimulation, EVs from mature DCs, activated by IFN-$\gamma$, were used in a new phase II clinical trial for NSCLC [382]. Again the treatment did not induce detectable $CD4^{+}$ or $CD8^{+}$ adaptive T cell responses in the treated patients, even though in some of them a boost in the NK cell antitumor immunity was reported [382]. EVs can also be modified to interfere with the immune checkpoint blockade exploited by cancer cells to evade more efficiently immune response. For instance, EVs have been engineered to present PD-1 receptors on their membranes and loaded with Indoleamine 2,3-dioxygenase (IDO), an immunosuppressive molecule overexpressed by both tumor and dendritic cells to limit the proliferation and function of effector T cells. These vesicles are able to bind the PD-L1 on cancer cells and simultaneously block the PD-1/PD-L1 binding, overcoming the inhibitory effects of IDO on effector T cells within the TME [383]. An interesting genetically manipulation of EV membrane has been performed by Chen and coworkers [384]. The authors constructed the synthetic multivalent antibodies retargeted exosomes (SMART-Exos), which express two distinct types of antibodies on their surface: the monoclonal antibodies specific for T cell CD3 co-receptor and for cancer-associated EGFR. These vesicles elicit a potent antitumor immunity *in vitro* and *in vivo*, inducing the cross-linking of T cells and EGFR-expressing breast cancer cells. This strategy allows the recruitment and activation of endogenous cytotoxic effector cells toward target cells for destruction [384].

Finally, an interesting strategy was developed to specifically remove TDEVs from the entire circulatory system. Using a therapeutic hemofiltration approach called ADAPT™ (adaptive dialysis-like affinity platform technology), TDEVs are captured on an antibody-coated matrix during extracorporeal dialysis [385].

### 5.2.3 EVs as nanocarriers for drug delivery

EVs do possess important characteristics that render them extremely suitable for the delivery of therapeutic molecules and very competitive respect to synthetic nanoparticles. First of all, they are well tolerated and do not elicit immune response in the host, being able, on the contrary, to exert an immunosuppressive effect, as also described in the previous paragraph. The biocompatibility was first described in mice [386,387] and then also in patients with graft versus host diseases, where vesicles derived from MSCs did not cause significant side effects and resulted in a better patient response than MSCs themselves [388,389]. Moreover, EVs display long-circulating half-life, protecting their content from degradation, the ability to target specific types of cells and also to cross the blood-brain barrier [390]. Native and drug-loaded mammalian cell-derived EVs have recently been developed and are contributing to the expanding research field known as "cell-free therapy". The finding of pro-regenerative effects of MSC-derived EVs in several pathological conditions [389,391,392] has led to explore their use in clinical trials, especially to treat chronic diseases [393], immunity diseases (NCT02138331; NCT03437759), and acute ischemic stroke (NCT03384433), for which the reports are not still available. The only phase I clinical trials involving MSC-derived vesicles applied in cancer started in January 2020 (NCT03608631). In this study, aimed at evaluating the best dose and possible side effects, MSC-derived exosomes that encapsulate K-RASG12D siRNA (iExosomes) will be used to treat patients with metastatic pancreatic cancer carrying K-RASG12D mutation [43].

The simplest method of drug loading is incubation of hydrophobic drugs with EVs. This strategy has been used by Sun and colleagues [394] to entrap curcumin in EVs. Through the hydrophobic interaction between the hydrophobic tails and hydrophobic drug, curcumin can be self-assembled into the lipid bilayer of vesicles and this may protect curcumin from degradation. The encapsulation of curcumin into EVs can increase solubility, stability, and bioavailability of curcumin [394]. However, this method shows an inefficient loading capacity and can be used mainly for hydrophobic compounds. Thus, the most employed approach to achieve the effective internalization of molecules in EVs has been electroporation. This strategy has been used for the encapsulation of the chemotherapeutic agent doxorubicin, commonly used to treat a broad range of cancers. The study has shown that vesicular doxorubicin is taken up faster into recipient cells than free or liposomal doxorubicin, inducing mitochondrial swelling followed by apoptotic cell death within hours of cell treatment. Interestingly, in contrast to other delivery methods for doxorubicin, EVs do not accumulate in the heart, thereby providing potential for limiting the cardiac side effects [395]. An alternative strategy to improve therapeutic functions of EVs involves the bioengineering of vesicles in order to regulate the expression of specific molecules on the surface or inside EV lumen. Morishita and colleagues genetically modified murine melanoma B16BL6 cells to produce EVs containing endogenous tumor antigens (gp100 and TRP2) and immunostimulatory CpG DNA. These EVs exhibited stronger *in vivo* antitumor effects in B16BL6 tumor-bearing mice than simple co-administration of exosomes and CpG DNA [396]. Moreover, the most employed method to enhance cargo protein loading on EV surface is the genetic fusion of cargo protein with the additional domain of protein that natively localizes into EV membrane [397,398]. Bioengineering has been also used to regulate the specificity and the efficiency in targeting recipient cells. For instance, expression of the gp350 peptide in the human embryonic kidney HEK293T cells leads to its incorporation into EVs, which B cells internalize more effectively [399]. In another study, specific targeting was achieved by engineering DCs to produce Lamp2b, an EV membrane protein, fused to the neuron-specific rabies virus glycoprotein (RVG) peptide. The recombinant protein promotes EV transport across the blood-brain barrier and release of functional cargo (siRNA) in the brain [400].

Translational applications are currently hampered by the need of biosources able to release great amounts of EVs as well as by the lack of suitable processes for their isolation, characterization and functionalization. The finding that vesicles are not bound to intraspecies interactions, but are also capable of interkingdom communication, has once again revolutionized the EV field. Among the natural sources alternative to mammalian cells that have been currently investigated there are plants, bovine milk, bacteria, fungus and parasites, that could constitute a more economically viable and sustainable source of EVs [401,402]. The scientific community is strongly committed in the set-up, standardization and validation of protocols for the efficient production, analysis and functionalization of vesicles. Definition of common Good Manufacturing Practices (GMPs) is also needed in order to improve reproducibility in the quantity and quality of EV preparations to be used in clinical trials. In this regard, the emerging role of plant-derived vesicles should be mentioned. To date, three studies involving plants, namely grape (NCT01668849), ginger or aloe (NCT03493984) and unspecified plants (NCT01294072) have been registered in clinical trials for the treatment of radiation- and chemotherapy-induced oral mucositis, polycystic ovary syndrome and colon cancer, respectively. However, all of them are still in the early stage [403].

## 6. Concluding remarks

Even though oncogenic genetic and epigenetic events are regarded as the ultimate drivers of tumor development and establishment, their role in non-cell autonomous processes emerge as an important complementary factor to their extensively studied cell autonomous functions. In the last decade, increasing evidence pointed to EV-mediated cancer cell communication as a key aspect of this complexity.

As discussed along the chapter, oncogenic transformation impacts on vesiculation, in terms of both vesicle number and content. Cancer EVs are able to mediate communication inside tumor mass, within its microenvironment and in a systemic way, up to reach distant organs where a pre-metastatic niche can be induced. TDEVs possess a plethora of functions, being able to induce conversion of non-stem cancer cells into CSCs, to activate fibroblasts, to induce angiogenesis, EMT, metastasis formation and immune response suppression. Cancer EVs are also involved in the development of undesirable chemoresistance. A further level of complexity is introduced by the documented ability of TME cells, such as stromal, endothelial and immune cells, to release vesicles that can influence cancer cell phenotype. This bilateral communication is fundamental for tumor progression and invasion. However, some interesting aspects still to be elucidated are: (i) the persistence of the EV-induced cellular effects, in spite of molecular degradation and dilution with cell division, and the underlying mechanisms involved; (ii) the *in vivo* relevance of EV-mediated effects in multicellular organisms, considering that most studies have been conducted in cell culture and using supra-physiological number of vesicles.

EV-based cell communication has become an extremely intriguing mechanism that attracted a lot of scientists for its great potential in basic research, to better understand tumor biology and its interaction with immune system. Furthermore, applied research widely recognized in EVs a promising tool for cancer diagnostics and therapeutics. As described in this chapter, cancer EVs are enriched in specific cargo with many different tumorigenic properties, are stable and potentially detectable in liquid biopsies, as they are expected to be at higher concentration than circulating tumor cells or free nucleic acids. Due to these features, cancer EVs might provide extremely powerful biomarkers to be used in cancer diagnosis, prognosis and treatment. Moreover, given that all EVs are very stable, extremely well tolerated in the body, able to target specific cells and to cross species boundaries, they definitely represent a valid alternative to synthetic nanocarriers for drug delivery and for the development of new pioneering therapeutics approaches.

As most emerging fields, EV studies have also to face important challenges. First of all, the standardization of the procedures of EV isolation, characterization, functionalization up to large-scale production, that is fundamental to ensure reliability, reproducibility and exploitation of the huge amount of data that have been collecting. EVs are very heterogenous in size, content, biochemical/biophysical and functional properties. Standardization and sharing of common validated procedures are fundamental to deal with all this heterogeneity and univocally unravel EV specific molecular and functional signatures. In the last years, a great effort has been made by the international scientific community to implement and standardize methodologies and technologies for the collection, isolation/purification, characterization and functional analysis of EVs. In 2012 ISEV was founded and provided researchers with minimal information for studies of EVs that is continually revised and updated. All the issues regarding the isolation and characterization of EVs have to be considered with a special care before making strong conclusions on their functional role in specific pathologies or their use as providers of biological markers. Nowadays, several national societies for studying EVs are also being formed, among them the Italian Society, named EVIta, was founded and organized its first scientific meeting in 2019. Dedicated EV databases, such as Vesiclepedia, EVpedia and exRNA Atlas, were also created to collect specific data on EV cargo content and related features as well as on used procedures. More recently, the EV-TRACK knowledgebase [404] has been developed for recording the experimental parameters of EV-related studies. The European Innovation Council, also, through the Horizon 2020-Featuring and Emerging Technologies (FET) Open Work program has been strongly investing on EV research, funding five projects focused on the development and integration of novel technological breaktroughs in EV field opening to new promising clinical approaches. The projects are INDEX (www.indexproject.eu), VES4US (ves4us.eu), evFOUNDRY (www.evfoundry.eu), GLADIATOR (www.fet-gladiator.eu) and MINDGAP (www.biomark.isep.ipp.pt./mindgap), the last two specifically focusing on applications in cancer therapy. All five projects started an active networking, trough clustering events and the elaboration on an EVmanifesto aiming at highlighting urgency of introducing and securing more firmly the EV-related theme as a cornerstone in the research agenda in European Union. The actively emerging knowledge on EVs might have a strong impact on the biomedical landscape of the future. The development of multidisciplinary and highly innovative approaches aiming at integrating the expertize and knowledge matured in both tumor biology and EV field will be fundamental to deliver highly effective clinical tools against cancer.

## Acknowledgements

The authors apologize to researchers whose original studies were not cited due to space limitations.

The authors acknowledge Mariarosaria Aletta and Sergio D'Antonio for bibliographic support and support by European Union's Horizon 2020 research and innovation program (Grant no. 801338) to GLL and by PO FESR 2014–2020 "SATIN" from Regione Campania to GLL and MPS.

## References

[1] D.F. Quail, J.A. Joyce, Microenvironmental regulation of tumor progression and metastasis, Nat. Med. (2013), https://doi.org/10.1038/nm.3394.

[2] S.K. Gopal, D.W. Greening, A. Rai, M. Chen, R. Xu, A. Shafiq, R.A. Mathias, H.J. Zhu, R.J. Simpson, Extracellular vesicles: their role in cancer biology and epithelial-mesenchymal transition, Biochem. J. (2017), https://doi.org/10.1042/BCJ20160006.

[3] J. Maia, S. Caja, M.C. Strano Moraes, N. Couto, B. Costa-Silva, Exosome-based cell-cell communication in the tumor microenvironment, Front. Cell Dev. Biol. 6 (2018) 1–19.

[4] G. Bazzoni, E. Dejana, Endothelial cell-to-cell junctions: molecular organization and role in vascular homeostasis, Physiol. Rev. (2004), https://doi.org/10.1152/physrev.00035.2003.

[5] B. Mattes, S. Scholpp, Emerging role of contact-mediated cell communication in tissue development and diseases, Histochem. Cell Biol. (2018), https://doi.org/10.1007/s00418-018-1732-3.

[6] F.A. Ramírez-Weber, T.B. Kornberg, Cytonemes: cellular processes that project to the principal signaling center in Drosophila imaginal discs, Cell (1999), https://doi.org/10.1016/S0092-8674(00)80771-0.

[7] T.B. Kornberg, S. Roy, Cytonemes as specialized signaling filopodia, Development (2014), https://doi.org/10.1242/dev.086223.

[8] A. Rustom, The missing link: does tunnelling nanotube-based supercellularity provide a new understanding of chronic and lifestyle diseases? Open Biol (2016), https://doi.org/10.1098/rsob.160057.
[9] I. Parolini, C. Federici, C. Raggi, L. Lugini, S. Palleschi, A. De Milito, C. Coscia, E. Iessi, M. Logozzi, A. Molinari, M. Colone, M. Tatti, M. Sargiacomo, S. Fais, Microenvironmental pH is a key factor for exosome traffic in tumor cells, J. Biol. Chem. (2009), https://doi.org/10.1074/jbc.M109.041152.
[10] P. Kucharzewska, M. Belting, Emerging roles of extracellular vesicles in the adaptive response of tumour cells to microenvironmental stress, J. Extracell. Vesicles (2013), https://doi.org/10.3402/jev.v2i0.20304.
[11] P. Kucharzewska, H.C. Christianson, M. Belting, Global profiling of metabolic adaptation to hypoxic stress in human glioblastoma cells, PLoS One (2015), https://doi.org/10.1371/journal.pone.0116740.
[12] V.R. Minciacchi, M.R. Freeman, D. Di Vizio, Extracellular vesicles in cancer: exosomes, microvesicles and the emerging role of large oncosomes, Semin. Cell Dev. Biol. (2015), https://doi.org/10.1016/j.semcdb.2015.02.010.
[13] K.W. Witwer, E.I. Buzás, L.T. Bemis, A. Bora, C. Lässer, J. Lötvall, E.N. Nolte-'t Hoen, M.G. Piper, S. Sivaraman, J. Skog, C. Théry, M.H. Wauben, F. Hochberg, Standardization of sample collection, isolation and analysis methods in extracellular vesicle research, J. Extracell. Vesicles (2013), https://doi.org/10.3402/jev.v2i0.20360.
[14] D. Charles, The Variation of Animals and Plants Under Domestication, 2010.
[15] L. Margolis, Y. Sadovsky, The biology of extracellular vesicles: the known unknowns, PLoS Biol. (2019), https://doi.org/10.1371/journal.pbio.3000363.
[16] H.C. Anderson, Vesicles associated with calcification in the matrix of epiphyseal cartilage, J. Cell Biol. (1969), https://doi.org/10.1083/jcb.41.1.59.
[17] N. Crawford, The presence of contractile proteins in platelet microparticles isolated from human and animal platelet-free plasma, Br. J. Haematol. (1971), https://doi.org/10.1111/j.1365-2141.1971.tb03416.x.
[18] B. Stegmayr, G. Ronquist, Promotive effect on human sperm progressive motility by prostasomes, Urol. Res. (1982), https://doi.org/10.1007/BF00255932.
[19] E.G. Trams, C.J. Lauter, J. Norman Salem, U. Heine, Exfoliation of membrane ecto-enzymes in the form of micro-vesicles, BBA - Biomembr (1981), https://doi.org/10.1016/0005-2736(81)90512-5.
[20] H.F. Dvorak, S.C. Quay, N.S. Orenstein, A.M. Dvorak, P. Hahn, A.M. Bitzer, A.C. Carvalho, Tumor shedding and coagulation, Science (1981), https://doi.org/10.1126/science.7195067.
[21] J.N. George, L.L. Thoi, L.M. McManus, T.A. Reimann, Isolation of human platelet membrane microparticles from plasma and serum, Blood (1982), https://doi.org/10.1182/blood.v60.4.834.bloodjournal604834.
[22] N. Satta, F. Toti, O. Feugeas, A. Bohbot, J. Dachary-Prigent, V. Eschwège, H. Hedman, J.M. Freyssinet, Monocyte vesiculation is a possible mechanism for dissemination of membrane-associated procoagulant activities and adhesion molecules after stimulation by lipopolysaccharide, J. Immunol. 153 (7) (1994) 3245–3255.
[23] P.J. Sims, T. Wiedmer, C.T. Esmon, H.J. Weiss, S.J. Shattil, Assembly of the platelet prothrombinase complex is linked to vesiculation of the platelet plasma membrane. Studies in Scott syndrome: an isolated defect in platelet procoagulant activity, J. Biol. Chem. 264 (29) (1989) 17049–17057.
[24] C. Hess, S. Sadallah, A. Hefti, R. Landmann, J.A. Schifferdi, Ectosomes released by human neutrophils are specialized functional units, Mol. Immunol. (1998), https://doi.org/10.1016/s0161-5890(98)90627-7.

[25] C. Harding, J. Heuser, P. Stahl, Receptor-mediated endocytosis of transferrin and recycling of the transferrin receptor in rat reticulocytes, J. Cell Biol. (1983), https://doi.org/10.1083/jcb.97.2.329.
[26] B.T. Pan, R.M. Johnstone, Fate of the transferrin receptor during maturation of sheep reticulocytes in vitro: selective externalization of the receptor, Cell (1983), https://doi.org/10.1016/0092-8674(83)90040-5.
[27] R.M. Johnstone, M. Adam, J.R. Hammonds, C. Turbide, Vesicle formation during reticulocyte maturation, J. Biol. Chem. 262 (19) (1987) 9412–9420.
[28] M. Colombo, G. Raposo, C. Théry, Biogenesis, secretion, and intercellular interactions of exosomes and other extracellular vesicles, Annu. Rev. Cell Dev. Biol. 30 (2014) 255–289.
[29] E. Cocucci, J. Meldolesi, J. Lu, A.G. Clark, L.J. Luense, M.Z. Carletti, L.K. Christenson, S. Hilz, A.J. Modzelewski, P.E. Cohen, A. Grimson, H.J.G.G. Raposo, H.W. Nijman, W. Stoorvogel, R. Liejendekker, C.V. Harding, C.J. Melief, L. Wang, C. Xu, B lymphocytes secrete antigen-presenting vesicles, J. Exp. Med. (2015), https://doi.org/10.1101/gr.132514.111.22.
[30] L. Zitvogel, A. Regnault, A. Lozier, J. Wolfers, C. Flament, D. Tenza, P. Ricciardi-Castagnoli, G. Raposo, S. Amigorena, Eradication of established murine tumors using a novel cell-free vaccine: dendritic cell-derived exosomes, Nat. Med. (1998), https://doi.org/10.1038/nm0598-594.
[31] M. Mesri, D.C. Altieri, Endothelial cell activation by leukocyte microparticles, J. Immunol. 161 (8) (1998) 4382–4387.
[32] O. Gasser, J.A. Schifferli, Activated polymorphonuclear neutrophils disseminate anti-inflammatory microparticles by ectocytosis, Blood (2004), https://doi.org/10.1182/blood-2004-01-0361.
[33] M. Baj-Krzyworzeka, M. Majka, D. Pratico, J. Ratajczak, G. Vilaire, J. Kijowski, R. Reca, A. Janowska-Wieczorek, M.Z. Ratajczak, Platelet-derived microparticles stimulate proliferation, survival, adhesion, and chemotaxis of hematopoietic cells, Exp. Hematol. (2002), https://doi.org/10.1016/S0301-472X(02)00791-9.
[34] A. Bobrie, M. Colombo, G. Raposo, C. Théry, Exosome secretion: molecular mechanisms and roles in immune responses, Traffic (2011), https://doi.org/10.1111/j.1600-0854.2011.01225.x.
[35] N. Chaput, C. Théry, Exosomes: immune properties and potential clinical implementations, Semin. Immunopathol. (2011), https://doi.org/10.1007/s00281-010-0233-9.
[36] C. Simon, D.W. Greening, D. Bolumar, N. Balaguer, L.A. Salamonsen, F. Vilella, Extracellular vesicles in human reproduction in health and disease, Endocr. Rev. (2018), https://doi.org/10.1210/er.2017-00229.
[37] I.M. Saadeldin, H.J. Oh, B.C. Lee, Embryonic–maternal cross-talk via exosomes: potential implications, Stem Cell. Clon Adv. Appl. (2015), https://doi.org/10.2147/SCCAA.S84991.
[38] S. Atay, C. Gercel-Taylor, M. Kesimer, D.D. Taylor, Morphologic and proteomic characterization of exosomes released by cultured extravillous trophoblast cells, Exp. Cell Res. (2011), https://doi.org/10.1016/j.yexcr.2011.01.014.
[39] S.S. Sidhu, A.T. Mengistab, A.N. Tauscher, J. LaVail, C. Basbaum, The microvesicle as a vehicle for EMMPRin in tumor-stromal interactions, Oncogene (2004), https://doi.org/10.1038/sj.onc.1207070.
[40] A.C. Gradilla, E. González, I. Seijo, G. Andrés, M. Bischoff, L. González-Mendez, V. Sánchez, A. Callejo, C. Ibáñez, M. Guerra, J.R. Ortigão-Farias, J.D. Sutherland, M. González, R. Barrio, J.M. Falcón-Pérez, I. Guerrero, Exosomes as Hedgehog carriers in cytoneme-mediated transport and secretion, Nat. Commun. (2014), https://doi.org/10.1038/ncomms6649.

[41] I.J. McGough, J.P. Vincent, Exosomes in developmental signalling, Development (2016), https://doi.org/10.1242/dev.126516.
[42] A. Becker, B.K. Thakur, J.M. Weiss, H.S. Kim, H. Peinado, D. Lyden, Extracellular vesicles in cancer: cell-to-cell mediators of metastasis, Canc. Cell 30 (2016) 836–848.
[43] R. Kalluri, V.S. LeBleu, The biology, function, and biomedical applications of exosomes, Science 367 (2020), https://doi.org/10.1126/science.aau6977.
[44] V. Budnik, C. Ruiz-Cañada, F. Wendler, Extracellular vesicles round off communication in the nervous system, Nat. Rev. Neurosci. (2016), https://doi.org/10.1038/nrn.2015.29.
[45] J. Howitt, A.F. Hill, Exosomes in the pathology of neurodegenerative diseases, J. Biol. Chem. (2016), https://doi.org/10.1074/jbc.R116.757955.
[46] S.V. Chitti, P. Fonseka, S. Mathivanan, Emerging role of extracellular vesicles in mediating cancer cachexia, Biochem. Soc. Trans. (2018), https://doi.org/10.1042/BST20180213.
[47] L. Yuan, J.Y. Li, Exosomes in Parkinson's disease: current perspectives and future challenges, ACS Chem. Neurosci. (2019), https://doi.org/10.1021/acschemneuro.8b00469.
[48] R. Kalluri, The biology and function of exosomes in cancer, J. Clin. Invest. (2016), https://doi.org/10.1172/JCI81135.
[49] D. Hanahan, R.A. Weinberg, Hallmarks of cancer: the next generation, Cell (2011), https://doi.org/10.1016/j.cell.2011.02.013.
[50] E. Cocucci, J. Meldolesi, Ectosomes, Curr. Biol. (2011), https://doi.org/10.1016/j.cub.2011.10.011.
[51] J. Meldolesi, Exosomes and ectosomes in intercellular communication, Curr. Biol. 28 (2018) R435–R444.
[52] V.R. Minciacchi, M.R. Freeman, D. Di Vizio, B. Sciences, L. Angeles, U. Diseases, B. Children, the emerging role of large oncosomes, Semin. Cell Dev. Biol. (2016) 41–51.
[53] R. Xu, A. Rai, M. Chen, W. Suwakulsiri, D.W. Greening, R.J. Simpson, Extracellular vesicles in cancer — implications for future improvements in cancer care, Nat. Rev. Clin. Oncol. (2018), https://doi.org/10.1038/s41571-018-0036-9.
[54] E. Willms, C. Cabañas, I. Mäger, M.J.A. Wood, P. Vader, Extracellular vesicle heterogeneity: subpopulations, isolation techniques, and diverse functions in cancer progression, Front. Immunol. (2018), https://doi.org/10.3389/fimmu.2018.00738.
[55] C. Théry, K.W. Witwer, E. Aikawa, M.J. Alcaraz, J.D. Anderson, R. Andriantsitohaina, A. Antoniou, T. Arab, F. Archer, G.K. Atkin-Smith, D.C. Ayre, J.M. Bach, D. Bachurski, H. Baharvand, L. Balaj, S. Baldacchino, N.N. Bauer, A.A. Baxter, M. Bebawy, C. Beckham, A. Bedina Zavec, A. Benmoussa, A.C. Berardi, P. Bergese, E. Bielska, C. Blenkiron, S. Bobis-Wozowicz, E. Boilard, W. Boireau, A. Bongiovanni, F.E. Borràs, S. Bosch, C.M. Boulanger, X. Breakefield, A.M. Breglio, M. Brennan, D.R. Brigstock, A. Brisson, M.L.D. Broekman, J.F. Bromberg, P. Bryl-Górecka, S. Buch, A.H. Buck, D. Burger, S. Busatto, D. Buschmann, B. Bussolati, E.I. Buzás, J.B. Byrd, G. Camussi, D.R.F. Carter, S. Caruso, L.W. Chamley, Y.T. Chang, A.D. Chaudhuri, C. Chen, S. Chen, L. Cheng, A.R. Chin, A. Clayton, S.P. Clerici, A. Cocks, E. Cocucci, R.J. Coffey, A. Cordeiro-da-Silva, Y. Couch, F.A.W. Coumans, B. Coyle, R. Crescitelli, M.F. Criado, C. D'Souza-Schorey, S. Das, P. de Candia, E.F. De Santana, O. De Wever, H.A. del Portillo, T. Demaret, S. Deville, A. Devitt, B. Dhondt, D. Di Vizio, L.C. Dieterich, V. Dolo, A.P. Dominguez Rubio, M. Dominici, M.R. Dourado, T.A.P. Driedonks, F.V. Duarte, H.M. Duncan, R.M. Eichenberger, K. Ekström, S. EL Andaloussi, C. Elie-Caille, U. Erdbrügger, J.M. Falcón-Pérez, F. Fatima,

J.E. Fish, M. Flores-Bellver, A. Försönits, A. Frelet-Barrand, F. Fricke, G. Fuhrmann, S. Gabrielsson, A. Gámez-Valero, C. Gardiner, K. Gärtner, R. Gaudin, Y.S. Gho, B. Giebel, C. Gilbert, M. Gimona, I. Giusti, D.C.I. Goberdhan, A. Görgens, S.M. Gorski, D.W. Greening, J.C. Gross, A. Gualerzi, G.N. Gupta, D. Gustafson, A. Handberg, R.A. Haraszti, P. Harrison, H. Hegyesi, A. Hendrix, A.F. Hill, F.H. Hochberg, K.F. Hoffmann, B. Holder, H. Holthofer, B. Hosseinkhani, G. Hu, Y. Huang, V. Huber, S. Hunt, A.G.E. Ibrahim, T. Ikezu, J.M. Inal, M. Isin, A. Ivanova, H.K. Jackson, S. Jacobsen, S.M. Jay, M. Jayachandran, G. Jenster, L. Jiang, S.M. Johnson, J.C. Jones, A. Jong, T. Jovanovic-Talisman, S. Jung, R. Kalluri, S. ichi Kano, S. Kaur, Y. Kawamura, E.T. Keller, D. Khamari, E. Khomyakova, A. Khvorova, P. Kierulf, K.P. Kim, T. Kislinger, M. Klingeborn, D.J. Klinke, M. Kornek, M.M. Kosanović, Á.F. Kovács, E.M. Krämer-Albers, S. Krasemann, M. Krause, I.V. Kurochkin, G.D. Kusuma, S. Kuypers, S. Laitinen, S.M. Langevin, L.R. Languino, J. Lannigan, C. Lässer, L.C. Laurent, G. Lavieu, E. Lázaro-Ibáñez, S. Le Lay, M.S. Lee, Y.X.F. Lee, D.S. Lemos, M. Lenassi, A. Leszczynska, I.T.S. Li, K. Liao, S.F. Libregts, E. Ligeti, R. Lim, S.K. Lim, A. Linē, K. Linnemannstöns, A. Llorente, C.A. Lombard, M.J. Lorenowicz, Á.M. Lörincz, J. Lötvall, J. Lovett, M.C. Lowry, X. Loyer, Q. Lu, B. Lukomska, T.R. Lunavat, S.L.N. Maas, H. Malhi, A. Marcilla, J. Mariani, J. Mariscal, E.S. Martens-Uzunova, L. Martin-Jaular, M.C. Martinez, V.R. Martins, M. Mathieu, S. Mathivanan, M. Maugeri, L.K. McGinnis, M.J. McVey, D.G. Meckes, K.L. Meehan, I. Mertens, V.R. Minciacchi, A. Möller, M. Møller Jørgensen, A. Morales-Kastresana, J. Morhayim, F. Mullier, M. Muraca, L. Musante, V. Mussack, D.C. Muth, K.H. Myburgh, T. Najrana, M. Nawaz, I. Nazarenko, P. Nejsum, C. Neri, T. Neri, R. Nieuwland, L. Nimrichter, J.P. Nolan, E.N.M. Nolte-'t Hoen, N. Noren Hooten, L. O'Driscoll, T. O'Grady, A. O'Loghlen, T. Ochiya, M. Olivier, A. Ortiz, L.A. Ortiz, X. Osteikoetxea, O. Ostegaard, M. Ostrowski, J. Park, D.M. Pegtel, H. Peinado, F. Perut, M.W. Pfaffl, D.G. Phinney, B.C.H. Pieters, R.C. Pink, D.S. Pisetsky, E. Pogge von Strandmann, I. Polakovicova, I.K.H. Poon, B.H. Powell, I. Prada, L. Pulliam, P. Quesenberry, A. Radeghieri, R.L. Raffai, S. Raimondo, J. Rak, M.I. Ramirez, G. Raposo, M.S. Rayyan, N. Regev-Rudzki, F.L. Ricklefs, P.D. Robbins, D.D. Roberts, S.C. Rodrigues, E. Rohde, S. Rome, K.M.A. Rouschop, A. Rughetti, A.E. Russell, P. Saá, S. Sahoo, E. Salas-Huenuleo, C. Sánchez, J.A. Saugstad, M.J. Saul, R.M. Schiffelers, R. Schneider, T.H. Schøyen, A. Scott, E. Shahaj, S. Sharma, O. Shatnyeva, F. Shekari, G.V. Shelke, A.K. Shetty, K. Shiba, P.R.M. Siljander, A.M. Silva, A. Skowronek, O.L. Snyder, R.P. Soares, B.W. Sódar, C. Soekmadji, J. Sotillo, P.D. Stahl, W. Stoorvogel, S.L. Stott, E.F. Strasser, S. Swift, H. Tahara, M. Tewari, K. Timms, S. Tiwari, R. Tixeira, M. Tkach, W.S. Toh, R. Tomasini, A.C. Torrecilhas, J.P. Tosar, V. Toxavidis, L. Urbanelli, P. Vader, B.W.M. van Balkom, S.G. van der Grein, J. Van Deun, M.J.C. van Herwijnen, K. Van Keuren-Jensen, G. van Niel, M.E. van Royen, A.J. van Wijnen, M.H. Vasconcelos, I.J. Vechetti, T.D. Veit, L.J. Vella, É. Velot, F.J. Verweij, B. Vestad, J.L. Viñas, T. Visnovitz, K.V. Vukman, J. Wahlgren, D.C. Watson, M.H.M. Wauben, A. Weaver, J.P. Webber, V. Weber, A.M. Wehman, D.J. Weiss, J.A. Welsh, S. Wendt, A.M. Wheelock, Z. Wiener, L. Witte, J. Wolfram, A. Xagorari, P. Xander, J. Xu, X. Yan, M. Yáñez-Mó, H. Yin, Y. Yuana, V. Zappulli, J. Zarubova, V. Žėkas, J. ye Zhang, Z. Zhao, L. Zheng, A.R. Zheutlin, A.M. Zickler, P. Zimmermann, A.M. Zivkovic, D. Zocco, E.K. Zuba-Surma, Minimal information for studies of extracellular vesicles 2018 (MISEV2018): a position statement of the International Society for Extracellular

Vesicles and update of the MISEV2014 guidelines, J. Extracell. Vesicles (2018), https://doi.org/10.1080/20013078.2018.1535750.
[56] S.J. Gould, G. Raposo, As we wait: coping with an imperfect nomenclature for extracellular vesicles, J. Extracell. Vesicles 2 (2013).
[57] J. Lötvall, A.F. Hill, F. Hochberg, E.I. Buzás, D. Di Vizio, C. Gardiner, Y.S. Gho, I.V. Kurochkin, S. Mathivanan, P. Quesenberry, S. Sahoo, H. Tahara, M.H. Wauben, K.W. Witwer, C. Théry, Minimal experimental requirements for definition of extracellular vesicles and their functions: a position statement from the International Society for Extracellular Vesicles, J. Extracell. Vesicles 3 (2014).
[58] K.W. Witwer, C. Soekmadji, A.F. Hill, M.H. Wauben, E.I. Buzás, D. Di Vizio, J.M. Falcon-Perez, C. Gardiner, F. Hochberg, I.V. Kurochkin, J. Lötvall, S. Mathivanan, R. Nieuwland, S. Sahoo, H. Tahara, A.C. Torrecilhas, A.M. Weaver, H. Yin, L. Zheng, Y.S. Gho, P. Quesenberry, C. Théry, Updating the MISEV minimal requirements for extracellular vesicle studies: building bridges to reproducibility, J. Extracell. Vesicles 6 (2017).
[59] M.P. Bebelman, M.J. Smit, D.M. Pegtel, S.R. Baglio, Biogenesis and function of extracellular vesicles in cancer, Pharmacol. Ther. 188 (2018) 1−11.
[60] G. Raposo, W. Stoorvogel, Extracellular vesicles: exosomes, microvesicles, and friends, J. Cell Biol. (2013), https://doi.org/10.1083/jcb.201211138.
[61] G. Van Niel, G. D'Angelo, G. Raposo, Shedding light on the cell biology of extracellular vesicles, Nat. Rev. Mol. Cell Biol. 19 (2018) 213−228.
[62] K. Al-Nedawi, B. Meehan, J. Micallef, V. Lhotak, L. May, A. Guha, J. Rak, Intercellular transfer of the oncogenic receptor EGFRvIII by microvesicles derived from tumour cells, Nat, Cell Biol. 10 (2008) 619−624.
[63] P. Vader, X.O. Breakefield, M.J.A. Wood, Extracellular vesicles: emerging targets for cancer therapy, Trends Mol. Med. 20 (2014) 385−393.
[64] K.M. McAndrews, R. Kalluri, Mechanisms associated with biogenesis of exosomes in cancer, Mol. Canc. (2019), https://doi.org/10.1186/s12943-019-0963-9.
[65] M. Mathieu, L. Martin-Jaular, G. Lavieu, C. Théry, Specificities of secretion and uptake of exosomes and other extracellular vesicles for cell-to-cell communication, Nat. Cell Biol. (2019), https://doi.org/10.1038/s41556-018-0250-9.
[66] C. Kahlert, R. Kalluri, Exosomes in tumor microenvironment influence cancer progression and metastasis, J. Mol. Med. (2013), https://doi.org/10.1007/s00109-013-1020-6.
[67] W.M. Henne, N.J. Buchkovich, S.D. Emr, The ESCRT pathway, Dev. Cell (2011), https://doi.org/10.1016/j.devcel.2011.05.015.
[68] K. Trajkovic, C. Hsu, S. Chiantia, L. Rajendran, D. Wenzel, F. Wieland, P. Schwille, B. Brügger, M. Simons, Ceramide triggers budding of exosome vesicles into multivesicular endosomes, Science (2008).
[69] P. Urbina, M. Flores-Díaz, A. Alape-Girón, A. Alonso, F.M. Goni, Phospholipase C and sphingomyelinase activities of the *Clostridium perfringens* α-toxin, Chem. Phys. Lipids (2009), https://doi.org/10.1016/j.chemphyslip.2009.02.007.
[70] V. Muralidharan-Chari, J. Clancy, C. Plou, M. Romao, P. Chavrier, G. Raposo, C. D'Souza-Schorey, ARF6-Regulated shedding of tumor cell-derived plasma membrane microvesicles, Curr. Biol. (2009), https://doi.org/10.1016/j.cub.2009.09.059.
[71] S. Schlienger, S. Campbell, A. Claing, ARF1 regulates the Rho/MLC pathway to control EGF-dependent breast cancer cell invasion, Mol. Biol. Cell (2014), https://doi.org/10.1091/mbc.E13-06-0335.
[72] B. Li, M.A. Antonyak, J. Zhang, R.A. Cerione, RhoA triggers a specific signaling pathway that generates transforming microvesicles in cancer cells, Oncogene 31 (2012) 4740−4749.

[73] F. Bianco, E. Pravettoni, A. Colombo, U. Schenk, T. Möller, M. Matteoli, C. Verderio, Astrocyte-derived ATP induces vesicle shedding and IL-1β release from microglia, J. Immunol. (2005), https://doi.org/10.4049/jimmunol.174.11.7268.
[74] Y. Gutiérrez-Martín, D. Bustillo, R. Gómez-Villafuertes, J. Sánchez-Nogueiro, C. Torregrosa-Hetland, T. Binz, L.M. Gutiérrez, M.T. Miras-Portugal, A.R. Artalejo, P2X7 receptors trigger ATP exocytosis and modify secretory vesicle dynamics in neuroblastoma cells, J. Biol. Chem. (2011), https://doi.org/10.1074/jbc.M110.139410.
[75] M.J. Lukey, K.F. Wilson, R.A. Cerione, Therapeutic strategies impacting cancer cell glutamine metabolism, Future Med. Chem. (2013), https://doi.org/10.4155/fmc.13.130.
[76] O.T. Fackler, R. Grosse, Cell motility through plasma membrane blebbing, J. Cell Biol. (2008), https://doi.org/10.1083/jcb.200802081.
[77] A.E. Sedgwick, J.W. Clancy, M. Olivia Balmert, C. D'Souza-Schorey, Extracellular microvesicles and invadopodia mediate non-overlapping modes of tumor cell invasion, Sci. Rep. (2015), https://doi.org/10.1038/srep14748.
[78] J.W. Clancy, A. Sedgwick, C. Rosse, V. Muralidharan-Chari, G. Raposo, M. Method, P. Chavrier, C. D'Souza-Schorey, Regulated delivery of molecular cargo to invasive tumour-derived microvesicles, Nat. Commun. (2015), https://doi.org/10.1038/ncomms7919.
[79] J. Kim, S. Morley, M. Le, D. Bedoret, D.T. Umetsu, D. Di Vizio, M.R. Freeman, Enhanced shedding of extracellular vesicles from amoeboid prostate cancer cells Potential effects on the tumor microenvironment, Canc. Biol. Ther. (2014), https://doi.org/10.4161/cbt.27627.
[80] C.D. Paul, P. Mistriotis, K. Konstantopoulos, Cancer cell motility: lessons from migration in confined spaces, Nat. Rev. Canc. (2017), https://doi.org/10.1038/nrc.2016.123.
[81] M. Lokar, Š. Perutková, V. Kralj-Iglič, A. Iglič, P. Veranič, Chapter 3 membrane nanotubes in urothelial cell line T24, Adv. Planar Lipid Bilayers Liposomes (2009), https://doi.org/10.1016/S1554-4516(09)10003-0.
[82] P. Veranič, M. Lokar, G.J. Schütz, J. Weghuber, S. Wieser, H. Hägerstrand, V. Kralj-Iglič, A. Iglič, Different types of cell-to-cell connections mediated by nanotubular structures, Biophys. J. (2008), https://doi.org/10.1529/biophysj.108.131375.
[83] A. De Gassart, C. Géminard, B. Février, G. Raposo, M. Vidal, Lipid raft-associated protein sorting in exosomes, Blood (2003), https://doi.org/10.1182/blood-2003-03-0871.
[84] D. Mazurov, L. Barbashova, A. Filatov, Tetraspanin protein CD9 interacts with metalloprotease CD10 and enhances its release via exosomes, FEBS J. (2013), https://doi.org/10.1111/febs.12110.
[85] D. Perez-Hernandez, C. Gutiérrez-Vázquez, I. Jorge, S. López-Martín, A. Ursa, F. Sánchez-Madrid, J. Vázquez, M. Yañez-Mó, The intracellular interactome of tetraspanin-enriched microdomains reveals their function as sorting machineries toward exosomes, J. Biol. Chem. (2013), https://doi.org/10.1074/jbc.M112.445304.
[86] G. van Niel, S. Charrin, S. Simoes, M. Romao, L. Rochin, P. Saftig, M.S. Marks, E. Rubinstein, G. Raposo, The tetraspanin CD63 regulates ESCRT-independent and -dependent endosomal sorting during melanogenesis, Dev. Cell (2011), https://doi.org/10.1016/j.devcel.2011.08.019.
[87] C. Théry, Exosomes: secreted vesicles and intercellular communications, F1000 Biol. Rep. (2011), https://doi.org/10.3410/B3-15.
[88] C. Géminard, A. de Gassart, L. Blanc, M. Vidal, Degradation of AP2 during reticulocyte maturation enhances binding of hsc70 and Alix to a common site on TfR for sorting in exosomes, Traffic (2004), https://doi.org/10.1111/j.1600-0854.2004.0167.x.

[89] S.I. Buschow, J.M.P. Liefhebber, R. Wubbolts, W. Stoorvogel, Exosomes contain ubiquitinated proteins, Blood Cells, Mol. Dis. (2005), https://doi.org/10.1016/j.bcmd.2005.08.005.
[90] N. Luhtala, A. Aslanian, J.R. Yates, T. Hunter, Secreted glioblastoma nanovesicles contain intracellular signaling proteins and active ras incorporated in a farnesylation-dependent manner, J. Biol. Chem. (2017), https://doi.org/10.1074/jbc.M116.747618.
[91] B. Shen, N. Wu, M. Yang, S.J. Gould, Protein targeting to exosomes/microvesicles by plasma membrane anchors, J. Biol. Chem. (2011), https://doi.org/10.1074/jbc.M110.208660.
[92] J.M. Yang, S.J. Gould, The cis-acting signals that target proteins to exosomes and microvesicles, Biochem. Soc. Trans. (2013), https://doi.org/10.1042/BST20120275.
[93] A. Elsherbini, E. Bieberich, Ceramide and exosomes: a novel target in cancer biology and therapy, Adv. Canc. Res. (2018), https://doi.org/10.1016/bs.acr.2018.05.004.
[94] R. Ghossoub, F. Lembo, A. Rubio, C.B. Gaillard, J. Bouchet, N. Vitale, J. Slavík, M. Machala, P. Zimmermann, Syntenin-ALIX exosome biogenesis and budding into multivesicular bodies are controlled by ARF6 and PLD2, Nat. Commun. (2014), https://doi.org/10.1038/ncomms4477.
[95] A.L. Egea-Jimenez, P. Zimmermann, Phospholipase D and phosphatidic acid in the biogenesis and cargo loading of extracellular vesicles, J. Lipid Res. (2018), https://doi.org/10.1194/jlr.R083964.
[96] D. Allan, R.H. Michell, Calcium ion-dependent diacylglycerol accumulation in erythrocytes is associated with microvesiculation but not with efflux of potassium ions, Biochem. J. (1977), https://doi.org/10.1042/bj1660495.
[97] K.O. Schink, K.-W. Tan, H. Stenmark, Phosphoinositides in control of membrane dynamics, Annu. Rev. Cell Dev. Biol. (2016), https://doi.org/10.1146/annurev-cellbio-111315-125349.
[98] P. Mayinger, Phosphoinositides and vesicular membrane traffic, Biochim. Biophys. Acta Mol. Cell Biol. Lipids (2012), https://doi.org/10.1016/j.bbalip.2012.01.002.
[99] S.T. Yang, A.J.B. Kreutzberger, J. Lee, V. Kiessling, L.K. Tamm, The role of cholesterol in membrane fusion, Chem. Phys. Lipids (2016), https://doi.org/10.1016/j.chemphyslip.2016.05.003.
[100] T. Skotland, K. Sagini, K. Sandvig, A. Llorente, An emerging focus on lipids in extracellular vesicles, Adv. Drug Deliv. Rev. (2020), https://doi.org/10.1016/j.addr.2020.03.002.
[101] A. Khayrullin, P. Krishnan, L. Martinez-Nater, B. Mendhe, S. Fulzele, Y. Liu, J. Mattison, M. Hamrick, Very long-chain C24:1 ceramide is increased in serum extracellular vesicles with aging and can induce senescence in bone-derived mesenchymal stem cells, Cells (2019), https://doi.org/10.3390/cells8010037.
[102] M. Lin, W. Liao, M. Dong, R. Zhu, J. Xiao, T. Sun, Z. Chen, B. Wu, J. Jin, Exosomal neutral sphingomyelinase 1 suppresses hepatocellular carcinoma via decreasing the ratio of sphingomyelin/ceramide, FEBS J. (2018), https://doi.org/10.1111/febs.14635.
[103] N. Mukhamedova, A. Hoang, D. Dragoljevic, L. Dubrovsky, T. Pushkarsky, H. Low, M. Ditiatkovski, Y. Fu, R. Ohkawa, P.J. Meikle, A. Horvath, B. Brichacek, Y.I. Miller, A. Murphy, M. Bukrinsky, D. Sviridov, Exosomes containing HIV protein Nef reorganize lipid rafts potentiating inflammatory response in bystander cells, PLoS Pathog. (2019), https://doi.org/10.1371/journal.ppat.1007907.
[104] M. Record, K. Carayon, M. Poirot, S. Silvente-Poirot, Exosomes as new vesicular lipid transporters involved in cell-cell communication and various pathophysiologies, Biochim. Biophys. Acta Mol. Cell Biol. Lipids (2014), https://doi.org/10.1016/j.bbalip.2013.10.004.

[105] J. Skog, T. Würdinger, S. van Rijn, D.H. Meijer, L. Gainche, W.T. Curry, B.S. Carter, A.M. Krichevsky, X.O. Breakefield, Glioblastoma microvesicles transport RNA and proteins that promote tumour growth and provide diagnostic biomarkers, Nat. Cell Biol. (2008), https://doi.org/10.1038/ncb1800.

[106] H. Valadi, K. Ekström, A. Bossios, M. Sjöstrand, J.J. Lee, J.O. Lötvall, Exosome-mediated transfer of mRNAs and microRNAs is a novel mechanism of genetic exchange between cells, Nat. Cell Biol. (2007), https://doi.org/10.1038/ncb1596.

[107] M. Lu, H. Xing, Z. Xun, T. Yang, P. Ding, C. Cai, D. Wang, X. Zhao, Exosome-based small RNA delivery: progress and prospects, Asian J. Pharm. Sci. (2018), https://doi.org/10.1016/j.ajps.2017.07.008.

[108] Y. Teng, Y. Ren, X. Hu, J. Mu, A. Samykutty, X. Zhuang, Z. Deng, A. Kumar, L. Zhang, M.L. Merchant, J. Yan, D.M. Miller, H.G. Zhang, MVP-mediated exosomal sorting of miR-193a promotes colon cancer progression, Nat. Commun. (2017), https://doi.org/10.1038/ncomms14448.

[109] C. Villarroya-Beltri, C. Gutiérrez-Vázquez, F. Sánchez-Cabo, D. Pérez-Hernández, J. Vázquez, N. Martin-Cofreces, D.J. Martinez-Herrera, A. Pascual-Montano, M. Mittelbrunn, F. Sánchez-Madrid, Sumoylated hnRNPA2B1 controls the sorting of miRNAs into exosomes through binding to specific motifs, Nat. Commun. (2013), https://doi.org/10.1038/ncomms3980.

[110] M.F. Bolukbasi, A. Mizrak, G.B. Ozdener, S. Madlener, T. Ströbel, E.P. Erkan, J.B. Fan, X.O. Breakefield, O. Saydam, MiR-1289 and "zipcode"-like sequence enrich mRNAs in microvesicles, Mol. Ther. Nucleic Acids (2012), https://doi.org/10.1038/mtna.2011.2.

[111] D. Koppers-Lalic, M. Hackenberg, I.V. Bijnsdorp, M.A.J. van Eijndhoven, P. Sadek, D. Sie, N. Zini, J.M. Middeldorp, B. Ylstra, R.X. de Menezes, T. Würdinger, G.A. Meijer, D.M. Pegtel, Nontemplated nucleotide additions distinguish the small RNA composition in cells from exosomes, Cell Rep. (2014), https://doi.org/10.1016/j.celrep.2014.08.027.

[112] M.L. Squadrito, C. Baer, F. Burdet, C. Maderna, G.D. Gilfillan, R. Lyle, M. Ibberson, M. De Palma, Endogenous RNAs modulate MicroRNA sorting to exosomes and transfer to acceptor cells, Cell Rep. (2014), https://doi.org/10.1016/j.celrep.2014.07.035.

[113] A.J. McKenzie, D. Hoshino, N.H. Hong, D.J. Cha, J.L. Franklin, R.J. Coffey, J.G. Patton, A.M. Weaver, KRAS-MEK signaling controls Ago2 sorting into exosomes, Cell Rep. (2016), https://doi.org/10.1016/j.celrep.2016.03.085.

[114] M.J. Shurtleff, M.M. Temoche-Diaz, K.V. Karfilis, S. Ri, R. Schekman, Y-box protein 1 is required to sort microRNAs into exosomes in cells and in a cell-free reaction, Elife (2016), https://doi.org/10.7554/eLife.19276.

[115] L. Balaj, R. Lessard, L. Dai, Y.J. Cho, S.L. Pomeroy, X.O. Breakefield, J. Skog, Tumour microvesicles contain retrotransposon elements and amplified oncogene sequences, Nat. Commun. (2011), https://doi.org/10.1038/ncomms1180.

[116] C. Kahlert, S.A. Melo, A. Protopopov, J. Tang, S. Seth, O. Koch, J. Zhang, J. Weitz, L. Chin, A. Futreal, R. Kalluri, Identification of doublestranded genomic dna spanning all chromosomes with mutated KRAS and P53 DNA in the serum exosomes of patients with pancreatic cancer, J. Biol. Chem. (2014), https://doi.org/10.1074/jbc.C113.532267.

[117] B.K. Thakur, H. Zhang, A. Becker, I. Matei, Y. Huang, B. Costa-Silva, Y. Zheng, A. Hoshino, H. Brazier, J. Xiang, C. Williams, R. Rodriguez-Barrueco, J.M. Silva, W. Zhang, S. Hearn, O. Elemento, N. Paknejad, K. Manova-Todorova, K. Welte, J. Bromberg, H. Peinado, D. Lyden, Double-stranded DNA in exosomes: a novel biomarker in cancer detection, Cell Res. (2014), https://doi.org/10.1038/cr.2014.44.

[118] M. Guescini, S. Genedani, V. Stocchi, L.F. Agnati, Astrocytes and Glioblastoma cells release exosomes carrying mtDNA, J. Neural. Transm. (2010), https://doi.org/10.1007/s00702-009-0288-8.

[119] P. Sansone, C. Savini, I. Kurelac, Q. Chang, L.B. Amato, A. Strillacci, A. Stepanova, L. Iommarini, C. Mastroleo, L. Daly, A. Galkin, B.K. Thakur, N. Soplop, K. Uryu, A. Hoshinob, L. Norton, M. Bonafé, M. Cricca, G. Gasparre, D. Lyden, J. Bromberg, Packaging and transfer of mitochondrial DNA via exosomes regulate escape from dormancy in hormonal therapy-resistant breast cancer, Proc. Natl. Acad. Sci. U.S.A. (2017), https://doi.org/10.1073/pnas.1704862114.

[120] X. Sisquella, Y. Ofir-Birin, M.A. Pimentel, L. Cheng, P. Abou Karam, N.G. Sampaio, J.S. Penington, D. Connolly, T. Giladi, B.J. Scicluna, R.A. Sharples, A. Waltmann, D. Avni, E. Schwartz, L. Schofield, Z. Porat, D.S. Hansen, A.T. Papenfuss, E.M. Eriksson, M. Gerlic, A.F. Hill, A.G. Bowie, N. Regev-Rudzki, Malaria parasite DNA-harbouring vesicles activate cytosolic immune sensors, Nat. Commun. (2017), https://doi.org/10.1038/s41467-017-02083-1.

[121] Y.Y. Lan, D. Londoño, R. Bouley, M.S. Rooney, N. Hacohen, Dnase2a deficiency uncovers lysosomal clearance of damaged nuclear DNA via autophagy, Cell Rep. (2014), https://doi.org/10.1016/j.celrep.2014.08.074.

[122] A. Takahashi, R. Okada, K. Nagao, Y. Kawamata, A. Hanyu, S. Yoshimoto, M. Takasugi, S. Watanabe, M.T. Kanemaki, C. Obuse, E. Hara, Exosomes maintain cellular homeostasis by excreting harmful DNA from cells, Nat. Commun. (2017), https://doi.org/10.1038/ncomms15287.

[123] H. Kalra, R.J. Simpson, H. Ji, E. Aikawa, P. Altevogt, P. Askenase, V.C. Bond, F.E. Borràs, X. Breakefield, V. Budnik, E. Buzas, G. Camussi, A. Clayton, E. Cocucci, J.M. Falcon-Perez, S. Gabrielsson, Y.S. Gho, D. Gupta, H.C. Harsha, A. Hendrix, A.F. Hill, J.M. Inal, G. Jenster, E.M. Krämer-Albers, S.K. Lim, A. Llorente, J. Lötvall, A. Marcilla, L. Mincheva-Nilsson, I. Nazarenko, R. Nieuwland, E.N.M. Nolte-'t Hoen, A. Pandey, T. Patel, M.G. Piper, S. Pluchino, T.S.K. Prasad, L. Rajendran, G. Raposo, M. Record, G.E. Reid, F. Sánchez-Madrid, R.M. Schiffelers, P. Siljander, A. Stensballe, W. Stoorvogel, D. Taylor, C. Thery, H. Valadi, B.W.M. van Balkom, J. Vázquez, M. Vidal, M.H.M. Wauben, M. Yáñez-Mó, M. Zoeller, S. Mathivanan, Vesiclepedia: a compendium for extracellular vesicles with continuous community annotation, PLoS Biol. (2012), https://doi.org/10.1371/journal.pbio.1001450.

[124] D.K. Kim, B. Kang, O.Y. Kim, D.S. Choi, J. Lee, S.R. Kim, G. Go, Y.J. Yoon, J.H. Kim, S.C. Jang, K.S. Park, E.J. Choi, K.P. Kim, D.M. Desiderio, Y.K. Kim, J. Lötvall, D. Hwang, Y.S. Gho, EVpedia: an integrated database of high-throughput data for systemic analyses of extracellular vesicles, J. Extracell. Vesicles (2013), https://doi.org/10.3402/jev.v2i0.20384.

[125] S.L. Subramanian, R.R. Kitchen, R. Alexander, B.S. Carter, K.H. Cheung, L.C. Laurent, A. Pico, L.R. Roberts, M.E. Roth, J.S. Rozowsky, A.I. Su, M.B. Gerstein, A. Milosavljevic, Integration of extracellular RNA profiling data using metadata, biomedical ontologies and Linked Data technologies, J. Extracell. Vesicles (2015), https://doi.org/10.3402/jev.v4.27497.

[126] M.N. Islam, S.R. Das, M.T. Emin, M. Wei, L. Sun, K. Westphalen, D.J. Rowlands, S.K. Quadri, S. Bhattacharya, J. Bhattacharya, Mitochondrial transfer from bone-marrow-derived stromal cells to pulmonary alveoli protects against acute lung injury, Nat. Med. (2012), https://doi.org/10.1038/nm.2736.

[127] T.J. Morrison, M.V. Jackson, E.K. Cunningham, A. Kissenpfennig, D.F. McAuley, C.M. O'Kane, A.D. Krasnodembskaya, Mesenchymal stromal cells modulate macrophages in clinically relevant lung injury models by extracellular vesicle mitochondrial

transfer, Am. J. Respir. Crit. Care Med. (2017), https://doi.org/10.1164/rccm.201701-0170OC.
[128] A.M. Rodriguez, J. Nakhle, E. Griessinger, M.L. Vignais, Intercellular mitochondria trafficking highlighting the dual role of mesenchymal stem cells as both sensors and rescuers of tissue injury, Cell Cycle (2018), https://doi.org/10.1080/15384101.2018.1445906.
[129] G. Taraboletti, S. D'Ascenzo, I. Giusti, D. Marchetti, P. Borsotti, D. Millimaggi, R. Giavazzi, A. Pavan, V. Dolo, Bioavailability of VEGF in tumor-shed vesicles depends on vesicle burst induced by acidic pH 1, Neoplasia 8 (2006) 96–103.
[130] S. Taverna, G. Ghersi, A. Ginestra, S. Rigogliuso, S. Pecorella, G. Alaimo, F. Saladino, V. Dolo, P. Dell'Era, A. Pavan, G. Pizzolanti, P. Mignatti, M. Presta, M.L. Vittorelli, Shedding of membrane vesicles mediates fibroblast growth factor-2 release from cells, J. Biol. Chem. (2003), https://doi.org/10.1074/jbc.M304192200.
[131] G. Schiera, P. Proia, C. Alberti, M. Mineo, G. Savettieri, I. Di Liegro, Neurons produce FGF2 and VEGF and secrete them at least in part by shedding extracellular vesicles, J. Cell Mol. Med. (2007), https://doi.org/10.1111/j.1582-4934.2007.00100.x.
[132] P. Proia, G. Schiera, M. Mineo, A.M.R. Ingrassia, G. Santoro, G. Savettieri, I. Di Liegro, Astrocytes shed extracellular vesicles that contain fibroblast growth factor-2 and vascular endothelial growth factor, Int. J. Mol. Med. (2008), https://doi.org/10.3892/ijmm.21.1.63.
[133] C. Wiesner, K. El Azzouzi, S. Linder, A specific subset of RabGTPases controls cell surface exposure of MT1-MMP, extracellular matrix degradation and three-dimensional invasion of macrophages, J. Cell Sci. (2013), https://doi.org/10.1242/jcs.122358.
[134] A. Matsumoto, Y. Takahashi, M. Nishikawa, K. Sano, M. Morishita, C. Charoenviriyakul, H. Saji, Y. Takakura, Accelerated growth of B16BL6 tumor in mice through efficient uptake of their own exosomes by B16BL6 cells, Canc. Sci. (2017), https://doi.org/10.1111/cas.13310.
[135] K. Denzer, M. van Eijk, M.J. Kleijmeer, E. Jakobson, C. de Groot, H.J. Geuze, Follicular dendritic cells carry MHC class II-expressing microvesicles at their surface, J. Immunol. (2000), https://doi.org/10.4049/jimmunol.165.3.1259.
[136] M. Gabrielli, N. Battista, L. Riganti, I. Prada, F. Antonucci, L. Cantone, M. Matteoli, M. Maccarrone, C. Verderio, Active endocannabinoids are secreted on extracellular membrane vesicles, EMBO Rep. (2015), https://doi.org/10.15252/embr.201439668.
[137] I. Prada, L. Amin, R. Furlan, G. Legname, C. Verderio, D. Cojoc, A new approach to follow a single extracellular vesicle–cell interaction using optical tweezers, Biotechniques (2016), https://doi.org/10.2144/000114371.
[138] W. Heusermann, J. Hean, D. Trojer, E. Steib, S. von Bueren, A. Graff-Meyer, C. Genoud, K. Martin, N. Pizzato, J. Voshol, D.V. Morrissey, S.E.L. Andaloussi, M.J. Wood, N.C. Meisner-Kober, Exosomes surf on filopodia to enter cells at endocytic hot spots, traffic within endosomes, and are targeted to the ER, J. Cell Biol. (2016), https://doi.org/10.1083/jcb.201506084.
[139] A.E. Morelli, A.T. Larregina, W.J. Shufesky, M.L.G. Sullivan, D.B. Stolz, G.D. Papworth, A.F. Zahorchak, A.J. Logar, Z. Wang, S.C. Watkins, L.D. Falo, A.W. Thomson, Endocytosis, intracellular sorting, and processing of exosomes by dendritic cells, Blood (2004), https://doi.org/10.1182/blood-2004-03-0824.
[140] T. Tian, Y.L. Zhu, Y.Y. Zhou, G.F. Liang, Y.Y. Wang, F.H. Hu, Z.D. Xiao, Exosome uptake through clathrin-mediated endocytosis and macropinocytosis and mediating miR-21 delivery, J. Biol. Chem. (2014), https://doi.org/10.1074/jbc.M114.588046.

[141] P. Lajoie, I.R. Nabi, Lipid rafts, caveolae, and their endocytosis, Int. Rev. Cell Mol. Biol. (2010), https://doi.org/10.1016/S1937-6448(10)82003-9.
[142] I. Nakase, N.B. Kobayashi, T. Takatani-Nakase, T. Yoshida, Active macropinocytosis induction by stimulation of epidermal growth factor receptor and oncogenic Ras expression potentiates cellular uptake efficacy of exosomes, Sci. Rep. (2015), https://doi.org/10.1038/srep10300.
[143] D. Feng, W.L. Zhao, Y.Y. Ye, X.C. Bai, R.Q. Liu, L.F. Chang, Q. Zhou, S.F. Sui, Cellular internalization of exosomes occurs through phagocytosis, Traffic (2010), https://doi.org/10.1111/j.1600-0854.2010.01041.x.
[144] T. Tian, Y. Wang, H. Wang, Z. Zhu, Z. Xiao, Visualizing of the cellular uptake and intracellular trafficking of exosomes by live-cell microscopy, J. Cell. Biochem. (2010), https://doi.org/10.1002/jcb.22733.
[145] C. Bissig, J. Gruenberg, ALIX and the multivesicular endosome: ALIX in Wonderland, Trends Cell Biol. (2014), https://doi.org/10.1016/j.tcb.2013.10.009.
[146] W. Lösche, T. Scholz, U. Temmler, V. Oberle, R.A. Claus, Platelet-derived microvesicles transfer tissue factor to monocytes but not to neutrophils, Platelets (2004), https://doi.org/10.1080/09537100310001649885.
[147] D. Fitzner, M. Schnaars, D. Van Rossum, G. Krishnamoorthy, P. Dibaj, M. Bakhti, T. Regen, U.K. Hanisch, M. Simons, Selective transfer of exosomes from oligodendrocytes to microglia by macropinocytosis, J. Cell Sci. (2011), https://doi.org/10.1242/jcs.074088.
[148] M. Chivet, C. Javalet, K. Laulagnier, B. Blot, F.J. Hemming, R. Sadoul, Exosomes secreted by cortical neurons upon glutamatergic synapse activation specifically interact with neurons, J. Extracell. Vesicles (2014), https://doi.org/10.3402/jev.v3.24722.
[149] J. Jiang, C.Y. Kao, E.T. Papoutsakis, How do megakaryocytic microparticles target and deliver cargo to alter the fate of hematopoietic stem cells? J. Contr. Release (2017) https://doi.org/10.1016/j.jconrel.2016.12.021.
[150] I. Nazarenko, S. Rana, A. Baumann, J. McAlear, A. Hellwig, M. Trendelenburg, G. Lochnit, K.T. Preissner, M. Zöller, Cell surface tetraspanin Tspan8 contributes to molecular pathways of exosome-induced endothelial cell activation, Canc. Res. (2010), https://doi.org/10.1158/0008-5472.CAN-09-2470.
[151] S. Bruno, C. Grange, M.C. Deregibus, R.A. Calogero, S. Saviozzi, F. Collino, L. Morando, A. Busca, M. Falda, B. Bussolati, C. Tetta, G. Camussi, Mesenchymal stem cell-derived microvesicles protect against acute tubular injury, J. Am. Soc. Nephrol. (2009), https://doi.org/10.1681/ASN.2008070798.
[152] S.A. Melo, L.B. Luecke, C. Kahlert, A.F. Fernandez, S.T. Gammon, J. Kaye, V.S. LeBleu, E.A. Mittendorf, J. Weitz, N. Rahbari, C. Reissfelder, C. Pilarsky, M.F. Fraga, D. Piwnica-Worms, R. Kalluri, Glypican-1 identifies cancer exosomes and detects early pancreatic cancer, Nature (2015), https://doi.org/10.1038/nature14581.
[153] A. Hoshino, B. Costa-Silva, T.L. Shen, G. Rodrigues, A. Hashimoto, M. Tesic Mark, H. Molina, S. Kohsaka, A. Di Giannatale, S. Ceder, S. Singh, C. Williams, N. Soplop, K. Uryu, L. Pharmer, T. King, L. Bojmar, A.E. Davies, Y. Ararso, T. Zhang, H. Zhang, J. Hernandez, J.M. Weiss, V.D. Dumont-Cole, K. Kramer, L.H. Wexler, A. Narendran, G.K. Schwartz, J.H. Healey, P. Sandstrom, K. Jørgen Labori, E.H. Kure, P.M. Grandgenett, M.A. Hollingsworth, M. De Sousa, S. Kaur, M. Jain, K. Mallya, S.K. Batra, W.R. Jarnagin, M.S. Brady, O. Fodstad, V. Muller, K. Pantel, A.J. Minn, M.J. Bissell, B.A. Garcia, Y. Kang, V.K. Rajasekhar, C.M. Ghajar, I. Matei, H. Peinado, J. Bromberg, D. Lyden, Tumour exosome integrins determine organotropic metastasis, Nature (2015), https://doi.org/10.1038/nature15756.

[154] W. Chen, X. Wen, M. Latzel, M. Heilmann, J. Yang, X. Dai, S. Huang, S. Shrestha, R. Patterson, S. Christiansen, G. Conibeer, Nanoscale characterization of carrier dynamic and surface passivation in InGaN/GaN multiple quantum wells on GaN nanorods, ACS Appl. Mater. Interfaces (2016), https://doi.org/10.1021/acsami.6b11675.

[155] T.H. Lee, S. Chennakrishnaiah, E. Audemard, L. Montermini, B. Meehan, J. Rak, Oncogenic ras-driven cancer cell vesiculation leads to emission of double-stranded DNA capable of interacting with target cells, Biochem. Biophys. Res. Commun. (2014), https://doi.org/10.1016/j.bbrc.2014.07.109.

[156] M. Takasugi, R. Okada, A. Takahashi, D. Virya Chen, S. Watanabe, E. Hara, Small extracellular vesicles secreted from senescent cells promote cancer cell proliferation through EphA2, Nat. Commun. (2017), https://doi.org/10.1038/ncomms15728.

[157] P. Kharaziha, S. Ceder, Q. Li, T. Panaretakis, Tumor cell-derived exosomes: a message in a bottle, Biochim. Biophys. Acta Rev. Canc 1826 (2012) 103–111.

[158] D. Choi, T.H. Lee, C. Spinelli, S. Chennakrishnaiah, E. D'Asti, J. Rak, Extracellular vesicle communication pathways as regulatory targets of oncogenic transformation, Semin. Cell Dev. Biol. 67 (2017) 11–22.

[159] M. Koivusalo, C. Welch, H. Hayashi, C.C. Scott, M. Kim, T. Alexander, N. Touret, K.M. Hahn, S. Grinstein, Amiloride inhibits macropinocytosis by lowering submembranous pH and preventing Rac1 and Cdc42 signaling, J. Cell Biol. (2010), https://doi.org/10.1083/jcb.200908086.

[160] T.H. Lee, S. Chennakrishnaiah, B. Meehan, L. Montermini, D. Garnier, E. D'Asti, W. Hou, N. Magnus, T. Gayden, N. Jabado, K. Eppert, L. Majewska, J. Rak, Barriers to horizontal cell transformation by extracellular vesicles containing oncogenic H-ras, Oncotarget (2016), https://doi.org/10.18632/oncotarget.10627.

[161] C.J. Tape, S. Ling, M. Dimitriadi, K.M. McMahon, J.D. Worboys, H.S. Leong, I.C. Norrie, C.J. Miller, G. Poulogiannis, D.A. Lauffenburger, C. Jørgensen, Oncogenic KRAS regulates tumor cell signaling via stromal reciprocation, Cell (2016), https://doi.org/10.1016/j.cell.2016.03.029.

[162] M.F. Baietti, Z. Zhang, E. Mortier, A. Melchior, G. Degeest, A. Geeraerts, Y. Ivarsson, F. Depoortere, C. Coomans, E. Vermeiren, P. Zimmermann, G. David, Syndecan-syntenin-ALIX regulates the biogenesis of exosomes, Nat. Cell Biol. (2012), https://doi.org/10.1038/ncb2502.

[163] F. Palacios, L. Price, J. Schweitzer, J.G. Collard, C. D'Souza-Schorey, An essential role for ARF6-regulated membrane traffic in adherens junction turnover and epithelial cell migration, EMBO J. (2001), https://doi.org/10.1093/emboj/20.17.4973.

[164] E. Makrodouli, E. Oikonomou, M. Koc, L. Andera, T. Sasazuki, S. Shirasawa, A. Pintzas, BRAF and RAS oncogenes regulate Rho GTPase pathways to mediate migration and invasion properties in human colon cancer cells: a comparative study, Mol. Canc. (2011), https://doi.org/10.1186/1476-4598-10-118.

[165] M.A. Antonyak, B. Li, L.K. Boroughs, J.L. Johnson, J.E. Druso, K.L. Bryant, D.A. Holowka, R.A. Cerione, Cancer cell-derived microvesicles induce transformation by transferring tissue transglutaminase and fibronectin to recipient cells, Proc. Natl. Acad. Sci. U.S.A. 108 (2011) 4852–4857.

[166] D.B. Wheeler, R. Zoncu, D.E. Root, D.M. Sabatini, C.L. Sawyers, Identification of an oncogenic RAB protein, Science (2015), https://doi.org/10.1126/science.aaa4903.

[167] M. Osswald, E. Jung, F. Sahm, G. Solecki, V. Venkataramani, J. Blaes, S. Weil, H. Horstmann, B. Wiestler, M. Syed, L. Huang, M. Ratliff, K. Karimian Jazi, F.T. Kurz, T. Schmenger, D. Lemke, M. Gömmel, M. Pauli, Y. Liao, P. Häring, S. Pusch, V. Herl, C. Steinhäuser, D. Krunic, M. Jarahian, H. Miletic, A.S. Berghoff, O. Griesbeck, G. Kalamakis, O. Garaschuk, M. Preusser, S. Weiss,

H. Liu, S. Heiland, M. Platten, P.E. Huber, T. Kuner, A. Von Deimling, W. Wick, F. Winkler, Brain tumour cells interconnect to a functional and resistant network, Nature (2015), https://doi.org/10.1038/nature16071.
[168] M. Logozzi, A. De Milito, L. Lugini, M. Borghi, L. Calabrò, M. Spada, M. Perdicchio, M.L. Marino, C. Federici, E. Iessi, D. Brambilla, G. Venturi, F. Lozupone, M. Santinami, V. Huber, M. Maio, L. Rivoltini, S. Fais, High levels of exosomes expressing CD63 and caveolin-1 in plasma of melanoma patients, PLoS One (2009), https://doi.org/10.1371/journal.pone.0005219.
[169] D.D. Taylor, C. Gercel-Taylor, MicroRNA signatures of tumor-derived exosomes as diagnostic biomarkers of ovarian cancer, Gynecol. Oncol. (2008), https://doi.org/10.1016/j.ygyno.2008.04.033.
[170] Y. Hisada, W. Alexander, R. Kasthuri, P. Voorhees, F. Mobarrez, A. Taylor, C. McNamara, H. Wallen, M. Witkowski, N.S. Key, U. Rauch, N. MacKman, Measurement of microparticle tissue factor activity in clinical samples: a summary of two tissue factor-dependent FXa generation assays, Thromb. Res. (2016), https://doi.org/10.1016/j.thromres.2016.01.011.
[171] D. Ribatti, A revisited concept: contact inhibition of growth. From cell biology to malignancy, Exp. Cell Res. 359 (2017) 17–19.
[172] J.D.T. Lapidot, C. Sirard, J. Vormoor, B. Murdoch, T. Hoang, J. Caceres-Cortes, M. Minden, B. Paterson, M.A. Caligiuri, A cell initiating human acute myeloid leukaemia after transplantation into SCID mice, Nature 367 (6464) (1994) 645–648.
[173] M.R. Atashzar, R. Baharlou, J. Karami, H. Abdollahi, R. Rezaei, F. Pourramezan, S.H. Zoljalali Moghaddam, Cancer stem cells: a review from origin to therapeutic implications, J. Cell. Physiol. 235 (2020) 790–803.
[174] V. Plaks, N. Kong, Z. Werb, The cancer stem cell niche: how essential is the niche in regulating stemness of tumor cells? Cell Stem Cell 16 (2015) 225–238.
[175] S. Cannito, E. Novo, A. Compagnone, L.V. di Bonzo, C. Busletta, E. Zamara, C. Paternostro, D. Povero, A. Bandino, F. Bozzo, C. Cravanzola, V. Bravoco, S. Colombatto, M. Parola, Redox mechanisms switch on hypoxia-dependent epithelial-mesenchymal transition in cancer cells, Carcinogenesis 29 (2008) 2267–2278.
[176] J. Anido, A. Sáez-Borderías, A. Gonzàlez-Juncà, L. Rodón, G. Folch, M.A. Carmona, R.M. Prieto-Sánchez, I. Barba, E. Martínez-Sáez, L. Prudkin, I. Cuartas, C. Raventós, F. Martínez-Ricarte, M.A. Poca, D. García-Dorado, M.M. Lahn, J.M. Yingling, J. Rodón, J. Sahuquillo, J. Baselga, J. Seoane, TGF-β receptor inhibitors target the CD44high/id1high glioma-initiating cell population in human glioblastoma, Canc. Cell 18 (2010) 655–668.
[177] J. Mao, S. Fan, W. Ma, P. Fan, B. Wang, J. Zhang, H. Wang, B. Tang, Q. Zhang, X. Yu, L. Wang, B. Song, L. Li, Roles of Wnt/β-catenin signaling in the gastric cancer stem cells proliferation and salinomycin treatment, Cell Death Dis. 5 (2014) 1–9.
[178] S. Hiratsuka, A. Watanabe, H. Aburatani, Y. Maru, Tumour-mediated upregulation of chemoattractants and recruitment of myeloid cells predetermines lung metastasis, Nat. Cell Biol. 8 (2006) 1369–1375.
[179] Z. Sun, L. Wang, L. Dong, X. Wang, Emerging role of exosome signalling in maintaining cancer stem cell dynamic equilibrium, J. Cell Mol. Med. 22 (2018) 3719–3728.
[180] Z. Sun, L. Wang, Y. Zhou, L. Dong, W. Ma, L. Lv, J. Zhang, X. Wang, Glioblastoma stem cell-derived exosomes enhance stemness and tumorigenicity of glioma cells by transferring Notch1 protein, Cell. Mol. Neurobiol. (2019), https://doi.org/10.1007/s10571-019-00771-8.
[181] R. Philip, S. Heiler, W. Mu, M.W. Büchler, M. Zöller, F. Thuma, Claudin-7 promotes the epithelial - mesenchymal transition in human colorectal cancer, Oncotarget 6 (2015) 2046–2063.

[182] D. Kyuno, K. Zhao, M. Schnölzer, J. Provaznik, T. Hackert, M. Zöller, Claudin7-dependent exosome-promoted reprogramming of nonmetastasizing tumor cells, Int. J. Canc. 145 (2019) 2182–2200.

[183] C. Grange, M. Tapparo, F. Collino, L. Vitillo, C. Damasco, M.C. Deregibus, C. Tetta, B. Bussolati, G. Camussi, Microvesicles released from human renal cancer stem cells stimulate angiogenesis and formation of lung premetastatic niche, Canc. Res. 71 (2011) 5346–5356.

[184] M. Najafi, K. Mortezaee, J. Majidpoor, Cancer stem cell (CSC) resistance drivers, Life Sci. 234 (2019).

[185] P. Samuel, M. Fabbri, D.R.F. Carter, Mechanisms of drug resistance in cancer: the role of extracellular vesicles, Proteomics 17 (2017) 1–10.

[186] J.C. Santos, N.D.S. Lima, L.O. Sarian, A. Matheu, M.L. Ribeiro, S.F.M. Derchain, Exosome-mediated breast cancer chemoresistance via miR-155 transfer, Sci. Rep. 8 (2018) 1–11.

[187] V. Ciravolo, V. Huber, G.C. Ghedini, E. Venturelli, F. Bianchi, M. Campiglio, D. Morelli, A. Villa, P. Della Mina, S. Menard, P. Filipazzi, L. Rivoltini, E. Tagliabue, S.M. Pupa, Potential role of HER2-overexpressing exosomes in countering trastuzumab-based therapy, J. Cell. Physiol. 227 (2012) 658–667.

[188] R. Safaei, B.J. Larson, T.C. Cheng, M.A. Gibson, S. Otani, W. Naerdemann, S.B. Howell, Abnormal lysosomal trafficking and enhanced exosomal export of cisplatin in drug-resistant human ovarian carcinoma cells, Mol. Canc. Therapeut. 4 (2005) 1595–1604.

[189] K.G. Chen, J.C. Valencia, B. Lai, G. Zhang, J.K. Paterson, F. Rouzaud, W. Berens, S.M. Wincovitch, S.H. Garfield, R.D. Leapman, V.J. Hearing, M.M. Gottesman, Melanosomal sequestration of cytotoxic drugs contributes to the intractability of malignant melanomas, Proc. Natl. Acad. Sci. U.S.A. 103 (2006) 9903–9907.

[190] E. Sahai, I. Astsaturov, E. Cukierman, D.G. DeNardo, M. Egeblad, R.M. Evans, D. Fearon, F.R. Greten, S.R. Hingorani, T. Hunter, R.O. Hynes, R.K. Jain, T. Janowitz, C. Jorgensen, A.C. Kimmelman, M.G. Kolonin, R.G. Maki, R.S. Powers, E. Puré, D.C. Ramirez, R. Scherz-Shouval, M.H. Sherman, S. Stewart, T.D. Tlsty, D.A. Tuveson, F.M. Watt, V. Weaver, A.T. Weeraratna, Z. Werb, A framework for advancing our understanding of cancer-associated fibroblasts, Nat. Rev. Canc. 20 (2020).

[191] F. Xing, J. Saidou, K. Watabe, Cancer associated fibroblasts (CAFs) in tumor microenvironment, Front. Biosci. 15 (2010) 166–179.

[192] W.R. Karnoub AE, A.B. Dash, A.P. Vo, A. Sullivan, M.W. Brooks, G.W. Bell, A.L. Richardson, K. Polyak, R. Tubo, Mesenchymal stem cells within tumour stroma promote breast cancer metastasis, Nature 449 (2007) 557–563.

[193] A.D. Rhim, P.E. Oberstein, D.H. Thomas, E.T. Mirek, F. Palermo, S.A. Sastra, E.N. Dekleva, T. Saunders, C.P. Becerra, I.W. Tattersall, C.B. Westphalen, J. Kitajewski, M.G. Fernandez-barrena, M.E. Fernandez-zapico, C. Iacobuzzio-donahue, K.P. Olive, B.Z. Stanger, Stromal elements act to retrain rather than support pdac, Cancel Cell 25 (2015) 735–747.

[194] M. Ranson, M.P. Stoppelli, A. De Vincenzo, S. Belli, P. Franco, M.L. Telesca, I. Iaccarino, G. Botti, M.V. Carriero, Paracrine recruitment and activation of fibroblasts by c-Myc expressing breast epithelial cells through the IGFs/IGF-1R axis, Int. J. Canc. 145 (2019) 2827–2839.

[195] J. Webber, R. Steadman, M.D. Mason, Z. Tabi, A. Clayton, Cancer exosomes trigger fibroblast to myofibroblast differentiation, Canc. Res. 70 (2010) 9621–9630.

[196] C.R. Goulet, G. Bernard, S. Tremblay, S. Chabaud, S. Bolduc, F. Pouliot, Exosomes induce fibroblast differentiation into cancer-associated fibroblasts through TGFb signaling, Mol. Canc. Res. 16 (2018) 1196–1204.

[197] K.W. Lee, J.A. Cho, H. Park, E.H. Lim, Exosomes from breast cancer cells can convert adipose tissue-derived mesenchymal stem cells into myofibroblast-like cells, Int. J. Oncol. 40 (2012) 130–138.
[198] M.Y. Fong, Z. Weiying, L. Liu, A.Y. Alontaga, M. Chandra, J. Ashby, A. Chow, S.T. O'Connor, S. Li, A.R. Chin, G. Somlo, Breast cancer-secreted miR-122 reprograms glucose metabolism in pre-metastatic niche to promote metastasis, Nat. Cell Biol. 17 (2015) 183–194.
[199] C.A. Sánchez, E.I. Andahur, R. Valenzuela, E.A. Castellón, J.A. Fullá, C.G. Ramos, J.C. Triviño, Exosomes from bulk and stem cells from human prostate cancer have a differential microRNA content that contributes cooperatively over local and pre-metastatic niche, Oncotarget 7 (2016) 3993–4008.
[200] W. Yan, X. Wu, W. Zhou, M.Y. Fong, M. Cao, J. Liu, X. Liu, C. Chen, O. Fadare, D.P. Pizzo, J. Wu, X. Liu, A.R. Chin, X. Ren, Y. Chen, J.W. Locasale, S.E. Wang, Cancer-cell-secreted exosomal miR-105 promotes tumour growth through the MYC-dependent metabolic reprogramming of stromal cells, Nat. Cell Biol 20 (2018) 597–609.
[201] V. Luga, J.L. Wrana, Tumor-stroma interaction: revealing fibroblast-secreted exosomes as potent regulators of Wnt-planar cell polarity signaling in cancer metastasis, Canc. Res. 73 (2013) 6843–6847.
[202] Y.B. Hu, C. Yan, L. Mu, Y.L. Mi, H. Zhao, H. Hu, X.L. Li, D.D. Tao, Y.Q. Wu, J.P. Gong, J.C. Qin, Exosomal Wnt-induced dedifferentiation of colorectal cancer cells contributes to chemotherapy resistance, Oncogene 38 (2019) 1951–1965.
[203] Y. Kojima, A. Acar, E.N. Eaton, K.T. Mellody, C. Scheel, I. Ben-Porath, T.T. Onder, Z.C. Wang, A.L. Richardson, R.A. Weinberg, A. Orimo, Autocrine TGF-β and stromal cell-derived factor-1 (SDF-1) signaling drives the evolution of tumor-promoting mammary stromal myofibroblasts, Proc. Natl. Acad. Sci. U.S.A. 107 (2010) 20009–20014.
[204] L.T. Vu, B. Peng, D.X. Zhang, V. Ma, C.A. Mathey-Andrews, C.K. Lam, T. Kiomourtzis, J. Jin, L. McReynolds, L. Huang, A. Grimson, W.C. Cho, J. Lieberman, M.T. Le, Tumor-secreted extracellular vesicles promote the activation of cancer-associated fibroblasts via the transfer of microRNA-125b, J. Extracell. Vesicles 8 (2019) 1–24.
[205] C.L. Au Yeung, N.N. Co, T. Tsuruga, T.L. Yeung, S.Y. Kwan, C.S. Leung, Y. Li, E.S. Lu, K. Kwan, K.K. Wong, R. Schmandt, K.H. Lu, S.C. Mok, Exosomal transfer of stroma-derived miR21 confers paclitaxel resistance in ovarian cancer cells through targeting APAF1, Nat. Commun. 7 (2016) 1–14.
[206] A. Santi, A. Caselli, F. Ranaldi, P. Paoli, C. Mugnaioni, E. Michelucci, P. Cirri, Cancer associated fibroblasts transfer lipids and proteins to cancer cells through cargo vesicles supporting tumor growth, Biochim. Biophys. Acta Mol. Cell Res. 1853 (2015) 3211–3223.
[207] P. Dauer, X. Zhao, V.K. Gupta, N. Sharma, K. Kesh, V. Dudeja, S.M. Vickers, S. Banerjee, A. Saluja, Inactivation of cancer-associated-fibroblasts (CAF) disrupts oncogenic signaling in pancreatic cancer cells and promotes its regression, Cancer Res. 78 (2018) 1321–1333.
[208] C. Jun, B.M. Lopes-Bastos, W.G. Jiang, Tumour-endothelial cell communications: important and indispensable mediators of tumour angiogenesis, Anticancer Res. 36 (2016) 1119–1126.
[209] C. Lu, T. Bonome, Y. Li, A.A. Kamat, L.Y. Han, R. Schmandt, R.L. Coleman, D.M. Gershenson, R.B. Jaffe, M.J. Birrer, A.K. Sood, Gene alterations identified by expression profiling in tumor-associated endothelial cells from invasive ovarian carcinoma, Canc. Res. 67 (2007) 1757–1768.

[210] S.Y. Ko, W.J. Lee, H.A. Kenny, L.H. Dang, L.M. Ellis, E. Jonasch, E. Lengyel, H. Naora, Cancer-derived small extracellular vesicles promote angiogenesis by heparin-bound, bevacizumab-insensitive VEGF, independent of vesicle uptake, Commun. Biol. 2 (2019) 1–17.

[211] L. Treps, R. Perret, S. Edmond, D. Ricard, J. Gavard, Glioblastoma stem-like cells secrete the pro-angiogenic VEGF-A factor in extracellular vesicles, J. Extracell. Vesicles 6 (2017).

[212] D. Millimaggi, M. Mari, S. D'Ascenzo, E. Carosa, E.A. Jannini, S. Zucker, G. Carta, A. Pavan, V. Dolo, Tumor vesicle-associated CD147 modulates the angiogenic capability of endothelial cells, Neoplasia 9 (2007) 349–357.

[213] S.K. Gopal, D.W. Greening, E.G. Hanssen, H.J. Zhu, R.J. Simpson, R.A. Mathias, Oncogenic epithelial cell-derived exosomes containing Rac1 and PAK2 induce angiogenesis in recipient endothelial cells, Oncotarget 7 (2016) 19709–19722.

[214] Y. Wang, L. Wang, C. Chen, X. Chu, New insights into the regulatory role of microRNA in tumor angiogenesis and clinical implications, Mol. Canc. 17 (2018) 1–10.

[215] N.O. Yamada, K. Heishima, Y. Akao, T. Senda, Extracellular vesicles containing microRNA-92a-3p facilitate partial endothelial-mesenchymal transition and angiogenesis in endothelial cells, Int. J. Mol. Sci. 20 (2019).

[216] C. Garnis, J. Lawson, C. Dickman, R. Towle, J. Jabalee, A. Javer, Extracellular vesicle secretion of miR-142-3p from lung adenocarcinoma cells induces tumor promoting changes in the stroma through cell-cell communication, Mol. Carcinog. 58 (2019) 376–387.

[217] Y.L. Hsu, J.Y. Hung, W.A. Chang, Y.S. Lin, Y.C. Pan, P.H. Tsai, C.Y. Wu, P.L. Kuo, Hypoxic lung cancer-secreted exosomal MIR-23a increased angiogenesis and vascular permeability by targeting prolyl hydroxylase and tight junction protein ZO-1, Oncogene 36 (2017) 4929–4942.

[218] W. Zhou, M.Y. Fong, Y. Min, G. Somlo, L. Liu, M.R. Palomares, Y. Yu, A. Chow, S.T.F. O'Connor, A.R. Chin, Y. Yen, Y. Wang, E.G. Marcusson, P. Chu, J. Wu, X. Wu, A.X. Li, Z. Li, H. Gao, X. Ren, M.P. Boldin, P.C. Lin, S.E. Wang, Cancer-Secreted miR-105 destroys vascular endothelial barriers to promote metastasis, Canc. Cell 25 (2014) 501–515.

[219] W. Mu, J. Provaznik, T. Hackert, M. Zöller, Endothelial Cell and Fibroblast Remodeling Relies On, 2020.

[220] J. Skog, T. Wurdinger, S. Van Rijn, D. Meijer, L. Gainche, M. Sena-esteves, W.T.C. Jr, R.S. Carter, A.M. Krichevsky, X.O. Breakefield, Glioblastoma microvesicles transport RNA and protein that promote tumor growth and provide diagnostic biomarkers, Nat. cell. Biol 10 (2012) 1470–1476.

[221] C. Gai, A. Carpanetto, M.C. Deregibus, G. Camussi, Extracellular vesicle-mediated modulation of angiogenesis, Histol. Histopathol. 31 (2016) 379–391.

[222] C.W. Kim, H.M. Lee, T.H. Lee, C. Kang, H.K. Kleinman, Y.S. Gho, Extracellular membrane vesicles from tumor cells promote angiogenesis via sphingomyelin, Canc. Res. 62 (2002) 6312–6317.

[223] G. Taraboletti, S. D'Ascenzo, P. Borsotti, R. Giavazzi, A. Pavan, V. Dolo, Shedding of the matrix metalloproteinases MMP-2, MMP-9, and MT1-MMP as membrane vesicle-associated components by endothelial cells, Am. J. Pathol. 160 (2002) 673–680.

[224] J.V. McCann, A. Liu, L. Musante, U. Erdbrügger, J. Lannigan, A.C. Dudley, A miRNA signature in endothelial cell-derived extracellular vesicles in tumor-bearing mice, Sci. Rep. 9 (2019) 1–8.

[225] G. Zhuang, X. Wu, Z. Jiang, I. Kasman, J. Yao, Y. Guan, J. Oeh, Z. Modrusan, C. Bais, D. Sampath, N. Ferrara, Tumour-secreted miR-9 promotes endothelial cell migration and angiogenesis by activating the JAK-STAT pathway, EMBO J. 31 (2012) 3513–3523.

[226] Y. Zeng, X. Yao, X. Liu, X. He, L. Li, X. Liu, Z. Yan, J. Wu, B.M. Fu, Anti-angiogenesis triggers exosomes release from endothelial cells to promote tumor vasculogenesis, J. Extracell. Vesicles 8 (2019).

[227] J.P. Thiery, H. Acloque, R.Y.J. Huang, M.A. Nieto, Epithelial-mesenchymal transitions in development and disease, Cell 139 (2009) 871–890.

[228] M.A. Nieto, R.Y.Y.J. Huang, R.A.A. Jackson, J.P.P. Thiery, Emt: 2016, Cell 166 (2016) 21–45.

[229] G. Greenburg, E.D. Hay, Epithelia suspended in collagen gels can lose polarity and express characteristics of migrating mesenchymal cells, J. Cell Biol. 95 (1982) 333–339.

[230] J. Yang, R.A. Weinberg, Epithelial-mesenchymal transition: at the crossroads of development and tumor metastasis, Dev. Cell (2008), https://doi.org/10.1016/j.devcel.2008.05.009.

[231] J. Lim, J.P. Thiery, Epithelial-mesenchymal transitions: insights from development, Development (2012), https://doi.org/10.1242/dev.071209.

[232] P.G. Santamaria, G. Moreno-Bueno, F. Portillo, A. Cano, EMT: present and future in clinical oncology, Mol. Oncol. (2017), https://doi.org/10.1002/1878-0261.12091.

[233] H. Peinado, D. Olmeda, A. Cano, Snail, ZEB and bHLH factors in tumour progression: an alliance against the epithelial phenotype? Nat. Rev. Canc. (2007) https://doi.org/10.1038/nrc2131.

[234] M.A. Nieto, The snail superfamily of zinc-finger transcription factors, Nat. Rev. Mol. Cell Biol. (2002), https://doi.org/10.1038/nrm757.

[235] J. Yang, S.A. Mani, J.L. Donaher, S. Ramaswamy, R.A. Itzykson, C. Come, P. Savagner, I. Gitelman, A. Richardson, R.A. Weinberg, Twist, a master regulator of morphogenesis, plays an essential role in tumor metastasis, Cell (2004), https://doi.org/10.1016/j.cell.2004.06.006.

[236] M. Rodriguez, E. Aladowicz, L. Lanfrancone, C.R. Goding, Tbx3 represses E-cadherin expression and enhances melanoma invasiveness, Canc. Res. (2008), https://doi.org/10.1158/0008-5472.CAN-08-0301.

[237] C. Battistelli, C. Cicchini, L. Santangelo, A. Tramontano, L. Grassi, F.J. Gonzalez, V. De Nonno, G. Grassi, L. Amicone, M. Tripodi, The Snail repressor recruits EZH2 to specific genomic sites through the enrollment of the lncRNA HOTAIR in epithelial-to-mesenchymal transition, Oncogene (2017), https://doi.org/10.1038/onc.2016.260.

[238] C. Battistelli, M. Tripodi, C. Cicchini, Targeting of polycombs to DNA in EMT, Oncotarget (2017), https://doi.org/10.18632/oncotarget.20211.

[239] A. Conigliaro, C. Cicchini, Exosome-mediated signaling in epithelial to mesenchymal transition and tumor progression, J. Clin. Med. 8 (2018) 26.

[240] S. Grotegut, D. Von Schweinitz, G. Christofori, F. Lehembre, Hepatocyte growth factor induces cell scattering through MAPK/Egr-1-mediated upregulation of Snail, EMBO J. (2006), https://doi.org/10.1038/sj.emboj.7601213.

[241] D. Kong, Z. Wang, S.H. Sarkar, Y. Li, S. Banerjee, A. Saliganan, H.-R.C. Kim, M.L. Cher, F.H. Sarkar, Platelet-derived growth factor-D overexpression contributes to epithelial-mesenchymal transition of PC3 prostate cancer cells, Stem Cell. (2008), https://doi.org/10.1634/stemcells.2007-1076.

[242] R. Derynck, Y.E. Zhang, Smad-dependent and Smad-independent pathways in TGF-β family signalling, Nature (2003), https://doi.org/10.1038/nature02006.

[243] M.G. Persico, G.L. Liguori, S. Parisi, D. D'Andrea, D.S. Salomon, G. Minchiotti, Cripto in tumors and embryo development, Biochim. Biophys. Acta Rev. Canc 1552 (2001).

[244] K.G. Leong, K. Niessen, I. Kulic, A. Raouf, C. Eaves, I. Pollet, A. Karsan, Jagged1-mediated Notch activation induces epithelial-to-mesenchymal transition through Slug-induced repression of E-cadherin, J. Exp. Med. (2007), https://doi.org/10.1084/jem.20071082.

[245] S.K. Gopal, D.W. Greening, A. Rai, M. Chen, R. Xu, A. Shafiq, R.A. Mathias, H.J. Zhu, R.J. Simpson, Extracellular vesicles: their role in cancer biology and epithelial-mesenchymal transition, Biochem. J. 474 (2017) 21–45.

[246] I. Pastushenko, C. Blanpain, EMT transition states during tumor progression and metastasis, Trends Cell Biol. 29 (2019) 212–226.

[247] F. Garibaldi, C. Cicchini, A. Conigliaro, L. Santangelo, A.M. Cozzolino, G. Grassi, A. Marchetti, M. Tripodi, L. Amicone, An epistatic mini-circuitry between the transcription factors Snail and HNF4α controls liver stem cell and hepatocyte features exhorting opposite regulation on stemness-inhibiting microRNAs, Cell Death Differ. (2012), https://doi.org/10.1038/cdd.2011.175.

[248] V. Costa, A. Lo Dico, A. Rizzo, F. Rajata, M. Tripodi, R. Alessandro, A. Conigliaro, MiR-675-5p supports hypoxia induced epithelial to mesenchymal transition in colon cancer cells, Oncotarget (2017), https://doi.org/10.18632/oncotarget.14464.

[249] A. Díaz-López, G. Moreno-Bueno, A. Cano, Role of microRNA in epithelial to mesenchymal transition and metastasis and clinical perspectives, Canc. Manag. Res. (2014), https://doi.org/10.2147/CMAR.S38156.

[250] D. Garnier, N. Magnus, T.H. Lee, V. Bentley, B. Meehan, C. Milsom, L. Montermini, T. Kislinger, J. Rak, Cancer cells induced to express mesenchymal phenotype release exosome-like extracellular vesicles carrying tissue factor, J. Biol. Chem. (2012), https://doi.org/10.1074/jbc.M112.401760.

[251] D. Garnier, N. Magnus, B. Meehan, T. Kislinger, J. Rak, Qualitative changes in the proteome of extracellular vesicles accompanying cancer cell transition to mesenchymal state, Exp. Cell Res. (2013), https://doi.org/10.1016/j.yexcr.2013.08.003.

[252] D.K. Jeppesen, A. Nawrocki, S.G. Jensen, K. Thorsen, B. Whitehead, K.A. Howard, L. Dyrskjøt, T.F. Ørntoft, M.R. Larsen, M.S. Ostenfeld, Quantitative proteomics of fractionated membrane and lumen exosome proteins from isogenic metastatic and nonmetastatic bladder cancer cells reveal differential expression of EMT factors, Proteomics (2014), https://doi.org/10.1002/pmic.201300452.

[253] B.J. Tauro, R.A. Mathias, D.W. Greening, S.K. Gopal, H. Ji, E.A. Kapp, B.M. Coleman, A.F. Hill, U. Kusebauch, J.L. Hallows, D. Shteynberg, R.L. Moritz, H.J. Zhu, R.J. Simpson, Oncogenic H-Ras reprograms madin-darby canine kidney (MDCK) cell-derived exosomal proteins following epithelial-mesenchymal transition, Mol. Cell. Proteomics (2013), https://doi.org/10.1074/mcp.M112.027086.

[254] S.K. Gopal, D.W. Greening, R.A. Mathias, H. Ji, A. Rai, M. Chen, H.J. Zhu, R.J. Simpson, YBX1/YB-1 induces partial EMT and tumourigenicity through secretion of angiogenic factors into the extracellular microenvironment, Oncotarget (2015), https://doi.org/10.18632/oncotarget.3764.

[255] O. Galindo-Hernandez, N. Serna-Marquez, R. Castillo-Sanchez, E.P. Salazar, Extracellular vesicles from MDA-MB-231 breast cancer cells stimulated with linoleic acid promote an EMT-like process in MCF10A cells, Prostagl. Leukot. Essent. Fat. Acids (2014), https://doi.org/10.1016/j.plefa.2014.09.002.

[256] C.A. Franzen, R.H. Blackwell, V. Todorovic, K.A. Greco, K.E. Foreman, R.C. Flanigan, P.C. Kuo, G.N. Gupta, Urothelial cells undergo epithelial-to-mesenchymal transition after exposure to muscle invasive bladder cancer exosomes, Oncogenesis (2015), https://doi.org/10.1038/oncsis.2015.21.

[257] J.N. Higginbotham, M. Demory Beckler, J.D. Gephart, J.L. Franklin, G. Bogatcheva, G.J. Kremers, D.W. Piston, G.D. Ayers, R.E. McConnell, M.J. Tyska, R.J. Coffey, Amphiregulin exosomes increase cancer cell invasion, Curr. Biol. (2011), https://doi.org/10.1016/j.cub.2011.03.043.

[258] L. Chen, P. Guo, Y. He, Z. Chen, L. Chen, Y. Luo, L. Qi, Y. Liu, Q. Wu, Y. Cui, F. Fang, X. Zhang, T. Song, H. Guo, HCC-derived exosomes elicit HCC progression and recurrence by epithelial-mesenchymal transition through MAPK/ERK signalling pathway article, Cell Death Dis. (2018), https://doi.org/10.1038/s41419-018-0534-9.

[259] M.T.N. Le, P. Hamar, C. Guo, E. Basar, R. Perdigão-Henriques, L. Balaj, J. Lieberman, MiR-200-containing extracellular vesicles promote breast cancer cell metastasis, J. Clin. Invest. (2014), https://doi.org/10.1172/JCI75695.

[260] A.M. Roccaro, A. Sacco, P. Maiso, A.K. Azab, Y.T. Tai, M. Reagan, F. Azab, L.M. Flores, F. Campigotto, E. Weller, K.C. Anderson, D.T. Scadden, I.M. Ghobrial, BM mesenchymal stromal cell-derived exosomes facilitate multiple myeloma progression, J. Clin. Invest. (2013), https://doi.org/10.1172/JCI66517.

[261] Z. Julier, A.J. Park, P.S. Briquez, M.M. Martino, Promoting tissue regeneration by modulating the immune system, Acta Biomater. 53 (2017) 13–28.

[262] H. Gonzalez, C. Hagerling, Z. Werb, Roles of the immune system in cancer: from tumor initiation to metastatic progression, Genes Dev. 32 (2018) 1267–1284.

[263] J.M. Kim, J.P. Rasmussen, A.Y. Rudensky, Regulatory T cells prevent catastrophic autoimmunity throughout the lifespan of mice, Nat. Immunol. 8 (2007) 191–197.

[264] K. Yanaba, J.D. Bouaziz, K.M. Haas, J.C. Poe, M. Fujimoto, T.F. Tedder, A regulatory B cell subset with a unique CD1dhiCD5+ phenotype controls T cell-dependent inflammatory responses, Immunity 28 (2008) 639–650.

[265] T.A. Wynn, K.M. Vannella, Macrophages in tissue repair, regeneration, and fibrosis, Immunity 44 (2016) 450–462.

[266] Y. Liu, X. Cao, Immunosuppressive cells in tumor immune escape and metastasis, J. Mol. Med. 94 (2016) 509–522.

[267] D.G. Jie Zhou, Y. Nefedova, A. Lei, Neutrophils and PMN-MDSCs: their biological role and interaction with stromal cells, Semin. Immunol. 35 (2018) 19–28.

[268] M. Swiecki, M. Colonna, Unraveling the functions of plasmacytoid dendritic cells during viral infections, autoimmunity, and tolerance, Immunol. Rev. 234 (2010) 142–162.

[269] P.G. Coulie, B.J. Van Den Eynde, P. Van Der Bruggen, T. Boon, Tumour antigens recognized by T lymphocytes: at the core of cancer immunotherapy, Nat. Rev. Canc. 14 (2014) 135–146.

[270] S. Aras, M. Raza Zaidi, TAMeless traitors: macrophages in cancer progression and metastasis, Br. J. Canc. 117 (2017) 1583–1591.

[271] L. Chiossone, P.Y. Dumas, M. Vienne, E. Vivier, Natural killer cells and other innate lymphoid cells in cancer, Nat. Rev. Immunol. 18 (2018) 671–688.

[272] B. Farhood, M. Najafi, K. Mortezaee, CD8+ cytotoxic T lymphocytes in cancer immunotherapy: a review, J. Cell. Physiol. 234 (2019) 8509–8521.

[273] A. Sarvaria, J.A. Madrigal, A. Saudemont, B cell regulation in cancer and anti-tumor immunity, Cell. Mol. Immunol. 14 (2017) 662–674.

[274] A.E. Chang, L. Qiao, H. Song, S. Teitz-Tennenbaum, E.J. Donald, M. Li, In vivo sensitized and in vitro activated B cells mediate tumor regression in cancer adoptive immunotherapy, J. Immunol. 183 (2009) 3195–3203.

[275] P. Tsou, H. Katayama, E.J. Ostrin, S.M. Hanash, The emerging role of b cells in tumor immunity, Canc. Res. 76 (2016) 5591–5601.

[276] M. Schwartz, Y. Zhang, J.D. Rosenblatt, B cell regulation of the anti-tumor response and role in carcinogenesis, J. Immunother. Cancer. 4 (2016) 1–15.

[277] A.B. Purevdorj, B. Olkhanud, B. Damdinsuren, M. Bodogai, R.E. Gress, R. Sen, K. Wejksza, E. Malchinkhuu, R.P. Wersto, Tumor-evoked regulatory B cells promote breast cancer metastasis by converting resting CD4+ T cells to T regulatory cells, Canc. Res. 71 (2011) 13505–13515.
[278] I. Algarra, A. García-Lora, T. Cabrera, F. Ruiz-Cabello, F. Garrido, The selection of tumor variants with altered expression of classical and nonclassical MHC class I molecules: implications for tumor immune escape, Cancer Immunol. Immunother. 53 (2004) 904–910.
[279] L. Zitvogel, A. Tesniere, G. Kroemer, Cancer despite immunosurveillance: immunoselection and immunosubversion, Nat. Rev. Immunol. 6 (2006) 715–727.
[280] D. Wolf, S. Sopper, A. Pircher, G. Gastl, A.M. Wolf, Treg(s) in cancer: friends or foe? J. Cell. Physiol. 230 (2015) 2598–2605.
[281] A.L. Mellor, D.H. Munn, Creating immune privilege: active local suppression that benefits friends, but protects foes, Nat. Rev. Immunol. 8 (2008) 161.
[282] J. Wolfers, A. Lozier, G. Raposo, A. Regnault, C. Théry, C. Masurier, C. Flament, S. Pouzieux, F. Faure, T. Tursz, E. Angevin, S. Amigorena, L. Zitvogel, Tumor-derived exosomes are a source of shared tumor rejection antigens for CTL cross-priming, Nat. Med. 7 (2001) 297–303.
[283] G.N. Barber, STING: infection, inflammation and cancer Glen, Nat. Rev. Immunol. 15 (2015) 760–770.
[284] L. Muller, P. Simms, C.S. Hong, M.I. Nishimura, E.K. Jackson, S.C. Watkins, T.L. Whiteside, Human tumor-derived exosomes (TEX) regulate Treg functions via cell surface signaling rather than uptake mechanisms, OncoImmunology 6 (2017) 1–9.
[285] F.M. Barros, F. Carneiro, J.C. Machado, S.A. Melo, Exosomes and immune response in cancer: friends or foes? Front. Immunol. 9 (2018).
[286] R. Gastpar, M. Gehrmann, M.A. Bausero, A. Asea, C. Gross, J.A. Schroeder, G. Multhoff, Heat shock protein 70 surface-positive tumor exosomes, October 65 (2005) 5238–5247.
[287] S. Yu, C. Liu, K. Su, J. Wang, Y. Liu, L. Zhang, C. Li, Y. Cong, R. Kimberly, W.E. Grizzle, C. Falkson, H.-G. Zhang, Tumor exosomes inhibit differentiation of bone marrow dendritic cells, J. Immunol. 178 (2007) 6867–6875.
[288] G. Berchem, M.Z. Noman, M. Bosseler, J. Paggetti, S. Baconnais, E. Le cam, A. Nanbakhsh, E. Moussay, F. Mami-Chouaib, B. Janji, S. Chouaib, Hypoxic tumor-derived microvesicles negatively regulate NK cell function by a mechanism involving TGF-β and miR23a transfer, OncoImmunology 5 (2016) 1–13.
[289] X. Wang, H. Shen, G. Zhangyuan, R. Huang, W. Zhang, Q. He, K. Jin, H. Zhuo, Z. Zhang, J. Wang, B. Sun, X. Lu, 14-3-3ζ delivered by hepatocellular carcinoma-derived exosomes impaired anti-tumor function of tumor-infiltrating T lymphocytes article, Cell Death Dis. 9 (2018).
[290] A. Clayton, J.P. Mitchell, J. Court, M.D. Mason, Z. Tabi, Human tumor-derived exosomes selectively impair lymphocyte responses to interleukin-2, Canc. Res. 67 (2007) 7458–7466.
[291] Y.Y. Wan, R.A. Flavell, "Yin-Yang" functions of transforming growth factor-β and T regulatory cells in immune regulation, Immunol. Rev. 220 (2007) 199–213.
[292] B. Zafirova, F.M. Wensveen, M. Gulin, B. Polić, Regulation of immune cell function and differentiation by the NKG2D receptor, Cell. Mol. Life Sci. 68 (2011) 3519–3529.
[293] Z.Z. Yang, D.M. Grote, S.C. Ziesmer, B. Xiu, N.R. Yates, F.J. Secreto, L.S. Hodge, T.E. Witzig, A.J. Novak, S.M. Ansell, Soluble and membrane-bound TGF-β-mediated regulation of intratumoral T cell differentiation and function in B-cell non-hodgkin lymphoma, PLoS One 8 (2013) 1–13.

[294] L. Ye, Q. Zhang, Y. Cheng, X. Chen, G. Wang, M. Shi, T. Zhang, Y. Cao, H. Pan, L. Zhang, G. Wang, Y. Deng, Y. Yang, G. Chen, Tumor-derived exosomal HMGB1 fosters hepatocellular carcinoma immune evasion by promoting TIM-1+ regulatory B cell expansion, J. Immunother. Cancer. 6 (2018) 1−15.

[295] G. Andreola, L. Rivoltini, C. Castelli, V. Huber, P. Perego, P. Deho, P. Squarcina, P. Accornero, F. Lozupone, L. Lugini, A. Stringaro, A. Molinari, G. Arancia, M. Gentile, G. Parmiani, S. Fais, Induction of lymphocyte apoptosis by tumor cell secretion of FasL-bearing microvesicles, J. Exp. Med. 195 (2002) 1303−1316.

[296] H.O. Alsaab, S. Sau, R. Alzhrani, K. Tatiparti, K. Bhise, S.K. Kashaw, A.K. Iyer, PD-1 and PD-L1 checkpoint signaling inhibition for cancer immunotherapy: mechanism, combinations, and clinical outcome, Front. Pharmacol. 8 (2017) 1−15.

[297] G. Chen, A.C. Huang, W. Zhang, G. Zhang, M. Wu, W. Xu, Z. Yu, J. Yang, B. Wang, H. Sun, H. Xia, Q. Man, W. Zhong, L.F. Antelo, B. Wu, X. Xiong, X. Liu, L. Guan, T. Li, S. Liu, R. Yang, Y. Lu, L. Dong, S. McGettigan, R. Somasundaram, R. Radhakrishnan, G. Mills, Y. Lu, J. Kim, Y.H. Chen, H. Dong, Y. Zhao, G.C. Karakousis, T.C. Mitchell, L.M. Schuchter, M. Herlyn, E.J. Wherry, X. Xu, W. Guo, Exosomal PD-L1 contributes to immunosuppression and is associated with anti-PD-1 response, Nature 560 (2018) 382−386.

[298] M. Hedlund, O. Nagaeva, D. Kargl, V. Baranov, L. Mincheva-Nilsson, Thermal- and oxidative stress causes enhanced release of NKG2D ligand-bearing immunosuppressive exosomes in leukemia/lymphoma T and B cells, PloS One 6 (2011).

[299] C. Battke, R. Ruiss, U. Welsch, P. Wimberger, S. Lang, S. Jochum, R. Zeidler, Tumour exosomes inhibit binding of tumour-reactive antibodies to tumour cells and reduce ADCC, Cancer Immunol, Immunotherapy 60 (2011) 639−648.

[300] X. Zhang, H. Shi, X. Yuan, P. Jiang, H. Qian, W. Xu, Tumor-derived exosomes induce N2 polarization of neutrophils to promote gastric cancer cell migration, Mol. Canc. 17 (2018) 1−16.

[301] Q. Han, H. Zhao, Y. Jiang, C. Yin, J. Zhang, HCC-derived exosomes: critical player and target for cancer immune escape, Cells 8 (2019) 558.

[302] J. Liu, L. Fan, H. Yu, J. Zhang, Y. He, D. Feng, F. Wang, X. Li, Q. Liu, Y. Li, Z. Guo, B. Gao, W. Wei, H. Wang, G. Sun, Endoplasmic reticulum stress causes liver cancer cells to release exosomal miR-23a-3p and up-regulate programmed death ligand 1 expression in macrophages, Hepatology 70 (2019) 241−258.

[303] K. Gabrusiewicz, X. Li, J. Wei, Y. Hashimoto, A.L. Marisetty, M. Ott, F. Wang, D. Hawke, J. Yu, L.M. Healy, A. Hossain, J.C. Akers, S.N. Maiti, S. Yamashita, Y. Shimizu, K. Dunner, M.A. Zal, J.K. Burks, J. Gumin, F. Nwajei, A. Rezavanian, S. Zhou, G. Rao, R. Sawaya, G.N. Fuller, J.T. Huse, J.P. Antel, S. Li, L. Cooper, E.P. Sulman, C. Chen, C. Geula, R. Kalluri, T. Zal, A.B. Heimberger, Glioblastoma stem cell-derived exosomes induce M2 macrophages and PD-L1 expression on human monocytes, OncoImmunology 7 (2018).

[304] X. Chen, J. Zhou, X. Li, X. Wang, Y. Lin, X. Wang, Exosomes derived from hypoxic epithelial ovarian cancer cells deliver microRNAs to macrophages and elicit a tumor-promoted phenotype, Canc. Lett. 435 (2018) 80−91.

[305] Y.L. Hsu, J.Y. Hung, W.A. Chang, S.F. Jian, Y.S. Lin, Y.C. Pan, C.Y. Wu, P.L. Kuo, Hypoxic lung-cancer-derived extracellular vesicle MicroRNA-103a increases the oncogenic effects of macrophages by targeting PTEN, Mol. Ther. 26 (2018) 568−581.

[306] M. Fabbri, A. Paone, F. Calore, R. Galli, E. Gaudio, R. Santhanam, F. Lovat, P. Fadda, C. Mao, G.J. Nuovo, N. Zanesi, M. Crawford, G.H. Ozer, D. Wernicke, H. Alder, M.A. Caligiuri, P. Nana-Sinkam, D. Perrotti, C.M. Croce, MicroRNAs bind to Toll-like receptors to induce prometastatic inflammatory response, Proc. Natl. Acad. Sci. U.S.A. 109 (2012).

[307] Y. Liu, Y. Gu, Y. Han, Q. Zhang, Z. Jiang, X. Zhang, B. Huang, X. Xu, J. Zheng, X. Cao, Tumor exosomal RNAs promote lung pre-metastatic niche formation by activating alveolar epithelial TLR3 to recruit neutrophils, Canc. Cell 30 (2016) 243–256.

[308] R.V. Sionov, Z.G. Fridlender, Z. Granot, The multifaceted roles neutrophils play in the tumor microenvironment, Cancer Microenviron. 8 (2015) 125–158.

[309] B. Costa-silva, N.M. Aiello, A.J. Ocean, S. Singh, B.K. Thakur, A. Becker, A. Hoshino, M.T. Mark, H. Molina, J. Xiang, T. Zhang, T. Theilen, G. García-santos, C. Williams, Y. Ararso, Y. Huang, G. Rodrigues, Pancreatic cancer exosomes initiate pre-metastatic niche formation in the liver Bruno, Nat. Cell Biol. 17 (2018) 816–826.

[310] M.J.P. Ferdinando Pucci, C. Garris, C.P. Lai, A. Newton, C. Pfirschke, C. Engblom, D. Alvarez, M. Sprachman, C. Evavold, A. Magnuson, U.H. von Andrian, K. Glatz, X. O Breakefield, T.R. Mempel, SCS macrophages suppress melanoma by restricting tumor-derived vesicle–B cell interactions, Science 352 (2016) 242–246.

[311] S. Hao, O. Bai, J. Yuan, M. Qureshi, J. Xiang, Dendritic cell-derived exosomes stimulate stronger CD8+ CTL responses and antitumor immunity than tumor cell-derived exosomes, Cell. Mol. Immunol. 3 (2006) 205–211.

[312] F. André, N. Chaput, N.E.C. Schartz, C. Flament, N. Aubert, J. Bernard, F. Lemonnier, G. Raposo, B. Escudier, D.-H. Hsu, T. Tursz, S. Amigorena, E. Angevin, L. Zitvogel, Exosomes as potent cell-free peptide-based vaccine. I. Dendritic cell-derived exosomes transfer functional MHC class I/peptide complexes to dendritic cells, J. Immunol. 172 (2004) 2126–2136.

[313] S. Bae, J. Brumbaugh, B. Bonavida, Exosomes derived from cancerous and non-cancerous cells regulate the anti-tumor response in the tumor microenvironment, Genes and Cancer 9 (2018) 87–100.

[314] L. Lugini, S. Cecchetti, V. Huber, F. Luciani, G. Macchia, F. Spadaro, L. Paris, L. Abalsamo, M. Colone, A. Molinari, F. Podo, L. Rivoltini, C. Ramoni, S. Fais, Immune surveillance properties of human NK cell-derived exosomes, J. Immunol. 189 (2012) 2833–2842.

[315] N. Seo, Y. Shirakura, Y. Tahara, F. Momose, N. Harada, H. Ikeda, K. Akiyoshi, H. Shiku, Activated CD8+ T cell extracellular vesicles prevent tumour progression by targeting of lesional mesenchymal cells, Nat. Commun. 9 (2018) 1–11.

[316] M. Yang, J. Chen, F. Su, B. Yu, F. Su, L. Lin, Y. Liu, J.D. Huang, E. Song, Microvesicles secreted by macrophages shuttle invasion-potentiating microRNAs into breast cancer cells, Mol. Canc. 10 (2011) 117.

[317] F. Chen, J. Chen, L. Yang, J. Liu, X. Zhang, Y. Zhang, Q. Tu, D. Yin, D. Lin, P.P. Wong, D. Huang, Y. Xing, J. Zhao, M. Li, Q. Liu, F. Su, S. Su, E. Song, Extracellular vesicle-packaged HIF-1$\alpha$-stabilizing lncRNA from tumour-associated macrophages regulates aerobic glycolysis of breast cancer cells, Nat. Cell Biol. 21 (2019) 498–510.

[318] N. Karimi, A. Cvjetkovic, S.C. Jang, R. Crescitelli, M.A. Hosseinpour Feizi, R. Nieuwland, J. Lötvall, C. Lässer, Detailed analysis of the plasma extracellular vesicle proteome after separation from lipoproteins, Cell. Mol. Life Sci. 75 (2018) 2873–2886.

[319] M.P. Caby, D. Lankar, C. Vincendeau-Scherrer, G. Raposo, C. Bonnerot, Exosomal-like vesicles are present in human blood plasma, Int. Immunol. (2005), https://doi.org/10.1093/intimm/dxh267.

[320] C. Lässer, V. Seyed Alikhani, K. Ekström, M. Eldh, P. Torregrosa Paredes, A. Bossios, M. Sjöstrand, S. Gabrielsson, J. Lötvall, H. Valadi, Human saliva, plasma and breast milk exosomes contain RNA: uptake by macrophages, J. Transl. Med. (2011), https://doi.org/10.1186/1479-5876-9-9.

[321] M. Li, E. Zeringer, T. Barta, J. Schageman, A. Cheng, A.V. Vlassov, Analysis of the RNA content of the exosomes derived from blood serum and urine and its potential as biomarkers, Philos. Trans. R. Soc. B Biol. Sci. (2014).
[322] D.A.A. Raj, I. Fiume, G. Capasso, G. Pocsfalvi, A multiplex quantitative proteomics strategy for protein biomarker studies in urinary exosomes, Kidney Int. (2012), https://doi.org/10.1038/ki.2012.25.
[323] R.C. Wiggins, A. Glatfelter, B. Kshirsagar, J. Brukman, Procoagulant activity in normal human urine associated with subcellular particles, Kidney Int. (1986), https://doi.org/10.1038/ki.1986.39.
[324] S. Keller, J. Ridinger, A.K. Rupp, J.W.G. Janssen, P. Altevogt, Body fluid derived exosomes as a novel template for clinical diagnostics, J. Transl. Med. (2011), https://doi.org/10.1186/1479-5876-9-86.
[325] Y. Ogawa, M. Kanai-Azuma, Y. Akimoto, H. Kawakami, R. Yanoshita, Exosome-like vesicles with dipeptidyl peptidase IV in human saliva, Biol. Pharm. Bull. (2008), https://doi.org/10.1248/bpb.31.1059.
[326] C. Lässer, S.E. O'Neil, L. Ekerljung, K. Ekström, M. Sjöstrand, J. Lötvall, RNA-containing exosomes in human nasal secretions, Am. J. Rhinol. Allergy. (2011), https://doi.org/10.2500/ajra.2011.25.3573.
[327] A.I. Masyuk, B.Q. Huang, C.J. Ward, S.A. Gradilone, J.M. Banales, T.V. Masyuk, B. Radtke, P.L. Splinter, N.F. LaRusso, Biliary exosomes influence cholangiocyte regulatory mechanisms and proliferation through interaction with primary cilia, Am. J. Physiol. Gastrointest. Liver Physiol. (2010), https://doi.org/10.1152/ajpgi.00093.2010.
[328] C. Admyre, S.M. Johansson, K.R. Qazi, J.-J. Filén, R. Lahesmaa, M. Norman, E.P.A. Neve, A. Scheynius, S. Gabrielsson, Exosomes with immune modulatory features are present in human breast milk, J. Immunol. (2007), https://doi.org/10.4049/jimmunol.179.3.1969.
[329] K.M. Perkumas, E.A. Hoffman, B.S. McKay, R.R. Allingham, W.D. Stamer, Myocilin-associated exosomes in human ocular samples, Exp. Eye Res. (2007), https://doi.org/10.1016/j.exer.2006.09.020.
[330] G. Ronquist, I. Brody, The prostasome: its secretion and function in man, BBA - Rev. Biomembr. (1985), https://doi.org/10.1016/0304-4157(85)90008-5.
[331] A. Asea, C. Jean-Pierre, P. Kaur, P. Rao, I.M. Linhares, D. Skupski, S.S. Witkin, Heat shock protein-containing exosomes in mid-trimester amniotic fluids, J. Reprod. Immunol. (2008), https://doi.org/10.1016/j.jri.2008.06.001.
[332] J.M. Street, P.E. Barran, C.L. Mackay, S. Weidt, C. Balmforth, T.S. Walsh, R.T.A. Chalmers, D.J. Webb, J.W. Dear, Identification and proteomic profiling of exosomes in human cerebrospinal fluid, J. Transl. Med. (2012), https://doi.org/10.1186/1479-5876-10-5.
[333] J.H.W. Distler, A. Jüngel, L.C. Huber, C.A. Seemayer, C.F. Reich, R.E. Gay, B.A. Michel, A. Fontana, S. Gay, D.S. Pisetsky, O. Distler, The induction of matrix metalloproteinase and cytokine expression in synovial fibroblasts stimulated with immune cell microparticles, Proc. Natl. Acad. Sci. U.S.A. (2005), https://doi.org/10.1073/pnas.0409781102.
[334] M.P. Bard, J.P. Hegmans, A. Hemmes, T.M. Luider, R. Willemsen, L.A.A. Severijnen, J.P. Van Meerbeeck, S.A. Burgers, H.C. Hoogsteden, B.N. Lambrecht, Proteomic analysis of exosomes isolated from human malignant pleural effusions, Am. J. Respir. Cell Mol. Biol. (2004), https://doi.org/10.1165/rcmb.2003-0238OC.
[335] B. Levänen, N.R. Bhakta, P. Torregrosa Paredes, R. Barbeau, S. Hiltbrunner, J.L. Pollack, C.M. Sköld, M. Svartengren, J. Grunewald, S. Gabrielsson, A. Eklund, B.M. Larsson, P.G. Woodruff, D.J. Erle, Å.M. Wheelock, Altered

microRNA profiles in bronchoalveolar lavage fluid exosomes in asthmatic patients, J. Allergy Clin. Immunol. (2013), https://doi.org/10.1016/j.jaci.2012.11.039.
[336] K.R. Qazi, P.T. Paredes, B. Dahlberg, J. Grunewald, A. Eklund, S. Gabrielsson, Proinflammatory exosomes in bronchoalveolar lavage fluid of patients with sarcoidosis, Thorax (2010), https://doi.org/10.1136/thx.2009.132027.
[337] V.R. Minciacchi, A. Zijlstra, M.A. Rubin, D. Di Vizio, Extracellular vesicles for liquid biopsy in prostate cancer: where are we and where are we headed? Prostate Cancer Prostatic Dis. 20 (2017) 251–258.
[338] M.S. Chiriacò, M. Bianco, A. Nigro, E. Primiceri, F. Ferrara, A. Romano, A. Quattrini, R. Furlan, V. Arima, G. Maruccio, Lab-on-chip for exosomes and microvesicles detection and characterization, Sensors (2018) 18.
[339] G. De Rubis, S. Rajeev Krishnan, M. Bebawy, Liquid biopsies in cancer diagnosis, monitoring, and prognosis, Trends Pharmacol. Sci. 40 (2019) 172–186.
[340] C. Théry, S. Amigorena, G. Raposo, A. Clayton, Isolation and characterization of exosomes from cell culture supernatants and biological fluids, Curr. Protoc. Cell Biol. (2006), https://doi.org/10.1002/0471143030.cb0322s30.
[341] C.A. Fitts, N. Ji, Y. Li, C. Tan, Exploiting exosomes in cancer liquid biopsies and drug delivery, Adv. Healthc. Mater. (2019), https://doi.org/10.1002/adhm.201801268.
[342] N. Garcia-Romero, S. Esteban-Rubio, G. Rackov, J. Carrión-Navarro, C. Belda-Iniesta, A. Ayuso-Sacido, Extracellular vesicles compartment in liquid biopsies: clinical application, Mol. Aspect. Med. (2018), https://doi.org/10.1016/j.mam.2017.11.009.
[343] F.L. Ricklefs, Q. Alayo, H. Krenzlin, A.B. Mahmoud, M.C. Speranza, H. Nakashima, J.L. Hayes, K. Lee, L. Balaj, C. Passaro, A.K. Rooj, S. Krasemann, B.S. Carter, C.C. Chen, T. Steed, J. Treiber, S. Rodig, K. Yang, I. Nakano, H. Lee, R. Weissleder, X.O. Breakefield, J. Godlewski, M. Westphal, K. Lamszus, G.J. Freeman, A. Bronisz, S.E. Lawler, E.A. Chiocca, Immune evasion mediated by PD-L1 on glioblastoma-derived extracellular vesicles, Sci. Adv. (2018), https://doi.org/10.1126/sciadv.aar2766.
[344] M.N. Theodoraki, S.S. Yerneni, T.K. Hoffmann, W.E. Gooding, T.L. Whiteside, Clinical significance of PD-L1 þ exosomes in plasma of head and neck cancer patients, Clin. Canc. Res. (2018), https://doi.org/10.1158/1078-0432.CCR-17-2664.
[345] G. Chen, A.C. Huang, W. Zhang, G. Zhang, M. Wu, W. Xu, Z. Yu, J. Yang, B. Wang, H. Sun, H. Xia, Q. Man, W. Zhong, L.F. Antelo, B. Wu, X. Xiong, X. Liu, L. Guan, T. Li, S. Liu, R. Yang, Y. Lu, L. Dong, S. McGettigan, R. Somasundaram, R. Radhakrishnan, G. Mills, Y. Lu, J. Kim, Y.H. Chen, H. Dong, Y. Zhao, G.C. Karakousis, T.C. Mitchell, L.M. Schuchter, M. Herlyn, E.J. Wherry, X. Xu, W. Guo, PDL1-exosome-melanoma_Nat2018, Nature (2018), https://doi.org/10.1038/s41586-018-0392-8.
[346] M. Del Re, R. Marconcini, G. Pasquini, E. Rofi, C. Vivaldi, F. Bloise, G. Restante, E. Arrigoni, C. Caparello, M. Grazia Bianco, S. Crucitta, I. Petrini, E. Vasile, A. Falcone, R. Danesi, PD-L1 mRNA expression in plasma-derived exosomes is associated with response to anti-PD-1 antibodies in melanoma and NSCLC, Br. J. Canc. (2018), https://doi.org/10.1038/bjc.2018.9.
[347] P.G. Moon, J.E. Lee, Y.E. Cho, S.J. Lee, J.H. Jung, Y.S. Chae, H.I. Bae, Y.B. Kim, I.S. Kim, H.Y. Park, M.C. Baek, Identification of developmental endothelial locus-1 on circulating extracellular vesicles as a novel biomarker for early breast cancer detection, Clin. Canc. Res. (2016), https://doi.org/10.1158/1078-0432.CCR-15-0654.

[348] B. Costa-Silva, N.M. Aiello, A.J. Ocean, S. Singh, H. Zhang, B.K. Thakur, A. Becker, A. Hoshino, M.T. Mark, H. Molina, J. Xiang, T. Zhang, T.M. Theilen, G. García-Santos, C. Williams, Y. Ararso, Y. Huang, G. Rodrigues, T.L. Shen, K.J. Labori, I.M.B. Lothe, E.H. Kure, J. Hernandez, A. Doussot, S.H. Ebbesen, P.M. Grandgenett, M.A. Hollingsworth, M. Jain, K. Mallya, S.K. Batra, W.R. Jarnagin, R.E. Schwartz, I. Matei, H. Peinado, B.Z. Stanger, J. Bromberg, D. Lyden, Pancreatic cancer exosomes initiate pre-metastatic niche formation in the liver, Nat. Cell Biol. (2015), https://doi.org/10.1038/ncb3169.

[349] S. Garinet, P. Laurent-Puig, H. Blons, J.-B. Oudart, Current and future molecular testing in NSCLC, what can we expect from new sequencing technologies? J. Clin. Med. (2018) https://doi.org/10.3390/jcm7060144.

[350] X. Lai, M. Wang, S.D. McElyea, S. Sherman, M. House, M. Korc, A microRNA signature in circulating exosomes is superior to exosomal glypican-1 levels for diagnosing pancreatic cancer, Canc. Lett. (2017), https://doi.org/10.1016/j.canlet.2017.02.019.

[351] A.E. Frampton, M.M. Prado, E. López-Jiménez, A.B. Fajardo-Puerta, Z.A.R. Jawad, P. Lawton, E. Giovannetti, N.A. Habib, L. Castellano, J. Stebbing, J. Krell, L.R. Jiao, Glypican-1 is enriched in circulating-exosomes in pancreatic cancer and correlates with tumor burden, Oncotarget (2018), https://doi.org/10.18632/oncotarget.24873.

[352] J.M. Lewis, A.D. Vyas, Y. Qiu, K.S. Messer, R. White, M.J. Heller, Integrated analysis of exosomal protein biomarkers on alternating current electrokinetic chips enables rapid detection of pancreatic cancer in patient blood, ACS Nano (2018), https://doi.org/10.1021/acsnano.7b08199.

[353] J.Y. Qian, Y.L. Tan, Y. Zhang, Y.F. Yang, X.Q. Li, Prognostic value of glypican-1 for patients with advanced pancreatic cancer following regional intra-arterial chemotherapy, Oncol. Lett. (2018), https://doi.org/10.3892/ol.2018.8701.

[354] Z. Zhao, J. Fan, Y.M.S. Hsu, C.J. Lyon, B. Ning, T.Y. Hu, Extracellular vesicles as cancer liquid biopsies: from discovery, validation, to clinical application, Lab Chip 19 (2019) 1114–1140.

[355] S. Yang, S.P.Y. Che, P. Kurywchak, J.L. Tavormina, L.B. Gansmo, P. Correa de Sampaio, M. Tachezy, M. Bockhorn, F. Gebauer, A.R. Haltom, S.A. Melo, V.S. LeBleu, R. Kalluri, Detection of mutant KRAS and TP53 DNA in circulating exosomes from healthy individuals and patients with pancreatic cancer, Canc. Biol. Ther. (2017), https://doi.org/10.1080/15384047.2017.1281499.

[356] R. Kalluri, V.S. Lebleu, Discovery of double-stranded genomic DNA in circulating exosomes, Cold Spring Harbor Symp. Quant. Biol. (2016), https://doi.org/10.1101/sqb.2016.81.030932.

[357] K. Allenson, J. Castillo, F.A. San Lucas, G. Scelo, D.U. Kim, V. Bernard, G. Davis, T. Kumar, M. Katz, M.J. Overman, L. Foretova, E. Fabianova, I. Holcatova, V. Janout, F. Meric-Bernstam, P. Gascoyne, I. Wistuba, G. Varadhachary, P. Brennan, S. Hanash, D. Li, A. Maitra, H. Alvarez, High prevalence of mutant KRAS in circulating exosome-derived DNA from early-stage pancreatic cancer patients, Ann. Oncol. (2017), https://doi.org/10.1093/annonc/mdx004.

[358] A.K. Krug, D. Enderle, C. Karlovich, T. Priewasser, S. Bentink, A. Spiel, K. Brinkmann, J. Emenegger, D.G. Grimm, E. Castellanos-Rizaldos, J.W. Goldman, L.V. Sequist, J.C. Soria, D.R. Camidge, S.M. Gadgeel, H.A. Wakelee, M. Raponi, M. Noerholm, J. Skog, Improved EGFR mutation detection using combined exosomal RNA and circulating tumor DNA in NSCLC patient plasma, Ann. Oncol. (2018), https://doi.org/10.1093/annonc/mdx765.

[359] F.A. San Lucas, K. Allenson, V. Bernard, J. Castillo, D.U. Kim, K. Ellis, E.A. Ehli, G.E. Davies, J.L. Petersen, D. Li, R. Wolff, M. Katz, G. Varadhachary, I. Wistuba, A. Maitra, H. Alvarez, Minimally invasive genomic and transcriptomic profiling of

visceral cancers by next-generation sequencing of circulating exosomes, Ann. Oncol. (2016), https://doi.org/10.1093/annonc/mdv604.
[360] J. Zhang, S.C. Liu, X.H. Luo, G.X. Tao, M. Guan, H. Yuan, D.K. Hu, Exosomal long noncoding RNAs are differentially expressed in the cervicovaginal lavage samples of cervical cancer patients, J. Clin. Lab. Anal. 30 (2016) 1116–1121.
[361] S. Fais, L. O'Driscoll, F.E. Borras, E. Buzas, G. Camussi, F. Cappello, J. Carvalho, A. Cordeiro Da Silva, H. Del Portillo, S. El Andaloussi, T. Ficko Trček, R. Furlan, A. Hendrix, I. Gursel, V. Kralj-Iglic, B. Kaeffer, M. Kosanovic, M.E. Lekka, G. Lipps, M. Logozzi, A. Marcilla, M. Sammar, A. Llorente, I. Nazarenko, C. Oliveira, G. Pocsfalvi, L. Rajendran, G. Raposo, E. Rohde, P. Siljander, G. Van Niel, M.H. Vasconcelos, M. Yáñez-Mó, M.L. Yliperttula, N. Zarovni, A.B. Zavec, B. Giebel, Evidence-based clinical use of nanoscale extracellular vesicles in nanomedicine, ACS Nano (2016), https://doi.org/10.1021/acsnano.5b08015.
[362] M. Salehi, M. Sharifi, Exosomal miRNAs as novel cancer biomarkers: challenges and opportunities, J. Cell. Physiol. 233 (2018) 6370–6380.
[363] C.C. Pritchard, H.H. Cheng, M. Tewari, MicroRNA profiling: approaches and considerations, Nat. Rev. Genet. (2012), https://doi.org/10.1038/nrg3198.
[364] I.H. Chen, L. Xue, C.C. Hsu, J.S.P. Paez, L. Panb, H. Andaluz, M.K. Wendt, A.B. Iliuk, J.K. Zhu, W.A. Tao, Phosphoproteins in extracellular vesicles as candidate markers for breast cancer, Proc. Natl. Acad. Sci. U.S.A. (2017), https://doi.org/10.1073/pnas.1618088114.
[365] J. Castillo, V. Bernard, F.A. San Lucas, K. Allenson, M. Capello, D.U. Kim, P. Gascoyne, F.C. Mulu, B.M. Stephens, J. Huang, H. Wang, A.A. Momin, R.O. Jacamo, M. Katz, R. Wolff, M. Javle, G. Varadhachary, I.I. Wistuba, S. Hanash, A. Maitra, H. Alvarez, Surfaceome profiling enables isolation of cancerspecific exosomal cargo in liquid biopsies from pancreatic cancer patients, Ann. Oncol. (2018), https://doi.org/10.1093/annonc/mdx542.
[366] M. Record, S. Silvente-Poirot, M. Poirot, M.J.O. Wakelam, Extracellular vesicles: lipids as key components of their biogenesis and functions, J. Lipid Res. 59 (2018) 1316–1324.
[367] T. Skotland, K. Ekroos, D. Kauhanen, H. Simolin, T. Seierstad, V. Berge, K. Sandvig, A. Llorente, Molecular lipid species in urinary exosomes as potential prostate cancer biomarkers, Eur. J. Canc. 70 (2017) 122–132.
[368] J.S. Yang, J.C. Lee, S.K. Byeon, K.H. Rha, M.H. Moon, Size dependent lipidomic analysis of urinary exosomes from patients with prostate cancer by flow field-flow fractionation and nanoflow liquid chromatography-tandem mass spectrometry, Anal. Chem. (2017), https://doi.org/10.1021/acs.analchem.6b04634.
[369] L. Tao, J. Zhou, C. Yuan, L. Zhang, D. Li, D. Si, D. Xiu, L. Zhong, Metabolomics identifies serum and exosomes metabolite markers of pancreatic cancer, Metabolomics (2019), https://doi.org/10.1007/s11306-019-1550-1.
[370] B. Burla, M. Arita, M. Arita, A.K. Bendt, A. Cazenave-Gassiot, E.A. Dennis, K. Ekroos, X. Han, K. Ikeda, G. Liebisch, M.K. Lin, T.P. Loh, P.J. Meikle, M. Orešič, O. Quehenberger, A. Shevchenko, F. Torta, M.J.O. Wakelam, C.E. Wheelock, M.R. Wenk, MS-based lipidomics of human blood plasma: a community-initiated position paper to develop accepted guidelines, J. Lipid Res. (2018), https://doi.org/10.1194/jlr.S087163.
[371] B. Mateescu, E.J.K. Kowal, B.W.M. van Balkom, S. Bartel, S.N. Bhattacharyya, E.I. Buzás, A.H. Buck, P. de Candia, F.W.N. Chow, S. Das, T.A.P. Driedonks, L. Fernández-Messina, F. Haderk, A.F. Hill, J.C. Jones, K.R. Van Keuren-Jensen, C.P. Lai, C. Lässer, I. di Liegro, T.R. Lunavat, M.J. Lorenowicz, S.L.N. Maas, I. Mäger, M. Mittelbrunn, S. Momma, K. Mukherjee, M. Nawaz, D.M. Pegtel, M.W. Pfaffl, R.M. Schiffelers, H. Tahara, C. Théry, J.P. Tosar, M.H.M. Wauben,

K.W. Witwer, E.N.M. Nolte-'t Hoen, Obstacles and opportunities in the functional analysis of extracellular vesicle RNA - an ISEV position paper, J. Extracell. Vesicles (2017), https://doi.org/10.1080/20013078.2017.1286095.

[372] A. Clayton, E. Boilard, E.I. Buzas, L. Cheng, J.M. Falcón-Perez, C. Gardiner, D. Gustafson, A. Gualerzi, A. Hendrix, A. Hoffman, J. Jones, C. Lässer, C. Lawson, M. Lenassi, I. Nazarenko, L. O'Driscoll, R. Pink, P.R.M. Siljander, C. Soekmadji, M. Wauben, J.A. Welsh, K. Witwer, L. Zheng, R. Nieuwland, Considerations towards a roadmap for collection, handling and storage of blood extracellular vesicles, J. Extracell. Vesicles 8 (2019).

[373] A. Savina, M. Furlán, M. Vidal, M.I. Colombo, Exosome release is regulated by a calcium-dependent mechanism in K562 cells, J. Biol. Chem. 278 (2003) 20083–20090.

[374] M. Simons, K. Trajkovic, C. Hsu, S. Chiantia, L. Rajendran, D. Wenzel, F. Wieland, P. Schwille, B. Brügger, Ceramide triggers budding of exosome vesicles into multivesicular endosomes, Science 319 (2008) 1244–1247.

[375] A. Bobrie, S. Krumeich, F. Reyal, C. Recchi, L.F. Moita, M.C. Seabra, M. Ostrowski, C. Théry, Rab27a supports exosome-dependent and -independent mechanisms that modify the tumor microenvironment and can promote tumor progression, Canc. Res. 72 (2012) 4920–4930.

[376] C.A. Franzen, P.E. Simms, A.F. Van Huis, K.E. Foreman, P.C. Kuo, G.N. Gupta, Characterization of uptake and internalization of exosomes by bladder cancer cells, BioMed Res. Int. 2014 (2014).

[377] J. Colino, C.M. Snapper, Exosomes from bone marrow dendritic cells pulsed with diphtheria toxoid preferentially induce type 1 antigen-specific IgG responses in naive recipients in the absence of free antigen, J. Immunol. 177 (2006) 3757–3762.

[378] L. Zitvogel, A. Regnault, A. Lozier, J. Wolfers, C. Flament, D. Tenza, P. Ricciardi-Castagnoli, G. Raposo, L. Zitvogel, A. Regnault, A. Lozier, J. Wolfers, C. Flament, D. Tenza, P. Ricciardi-Castagnoli, G. Raposo, S. Amigorena, Eradication of established murine tumors using a novel cell-free vaccine: dendritic cell-derived exosomes, Nat. Med. 4 (1998) 594–600.

[379] Y. Yao, C. Wang, W. Wei, C. Shen, X. Deng, L. Chen, L. Ma, S. Hao, Dendritic cells pulsed with leukemia cell-derived exosomes more efficiently induce antileukemic immunities, PLoS One 9 (2014) 2–8.

[380] B. Escudier, T. Dorval, N. Chaput, F. André, M.P. Caby, S. Novault, C. Flament, C. Leboulaire, C. Borg, S. Amigorena, C. Boccaccio, C. Bonnerot, O. Dhellin, M. Movassagh, S. Piperno, C. Robert, V. Serra, N. Valente, J.B. Le Pecq, A. Spatz, O. Lantz, T. Tursz, E. Angevin, L. Zitvogel, Vaccination of metastatic melanoma patients with autologous dendritic cell (DC) derived-exosomes: results of the first phase 1 clinical trial, J. Transl. Med. 3 (2005).

[381] M.A. Morse, J. Garst, T. Osada, S. Khan, A. Hobeika, T.M. Clay, N. Valente, R. Shreeniwas, M.A. Sutton, A. Delcayre, D.H. Hsu, J.B. Le Pecq, H.K. Lyerly, A phase I study of dexosome immunotherapy in patients with advanced non-small cell lung cancer, J. Transl. Med. 3 (2005).

[382] B. Besse, M. Charrier, V. Lapierre, E. Dansin, O. Lantz, D. Planchard, T. Le Chevalier, A. Livartoski, F. Barlesi, A. Laplanche, S. Ploix, N. Vimond, I. Peguillet, C. Théry, L. Lacroix, I. Zoernig, K. Dhodapkar, M. Dhodapkar, S. Viaud, J.C. Soria, K.S. Reiners, E. Pogge von Strandmann, F. Vély, S. Rusakiewicz, A. Eggermont, J.M. Pitt, L. Zitvogel, N. Chaput, Dendritic cell-derived exosomes as maintenance immunotherapy after first line chemotherapy in NSCLC, OncoImmunology 5 (2016).

[383] X. Zhang, C. Wang, J. Wang, Q. Hu, B. Langworthy, Y. Ye, W. Sun, J. Lin, T. Wang, J. Fine, H. Cheng, G. Dotti, P. Huang, Z. Gu, PD-1 blockade cellular vesicles for cancer immunotherapy, Adv. Mater. 30 (2018) 1–8.

[384] Y.Z. Qinqin Cheng, X. Shi, M. Han, G. Smbatyan, H.-J. Lenz, Reprogramming exosomes as nanoscale controllers of cellular immunity, J. Am. Chem. Soc. 140 (2018) 16413–16417.

[385] A.M. Marleau, C.S. Chen, J.A. Joyce, R.H. Tullis, Exosome removal as a therapeutic adjuvant in cancer, J. Transl. Med. 10 (2012) 1.

[386] M. Mendt, S. Kamerkar, H. Sugimoto, K.M. McAndrews, C.C. Wu, M. Gagea, S. Yang, E.V.R. Blanko, Q. Peng, X. Ma, J.R. Marszalek, A. Maitra, C. Yee, K. Rezvani, E. Shpall, V.S. LeBleu, R. Kalluri, Generation and testing of clinical-grade exosomes for pancreatic cancer, JCI Insight 3 (2018) 1–22.

[387] J. van R. Anoek Zomer, C. Maynard, F.J. Verweij, A. Kamermans, R. Schäfer, E. Beerling, R.M. Schiffelers, E. de Wit, J. Berenguer, S. Inge, J. Ellenbroek, T. Wurdinger, D.M. Pegtel, In vivo imaging reveals extracellular vesicle-mediated phenocopying of metastatic behavior, Cell 161 (2015).

[388] S. Lim, R.W. Yeo, R.C. Lai, B. Zhang, S.S. Tan, Y. Yin, B.J. Teh, Mesenchymal stem cell: an efficient mass producer of exosomes for drug delivery, Adv. Drug Deliv. Rev. 65 (2013) 336–341.

[389] L. Kordelas, V. Rebmann, A.K. Ludwig, S. Radtke, J. Ruesing, T.R. Doeppner, M. Epple, P.A. Horn, D.W. Beelen, B. Giebel, MSC-derived exosomes: a novel tool to treat therapy-refractory graft-versus-host disease, Leukemia 28 (2014) 970–973.

[390] J. Zhang, J. Matsumoto, T. Stewart, W.A. Banks, The transport mechanism of extracellular vesicles at the blood-brain barrier, Curr. Pharmaceut. Des. 23 (2017) 6206–6214.

[391] R.C. Lai, F. Arslan, M.M. Lee, N.S.K. Sze, A. Choo, T.S. Chen, M. Salto-Tellez, L. Timmers, C.N. Lee, R.M. El Oakley, G. Pasterkamp, D.P.V. de Kleijn, S.K. Lim, Exosome secreted by MSC reduces myocardial ischemia/reperfusion injury, Stem Cell Res. 4 (2010) 214–222.

[392] X. Zou, G. Zhang, Z. Cheng, D. Yin, T. Du, G. Ju, S. Miao, G. Liu, M. Lu, Y. Zhu, Microvesicles derived from human Wharton's Jelly mesenchymal stromal cells ameliorate renal ischemia-reperfusion injury in rats by suppressing CX3CL1, Stem Cell Res. Ther. 5 (2014).

[393] W. Nassar, M. El-Ansary, D. Sabry, M.A. Mostafa, T. Fayad, E. Kotb, M. Temraz, A.N. Saad, W. Essa, H. Adel, Umbilical cord mesenchymal stem cells derived extracellular vesicles can safely ameliorate the progression of chronic kidney diseases, Biomater. Res. 20 (2016).

[394] D. Sun, X. Zhuang, X. Xiang, Y. Liu, S. Zhang, C. Liu, S. Barnes, W. Grizzle, D. Miller, H.G. Zhang, A novel nanoparticle drug delivery system: the anti-inflammatory activity of curcumin is enhanced when encapsulated in exosomes, Mol. Ther. 18 (2010) 1606–1614.

[395] C. Schindler, A. Collinson, C. Matthews, A. Pointon, L. Jenkinson, R.R. Minter, T.J. Vaughan, N.J. Tigue, Exosomal delivery of doxorubicin enables rapid cell entry and enhanced in vitro potency, PLoS One 14 (2019) 1–19.

[396] T.Y. Morishita M, Y. Takahashi, A. Matsumoto, M. Nishikawa, Exosome-based tumor antigens-adjuvant co-delivery utilizing genetically engineered tumor cell-derived exosomes with immunostimulatory CpG DNA, Biomaterials 111 (2016) 55–65.

[397] B.M. Bell, I.D. Kirk, S. Hiltbrunner, S. Gabrielsson, J.J. Bultema, Designer exosomes as next-generation cancer immunotherapy, Nanomedicine Nanotechnology, Biol. Med. 12 (2016) 163–169.

[398] B. György, M.E. Hung, X.O. Breakefield, J.N. Leonard, Therapeutic applications of extracellular vesicles: clinical promise and open questions, Annu. Rev. Pharmacol. Toxicol. 55 (2015) 439–464.

[399] R. Ruiss, S. Jochum, R. Mocikat, W. Hammerschmidt, R. Zeidler, EBV-gp350 confers B-cell tropism to tailored exosomes is a neo-antigen in normal and malignant B cells-a new option for the treatment of B-CLL, PLoS One 6 (2011).

[400] L. Alvarez-Erviti, Y. Seow, H. Yin, C. Betts, S. Lakhal, M.J.A. Wood, Delivery of siRNA to the mouse brain by systemic injection of targeted exosomes, Nat. Biotechnol. 29 (2011) 341–345.

[401] R.C.G. Radha Munagala, F. Aqil, J. Jeyabalan, Bovine milk-derived exosomes for drug delivery Radha, Canc. Lett. 371 (2016) 48–61.

[402] B. Wang, X. Zhuang, Z. Bin Deng, H. Jiang, J. Mu, Q. Wang, X. Xiang, H. Guo, L. Zhang, G. Dryden, J. Yan, D. Miller, H.G. Zhang, Targeted drug delivery to intestinal macrophages by bioactive nanovesicles released from grapefruit, Mol. Ther. 22 (2014) 522–534.

[403] Y.S. Chen, E.Y. Lin, T.W. Chiou, H.J. Harn, Exosomes in clinical trial and their production in compliance with good manufacturing practice, Tzu Chi Med. J. 32 (2020) 113–120.

[404] J. Van Deun, P. Mestdagh, P. Agostinis, Ö. Akay, S. Anand, J. Anckaert, Z.A. Martinez, T. Baetens, E. Beghein, L. Bertier, G. Berx, J. Boere, S. Boukouris, M. Bremer, D. Buschmann, J.B. Byrd, C. Casert, L. Cheng, A. Cmoch, D. Daveloose, E. De Smedt, S. Demirsoy, V. Depoorter, B. Dhondt, T.A.P. Driedonks, A. Dudek, A. Elsharawy, I. Floris, A.D. Foers, K. Gärtner, A.D. Garg, E. Geeurickx, J. Gettemans, F. Ghazavi, B. Giebel, T.G. Kormelink, G. Hancock, H. Helsmoortel, A.F. Hill, V. Hyenne, H. Kalra, D. Kim, J. Kowal, S. Kraemer, P. Leidinger, C. Leonelli, Y. Liang, L. Lippens, S. Liu, A. Lo Cicero, S. Martin, S. Mathivanan, P. Mathiyalagan, T. Matusek, G. Milani, M. Monguió-Tortajada, L.M. Mus, D.C. Muth, A. Németh, E.N.M. Nolte-'T Hoen, L. O'Driscoll, R. Palmulli, M.W. Pfaffl, B. Primdal-Bengtson, E. Romano, Q. Rousseau, S. Sahoo, N. Sampaio, M. Samuel, B. Scicluna, B. Soen, A. Steels, J.V. Swinnen, M. Takatalo, S. Thaminy, C. Théry, J. Tulkens, I. Van Audenhove, S. Van Der Grein, A. Van Goethem, M.J. Van Herwijnen, G. Van Niel, N. Van Roy, A.R. Van Vliet, N. Vandamme, S. Vanhauwaert, G. Vergauwen, F. Verweij, A. Wallaert, M. Wauben, K.W. Witwer, M.I. Zonneveld, O. De Wever, J. Vandesompele, A. Hendrix, EV-TRACK: transparent reporting and centralizing knowledge in extracellular vesicle research, Nat. Methods 14 (2017) 228–232.

CHAPTER THREE

# Treatment with platelet- and extracellular vesicle-rich plasma in otorhinolaryngology-a review and future perspectives

**Domen Vozel[a,b,*], Darja Božič[c,d], Marko Jeran[c,d], Zala Jan[c], Manca Pajnič[c], Ljubiša Pađen[c], Bojana Uršič[e], Aleš Iglič[d,f,g], Veronika Kralj-Iglič[g,h] and Saba Battelino[a,b]**

[a]Department of Otorhinolaryngology and Cervicofacial Surgery, University Medical Centre Ljubljana, Ljubljana, Slovenia
[b]Faculty of Medicine, University of Ljubljana, Ljubljana, Slovenia
[c]Laboratory of Clinical Biophysics, Faculty of Health Sciences, University of Ljubljana, Ljubljana, Slovenia
[d]Laboratory of Physics, Faculty of Electrical Engineering, University of Ljubljana, Ljubljana, Slovenia
[e]Department of Urology, University Medical Centre Ljubljana, Ljubljana, Slovenia
[f]Laboratory of Clinical Biophysics, Faculty of Medicine, University of Ljubljana, Ljubljana, Slovenia
[g]Extracellular Vesicles and Mass Spectrometry Laboratory, Institute of Biosciences and BioResources, National Research Council of Italy, Naples, Italy
[h]University of Ljubljana, Faculty of Health Sciences, Laboratory of Clinical Biophysics, Ljubljana, Slovenia
*Corresponding author: E-mail: domen.vozel@kclj.si

## Contents



*Advances in Biomembranes and Lipid Self-Assembly*, Volume 33
ISSN 2451-9634
https://doi.org/10.1016/bs.abl.2020.05.003

## Abstract


Extracellular vesicles (EVs), also known as microvesicles, exosomes, nanovesicles, microparticles etc. are a heterogeneous group of nanometrically sized cell-derived membranous structures which can originate from any cell, including plant cell or bacteria. They were found in many body fluids including platelet-rich plasma (PRP). PRP is a blood-derived product with immune, hemostatic and regenerative effects. PRP is expected to contain important concentrations of EVs which could be important contributors to PRPs effects. For that reason, we call the preparation from PRP a platelet- and extracellular vesicle-rich plasma (PVRP). PVRP has been used in various medical fields including orthopedics, traumatology, wound care surgery, plastic, reconstructive and esthetic surgery, hand surgery, maxillofacial and oral surgery, gynecology, urology, ophthalmology, cardiothoracic surgery, neurosurgery and otorhinolaryngology. Otorhinolaryngology is a field in which PVRP has a great potential in the treatment of diseases of the face, neck, nose, paranasal sinuses, anterior skull base, lips, oral cavity, pharynx, esophagus, salivary glands, ear and lateral skull base. To present this potential and also hitherto gathered results in the field of otorhinolaryngology, this chapter describes the role of platelets and EVs in diagnostics and treatment with an emphasis on otorhinolaryngology. Also, we discuss future perspectives of EVs and PVRP in treatment of moist radical cavities after cholesteatoma surgery.


# 1. Extracellular vesicles

According to Théry et al. [1] International Society for Extracellular Vesicles endorses extracellular vesicle (EV) as a generic term for particles naturally released from the cell that are delimited by a lipid bilayer and cannot replicate, i.e. do not contain a functional nucleus. This definition encompasses any cell, including plant cell or bacteria which can present a source of EVs [1–4]. First, it was believed that EVs serve in transport of cell waste only [2], but later their vital role in intercellular communication in health and disease became apparent [4–7].

EVs have been classified as microvesicles and exosomes in the majority of reviewed studies, therefore this chapter involves this classification also [2]. However, there is no consensus for specific markers of EV subtypes, so far and EVs should not be described as microvesicles or exosomes, unless their reliable specific markers of subcellular origin can be established within an experimental system. According to the minimal requirements of MISEV2018 publication [1] EVs should be described based on their:

**(a)** physical characteristics:
   **i.** size: small EVs (sEVs) for $< 200$ nm and medium/large EVs (m/lEVs) for $> 200$ nm.
   **ii.** density: low, middle, high.
**(b)** biochemical composition.
**(c)** descriptions of conditions or cell of origin.

## 1.1 Classification and biogenesis of extracellular vesicles

As already mentioned, EVs have been classified as microvesicles and exosomes also. This classification is based on EV formation mechanisms, size, composition and source. Van Niel et al. [2] classifies microvesicles as 50–500 nm-sized membranous particles which are formed by ectocytosis (i.e. plasma membrane budding) and exosomes as 50–150 nm-sized membranous particles which are formed by exocytosis of intracellular compartments [2,4]. Some molecular mechanisms of EV formation have been identified, therefore formation can be arbitrarily divided into stages [2,7].

### *1.1.1 Microdomain formation*

An initial step of EV formation is the aggregation of proteins and lipids in the plasma membrane or multivesicular endosome (MVE) to form microdomains (i.e. aggregates of membranous molecules). Formation of microdomains on plasma membrane precedes the formation of microvesicles and formation of microdomains in MVE precedes the formation of intraluminal vesicles (ILV). Understandably, MVE is an intracellular compartment which contains vesicles formed by inward budding of MVE [2].

#### *1.1.2 Binding of cytosolic molecules*

Microdomains bind cytosolic molecules (e.g. proteins, RNA) which are later encapsulated within microvesicle or ILV [2].

#### *1.1.3 Budding*

Aggregates of microdomains and cytosolic molecules begin to bud outward (i.e. ectocytosis) from the plasma membrane in extracellular space to form microvesicle or inward of MVE to form ILV [2].

#### *1.1.4 Fission*

Microvesicles are finally formed by cleavage (i.e. fission) of plasma membrane bud and ILV by cleavage of MVE bud. On this stage, exosomes are encapsulated within MVE but called ILV [2].

#### *1.1.5 Exocytosis*

The final step of exosome formation is presumably exocytosis of MVE which releases ILV in extracellular space. On this stage ILV are named exosomes [2].

According to described mechanisms of EVs formation, it is expected that exosomes involve more steps of formation, are smaller and carry different cargo than microvesicles [2].

### 1.2 Mode of action of extracellular vesicles

EVs exert a physiological or pathological response when reaching a target cell. Target cell can be adjacent, distant or an origin cell. The latter implies an important role of EVs in autoregulation. There are three modes of action of EVs:

1. binding with plasma cell membranous receptors which induces cellular signaling:
2. the fusion of EV and plasma membrane which causes a cytosolic release of EV cargo
3. uptake of EV by plasma membrane with endocytosis [2].

### 1.3 Isolation of extracellular vesicles

Although the applicability of EVs is promising and mechanisms of their formation are thoroughly studied, there is presently no standard isolation protocol to harvest EVs and prepare EVs isolates so far. Different body fluids and cell-culture media have already been used to harvest and inspect EVs microscopically [8—10]. It is well-known that blood-derived EVs are a dynamic material formed by blood-cell fragments [11]. Additionally, blood-derived EV isolation protocols are based on red blood cell (RBC) and leukocyte elimination, therefore harvested pellet contains EVs in addition to platelet- and megakaryocyte-derived molecules [12]. However standard laboratory tests do not detect EVs exclusively, concentrations of EVs in body fluids, culture media or EVs isolates can be high [13].

### 1.4 Applicability of extracellular vesicles isolates in clinical practice

EVs contain different molecules which are important in intercellular communication, therefore they could be applied in diagnosis or therapy of numerous diseases. For that reason, it is appropriate to term them as theranostic tools. They could be used as biomarkers in the diagnosis of infectious, neurodegenerative, autoimmune diseases or tumors. Therapeutically, they could be used as vaccines to prevent infectious diseases or tumors, as immunosuppressive or regenerative therapy or as drug-delivery tools [6].

Although EV isolates still have not been used for therapy in daily clinical practice, they have been regularly used via platelet-rich plasma (PRP). PRP contains high concentrations of platelets and platelet-derived EVs (PDEVs), which possess PRP's main regenerative effects [12,14]. In fact EV isolates are expected to exert at least the same regeneration potential as PRP alone if not superior [12]. For that reason, the preparation can be appropriately named platelet- and extracellular vesicle-rich plasma (PVRP).

## 2. Platelet- and extracellular vesicle-rich plasma (PVRP)

The terminology of PVRP and its types is inconsistent (e.g. platelet-rich plasma, platelet-rich fibrin matrix, platelet-rich fibrin, platelet-rich growth factors, platelet concentrate, platelet gel) due to different observation methods, preparation and activation protocols [15]. Moreover, it was confirmed that PVRP's main regenerative functions are exhibited by PDEV derived from PVRP according to in-vitro and in-vivo animals studies [16]. An abundance of PDEVs and activated platelets prone to shed PDEVs in PVRP have been described and furthermore observed microscopically [14]. Fig. 1 shows a sample of PVRP, which contains large quantity of platelets (A) and sub-micron particles (B), believed to be PDEVs. Some red blood cells and leukocytes can usually be found as well. It can be seen that platelets are in activated state. It is known that upon activation, platelets are prone to

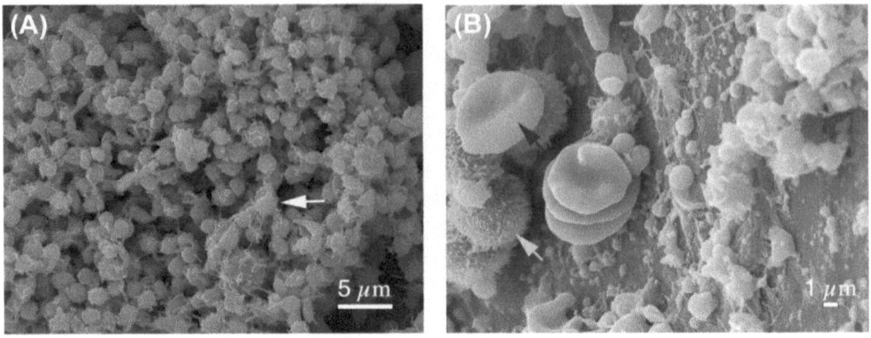


**Fig. 1** Scanning electron micrographs of platelet- and extracellular vesicle-rich plasma. Platelets (Panel A) and sub-micron particles (Panel B) are present in high concentrations. White arrow (Panel A) shows activated, shedding platelet which is prone to produce extracellular vesicles by shedding (i.e. ectocytosis), black arrow (Panel B) shows a red blood cell and gray arrow (Panel B) a leukocyte.

shed vesicles [12]. Roles of PDEVs in PVRP additionally interferes the PVRP terminology. However, it is reasonable to define PVRP as a fraction of blood plasma with platelet and EV concentrations higher than in peripheral blood [17].

Additionally, it is of utmost importance to thoroughly describe the preparation and administration protocol to enable further analyses. Briefly, PVRP can be prepared whether by isolation of plasma or buffy coat from blood sample via distinctive centrifugation-based protocols. The following subchapters describe blood fractions, platelet and EV functions to better understand preparation protocols and biological roles of PVRP.

## 2.1 Platelets and platelet-derived extracellular vesicles

Fig. 2 shows a sample of blood of a healthy subject that underwent centrifugation 1000–2000 g for 20–30 min. It can be seen that erythrocytes which are colored red gather at the bottom of the tube and normally occupy approximately 55% of blood volume (i.e. hematocrit). Plasma, a yellowish liquid above the erythrocytes, of a healthy person normally occupies approximately 45% of blood volume. Between plasma and hematocrit is a thin layer called buffy coat which is consisted of leukocytes and platelets. Buffy coat normally occupies less than 1% of blood volume. To sum up, platelets are abundant in the buffy coat. However, due to a heterogenous platelet size and shape which cause variable sedimentation rates, they can be found in plasma and hematocrit also. This should be taken into consideration when isolating platelets in PVRP preparation.

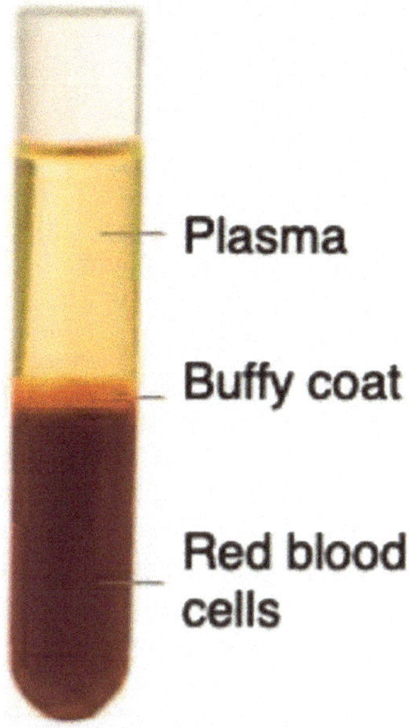


**Fig. 2** Blood fractions after centrifugation at 1000–2000 g for 20–30 min. Upper yellowish layer is plasma and lower dark red are red blood cells. The ratio between the volume of red blood cells and total volume of blood is called hematocrit. Buffy coat, which occupies less than 1% of blood volume in a normal human subject, is observed as a greyish layer between plasma and red blood cells. Plasma or buffy coat are harvested from whole blood sample after the first centrifugation to produce platelet- and extracellular vesicle-rich plasma.

Platelets and platelet-derived EVs carry the most important regenerative functions of PVRP [12,14,16]. Resting platelets are 2–3 μm sized disc-like blood cells without nuclei, formed by megakaryocyte cytoplasm fragmentation in the bone marrow or lungs. Their lifespan is relatively short - 7–10 days [18]. Their blood concentration in healthy human subject is 150–400 × $10^9$/L. Blood concentration qualifies them in the second place, after RBCs [19]. Platelets have an essential role in hemostasis, immune response and tissue regeneration. Their shape and size enable them to travel near the vessel wall and detect endothelial damage first [20]. They possess their functions via secretion of bioactive molecules contained in granules, cytoplasm or cellular organelles (i.e. extra-granular molecules) (Table 1). Alpha granules which are the largest and most abundant contain approximately 280 different types of proteins. Delta granules (i.e. dense granules) are smaller, less abundant and contain smaller molecules (Table 1). Lambda, the least abundant granules, are lysosomes, which contain mainly enzymes. T-granules have also been described

**Table 1** Some molecules from platelets and platelet-derived extracellular vesicles in platelet- and extracellular vesicle-rich plasma with their functional roles in hemostasis, immune response and regeneration [16,21,23].

| Molecule | Functions |
|---|---|
| ***Alpha granules*** | |
| bFGF | Chemotaxis, cell growth and differentiation, angiogenesis |
| CD63 | Transmembrane adapter protein, leukocyte recruitment |
| CD40L | TNF-α family: B-cell response, endothelial cell and antigen-presenting cell activation |
| Cyclophilin A | Smooth muscle cell growth factor |
| MIP-1α | Neutrophil, eosinophil activation, immunoglobulin formation |
| MMP-2, MMP-9 | Extracellular matrix disintegration, platelet-leukocyte aggregate formation |
| PDGF | Cell growth and differentiation, monocyte/macrophage differentiation |
| PF4 | Monocyte, neutrophil, T-cell recruitment and helper T-cell differentiation |
| Ppbp β-thromboglobulin NAP-2 | Neutrophil activation and recruitment, macrophage phagocytosis activation |
| Protein S | Thrombin production and its regulation, angiogenesis |
| P-selectin | Leukocyte adhesion, complement activation |
| SDF-1 | T-cell, monocyte and neutrophil chemotaxis |
| TGF-β | Cell proliferation, T-cell differentiation, B-cell and macrophage phenotype regulation |
| Thrombocidins | Bactericidal and fungicidal properties |
| Thrombospondins | Apoptosis, endothelial cell inflammation, platelet-macrophage aggregates formation |
| VEGF | Angiogenesis, adhesion molecule expression |
| vWF | Platelet adhesion, neutrophil extravasation |
| ***Delta granules*** | |
| ADP | Platelet, leukocyte and endothelial cell activation |
| Glutamate | T-cell regulation |
| Histamine | Vascular reactivity enhancement and degranulation, pro- and anti-inflammatory functions |
| Polyphosphates | Immune response stimulation |
| Serotonin | Dendritic cell and T-cell regulation, vasoconstriction and increased capillary permeability |

**Table 1** Some molecules from platelets and platelet-derived extracellular vesicles in platelet- and extracellular vesicle-rich plasma with their functional roles in hemostasis, immune response and regeneration [16,21,23].—cont'd

| Molecule | Functions |
|---|---|
| ***Extra-granular molecules*** | |
| GPIbα | Binding with leukocyte Mac-1 |
| IL-1β | Acute inflammatory response, leukocyte and endothelial cell activation |
| Nitric oxide | Anti-inflammatory and antithrombotic effect |
| Thromboxane | T-cell differentiation, monocyte activation |

*ADP*, adenosine diphosphate; *bFGF*, basic fibroblast growth factor; *CD40L*, CD40 ligand; *GPIbα*, glycoprotein Ibα; *IL*, interleukin; *MIP*, macrophage inhibitory protein; *MMP*, metalloproteinase; *NAP*, neutrophil activating peptide; *PDGF*, platelet-derived growth factor; *PF*, platelet factor 4; *Ppbp*, Pro-platelet basic protein; *SDF*, stromal cell-derived factor; *TGF*, transforming growth factor; *TNF*, tumor necrosis factor; *VEGF*, vascular endothelial growth factor; *vWF*, von Willebrand factor.

and contain molecules which are secreted via exocytosis or simply translocated to the extracellular surface of the plasma membrane [21]. Notably, some molecules from granules also provide antimicrobial effects of PVRP, identified in numerous in-vitro and in-vivo animal studies [22].

Understandably platelets are the most numerous blood immune cells due to a concentration greater than the blood concentration of leukocytes and important immunological and antimicrobial roles [19].

Platelet-derived extracellular vesicles (PDEVs) which are formed after platelet activation [18], possess cell functions of their origin, therefore they are important in hemostasis, immune response and tissue regeneration [19–21,24]. Additionally, platelet-derived exosomes isolated from PVRP possess a substantial role in the proliferation-, migration- and vessel formation-promoting effects of PVRP. These roles are mediated by composition of exosomes, which are rich in growth factors, especially bFGF, PDGF, TFG-β and VEGF (Table 1) [16].

## 2.2 Platelet- and extracellular vesicle-rich plasma preparation protocols

The knowledge of blood composition and characteristics of platelets described above enables us to prepare PVRP with different protocols. In fact, statistical review of 105 clinical studies describing the PVRP preparation protocols for later clinical application revealed that their repeatability is 10% [25].

At this point it is important to mention that platelets and leukocytes are also present in other blood fractions in addition to buffy coat. Therefore, it is impossible to completely separate platelets from other blood cells with centrifugation due to intra- and inter-subject blood components variability despite the protocol type. In general, there are 2 types of centrifugation-based PVRP preparation protocols: plasma-based and buffy coat-based protocols, differing in platelet yield, leukocyte concentration, volume of PVRP and consequent clinical applicability of prepared PVRP.

Additionally venepuncture technique (e.g. needle gauge, tourniquet use), materials used in PVRP preparation and staff skills are important in EV formation and therefore in PVRP preparation [11,26,27]. Blood donor and PVRP recipient are different in heterologous or the same in autologous in PVRP application.

### 2.2.1 Plasma-based protocols

These preparation protocols are based on the separation of blood plasma and platelets from leukocytes and erythrocytes. This is acquired by discarding hematocrit and buffy coat after the first centrifugation step (Fig. 3). Because buffy coat contains the majority of leukocytes and a considerable number of platelets due to their sedimentation rates, plasma-based protocols produce PVRP with low or minute leukocyte count at the expense of low platelet count. Yielded platelet concentrations are typically approximately 2–3 times higher ($300-500 \times 10^9$ platelets/L) than baseline whole blood platelet concentration [17]. To increase platelet count and to lower leukocyte count in PVRP, differential instead of single centrifugation can be utilized. Protocol shown in Fig. 3 is based on differential centrifugation of blood sample at low spin rate and short spin regimen (about 200–400 g for 4–6 min at 18 °C) to isolate RBCs initially, followed by centrifugation of harvested supernatant (i.e. plasma above buffy coat) at high spin rate and long spin regimen (about 600–800 g for 16–18 min at 18 °C) to concentrate platelets.

### 2.2.2 Buffy coat-based protocols

These preparation protocols focus on platelet isolation. To achieve this, buffy coat is isolated from blood sample by centrifugation at high spin rates and long spin regimens. Namely, the platelet count is high on the expense of high platelet count. Typically buffy coat-based preparation protocols yield 3–8 times higher ($500-1.500 \times 10^9$ platelets/L) than baseline whole blood platelet concentration [17].

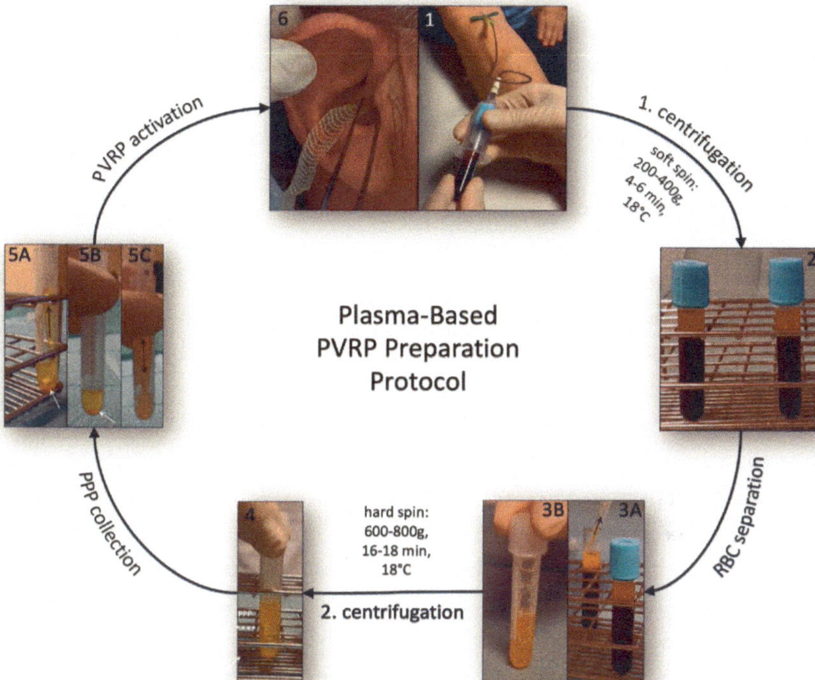


**Fig. 3** A scheme of the plasma-based platelet- and extracellular vesicle-rich plasma preparation protocol. Blood samples are collected into tubes with anticoagulant (Panel 1) and are centrifuged at several hundred g and temperature close to room temperature for some minutes to prepare a product with distinctive layers (Panels 2 and 3A): red blood cells, plasma and buffy coat. Plasma that was harvested into citrated tubes (*black arrow* in Panel 3A) is transferred with a sterile pipette into a single tube (Fig. 2, Panel 3B). Accuracy is required to prevent buffy coat and red blood cell harvesting. Plasma is then centrifuged again (at about 500—1000 g for about 20 min at the same temperature) to produce a preparation with distinctive hazy pellet (*white arrows* in Panels 4, 5A and 5B). The upper half of this product (*bordered by the white dashed line* in Panel 4) that is called platelet-poor plasma (PPP) is discarded (Panel 5A) without prior resuspension. The remaining preparation (Panel 5B) is resuspended in the same tube (Panel 5C). From that point on, the preparation which is optionally exogenously activated prior autologous administration (Panel 6), is called platelet- and extracellular vesicle-rich plasma.

## 2.3 Composition of platelet- and extracellular vesicle-rich plasma

Ideal average concentrations of PVRP have been suggested according to healing effects. However, it is postulated that platelet concentrations above $1250 \times 10^3/\mu L$ could exhibit inhibitory effects [17]. Additionally, the healing effects of other platelet concentrations in PVRP have been studied. PAW classification was published by Delong et al., which classifies PVRP according to platelet concentration (P), activation method (A) and leukocyte concentration (W). It provides guidance in PVRP preparation and standardization of data for subsequent meta-analyses of studies in various clinical settings. Delong et al., also proposed that these studies should include baseline whole blood platelet concentration [17].

### 2.3.1 Platelet concentration

Platelet concentration in PVRP is classified as low, moderate, high or super. Low means that platelet concentration in PVRP is lower than baseline and is termed as platelet-poor plasma. It has a minute effect; therefore, it is often used as a placebo. Moderate PVRP platelet concentrations are up to 4-times higher than baseline (i.e., up to $750 \times 10^9$/L). Majority of clinical studies that reported satisfactory effect used moderate platelet concentrations. If platelet concentration is 4- to 6-times higher than baseline (i.e., 750—$1250 \times 10^9$/L), it is classified as high. This concentration is usually achieved with buffy coat-based PVRP preparation protocols. Several in-vivo clinical and animal studies report a beneficial effect of the high platelet concentration. Platelet concentrations in PVRP higher than 6-times baseline (i.e., above $1250 \times 10^9$/L) could be detrimental and could induce harmful effects due to paradoxical inhibition as excessive concentrations of platelets and growth factors lead to platelet apoptosis, receptor downregulation and desensitization [17].

### 2.3.2 Leukocyte concentration

Leukocyte concentration depends on the PVRP preparation protocol. In contrast to plasma-based protocols, buffy coat-based protocols yield leukocyte concentrations higher than baseline. High leukocyte concentrations, particularly neutrophils, promote scar formation, for that reason leukocytes are undesirable for these cases. In the prevention of infections and treatment of open wounds, high concentrations may be required [17].

## 2.4 Roles of anticoagulants in blood sampling and platelet- and extracellular vesicle-rich plasma activation type

When PVRP is prepared it can be administered to target tissue unchanged or administered as a gel.

PVRP gel is a result of an in-vitro (i.e. exogenous) platelet activation which can be simplified by blood clot formation. It can be achieved by using various substances (i.e. activators), e.g. bovine thrombin, calcium chloride or both. Activators also dilute PVRP due to the increase of the volume. Side effects were observed after the application of the preparation; bovine thrombin was found to cause hypersensitivity reactions while calcium chloride induced pain or severe burning due to a decrease of pH in the preparation [17]. PVRP gel can be prepared also if blood is processed in a tube without anticoagulant. Indicatively, PVRP gel showed higher probability for adverse effects compared to native PVRP. Nomenclature of PVRP gel is inconsistent as it is often termed as platelet-rich fibrin, platelet-rich fibrin growth factors, platelet gel, platelet-rich gel etc. Therefore, classification of DeLong et al. suggests a description of activation type. Despite disadvantages, PVRP gel enables easier application in some cases due to mechanical characteristics (i.e. it can be sutured or manually applied to the target tissue). Additionally PVRP gel is supposed to release growth factors in a better controlled manner compared to native PVRP [17].

If blood was drawn into tubes without anticoagulant, a blood clot would form during the first centrifugation and the product is not defined as PVRP. Therefore, blood must be drawn into tubes containing anticoagulant [28]. The classification by Delong et al. does not consider the type of anticoagulant in the preparation of PVRP, although it is important [17]. Based on a study comparing three anticoagulants: sodium citrate, ACD (acid citrate dextrose solution) and EDTA (ethylenediaminetetraacetate) in the preparation of PVRP, do Amaral et al. [28] recommended the use of sodium citrate as an anticoagulant in venepunction. Sodium citrate enables the preparation of more concentrated PVRP and stimulates the proliferation of mesenchymal stromal cells without interfering with their phenotype [28].

Also, PVRP can be administered to target tissue without prior intentional activation (i.e. native PVRP). Then activation process happens in the target tissue (i.e. endogenously) which mimics the physiological platelet activation mechanism via interaction with type I collagen. Type I collagen is even more potent activator than thrombin. Endogenous activation prevents oversaturation of growth factor receptors due to natural growth factor release pattern. Additionally, native PVRP can be injected to the target tissue [17].

Although an activation type is important, PVRP is always activated exogenously and endogenously. Namely, platelets are imperatively exogenously activated during venepuncture and centrifugation and also endogenously by type 1 collagen [17,26]. Activation of plasma is targeted at transformation of platelets as regards the reactions that take place in the platelet membranes and their surroundings. However, as the local membrane composition is connected to the local curvature, changes in the platelet shape also take place. Activated platelets first exhibit tubular protrusions while the mother cell becomes globular. Eventually, the protrusions detach from the mother cell which is prone to fragment. Remodulation involving material from adjacent cells and/or fragments, including residual RBCs, may take place. Fragmented and reconstituted platelets and RBCs thereby form EVs in the preparation [14,19–21,24].

Basic knowledge of PVRP preparation protocols, PVRP composition, type of PVRP activation and desired PVRP administration technique allow a clinician to appropriately choose the most suitable PVRP for a patient. It is imperative to be aware of the importance of EVs in PVRP as mentioned above.

## 3. Treatment with platelet- and extracellular vesicle-rich plasma

Ten years after the first use of PVRP in hematology in 1970, it began to be used in maxillofacial surgery followed by the treatment of musculoskeletal injuries in athletes [15]. PVRP use has expanded to other fields of medicine especially to surgical treatment of various disorders (Table 2). Although PVRP's wide indication area, there is a deficit of systematic reviews or meta-analyses regarding its use for the majority of diseases.

There is a lack of meta-analyses and systematic reviews of PVRP application that would enable further validation of this treatment option for different diseases, especially in otorhinolaryngology [60]. However, the application of PVRP was thoroughly described according to an anatomical area in otorhinolaryngology due to its invaluable potential (Table 3).

### 3.1 Face, nose and paranasal sinuses

PVRP has been proved efficient in different methods of facial rejuvenation. It was administrated as a local injection on half of the face every 2 weeks for 3 months. The effect was compared to the contralateral half of the face treated with injected growth factors (i.e. mesotherapy). PVRP was more efficient in rejuvenation compared to mesotherapy according to patient-reported and objective measures [62]. Additionally, a recent systematic review

**Table 2** A survey on platelet- and extracellular vesicle-rich plasma clinical application in different fields of medicine (except otorhinolaryngology).

| Field of medicine | Medical indication, disease or surgical procedure associated with PVRP application |
|---|---|
| Orthopedics and traumatology | [a]Plantar fasciitis [29]<br>[a]Elbow epicondylitis [29]<br>[a]Knee osteoarthritis [30,31]<br>Hip osteoarthritis [32]<br>Achilles tendinopathy [14]<br>Patellar tendinopathy [14]<br>Bone fracture [31]<br>ACL rupture [31] |
| Plastic, reconstructive, esthetic and hand surgery | Reduction mammaplasty [33]<br>Abdominoplasty [33]<br>Carpal tunnel syndrome [34]<br>Alopecia [35,36]<br>Breast reconstruction with lipofilling [37] |
| Wound care surgery | [a]Diabetic ulcers [38]<br>Venous leg ulcers [39]<br>Arterial ulcers [40]<br>Pressure ulcers [39]<br>Traumatic wounds [40]<br>Chronic wounds [31,41] |
| Maxillofacial and oral surgery | [a]Periodontal defects [42]<br>[a]Periodontal disease [43,44]<br>[a]Temporomandibular joint osteoarthritis [45]<br>Odontogenic mandibular cysts surgery [46]<br>Mandibular fracture surgery [46,47]<br>Dental implant surgery [43]<br>Maxillary sinus lift [43]<br>Bisphosphonate related osteonecrosis of the mandible [43]<br>Orthognathic surgery [48]<br>Dental extraction [43] |
| Gynecology and urology | Caesarean section [37]<br>Cervical ectopia [37]<br>Vulvar dystrophy [37]<br>Cancer vulvectomy [37]<br>Vesicular, perianal, rectovaginal fistulae [37]<br>Erectile dysfunction [49]<br>Urinary incontinence [37,49]<br>Infertility [37]<br>Vaginal infections [37]<br>Vaginal aging [37]<br>Peyronie's disease [49] |

(*Continued*)

**Table 2** A survey on platelet- and extracellular vesicle-rich plasma clinical application in different fields of medicine (except otorhinolaryngology).—cont'd

| Field of medicine | Medical indication, disease or surgical procedure associated with PVRP application |
|---|---|
| Ophthalmology | Hyposecretory dry eye [50]<br>Dry eye syndrome after laser refractive surgery [51]<br>Sjögren's syndrome [52]<br>Corneal ulcers [53] |
| Cardiothoracic surgery | Sternotomy wounds [54,55]<br>Chylothorax post esophagectomy [56]<br>Mediastinitis after cardiac surgery [55] |
| Neurosurgery | Post-dural puncture headache [57]<br>Lumbar spine surgery [58]<br>CSF leakage after cranial and spinal surgery [59] |

*ACL*, anterior cruciate ligament; *CSF*, erebrospinal fluid; *PVRP*, platelet- and extracellular vesicle-rich plasma.
[a]Denotes systematic review or meta-analysis. Some fields of medicine share medical indications, diseases or surgical procedures listed in the table.

showed that PVRP injections are safe and may have modest benefit in the treatment of skin aging [61]. A meta-analysis of 7 studies showed that PVRP is efficient as a combination therapy with fractional carbon laser therapy or microneedling for atrophic acne scars [63]. PVRP has also been used in face and neck lift to accelerate surgical wound healing [33].

Wound healing was accelerated by PVPR also in rhinoplasty, especially in the early postoperative period [64]. PVRP was proved effective in increasing viability and keeping the form of inserted diced cartilage in nasal dorsum after rhinoplasty [65].

Administration of PVRP-soaked nasal packs reduced crusting, bleeding, complaints and accelerated healing after rhinosurgical procedures [66].

Additionally, PVRP was efficiently applied concomitantly with graft materials for anterior skull base cerebrospinal fluid leaks [70], frontal sinus obliteration after surgical treatment of chronic frontal sinusitis [68,69] and fibrous dysplasia [67].

## 3.2 Lips and oral cavity

PVRP is efficient also in the pediatric population, in reducing scar width after cleft lip reconstruction according to randomized controlled study [71]. It

**Table 3** A survey on treatment with platelet- and extracellular vesicle-rich plasma in distinct anatomical areas of otorhinolaryngology.

| An anatomical area in otorhinolaryngology | Medical indication, disease or surgical procedure associated with PVRP application |
|---|---|
| Face, neck, nose, paranasal sinuses and anterior skull base | Facial rejuvenation [61,62]<br>Neck rejuvenation [62]<br>Facial acne scars[a] [63]<br>Facelift [33]<br>Neck lift [33]<br>Rhinoplasty [64,65]<br>Postsurgical nasal packing [66]<br>Frontal sinus obliteration [67–69]<br>Anterior skull base cerebrospinal fluid leaks [70] |
| Lips and oral cavity | Cleft lip [71]<br>Oral mucosal lesions [72,73]<br>Oral pemphigus vulgaris [74]<br>Osteoradionecrosis of the mandible [75] |
| Pharynx and esophagus | Pharyngoplasty for obstructive sleep apnea [76]<br>Traumatic esophagocutaneous fistula [77] |
| Salivary glands | Suprafacial parotidectomy [78,79] |
| ***Ear and lateral skull base*** | |
| External ear | Auricle amputation [80] |
| Middle ear | Acute eardrum perforation [81]<br>Chronic eardrum perforation [82–91]<br>Reconstruction of the posterior external ear canal wall [92]<br>Mastoid reconstruction after CWD mastoidectomy [93,94] |

*CWD,* canal wall down; *PVRP*, platelet- and extracellular vesicle-rich plasma.
[a]Denotes systematic review or meta-analysis.

improves healing in the early postoperative period and significantly reduces pain when administered as a gel on oral defects after excision of potentially malignant lesions [72,73]. The efficiency of intralesional PVRP injections was comparable to corticosteroid injections for the treatment of oral pemphigus vulgaris, therefore PVPR was suggested as an alternative treatment of this disease [74]. Another disease treated with PVRP gel is mandibular osteoradionecrosis [75].

However, no systematic reviews or meta-analyses were found for the treatment of the diseases of lips and oral cavity with PVRP.

## 3.3 Pharynx and esophagus

In pharyngology, PVRP application was associated with pharyngoplasty for the treatment of obstructive sleep apnea. It was administered as a gel to the surgical wound before suturing. PVRP statistically significantly decreased the incidence of palatal wound breakdown, provided better pain management and reduced the time taken to return to the normal diet [76].

In esophagology, PVRP was used in the treatment of traumatic esophagocutaneous fistula in a single pediatric patient [77].

## 3.4 Salivary glands

PVRP enhanced healing process also when its application was associated with superficial parotidectomy due to benign parotid gland tumor. In a recent controlled trial, it was administered as a gel in the surgical wound which reduced complications rate, length of hospitalization and duration of drainage tube [78]. Another randomized controlled study reported a benefit for the same indication [79].

## 3.5 Ear and lateral skull base

The vast majority of studies in otology and lateral skull base surgery report the administration of PVRP for tympanic membrane perforations.

### *3.5.1 The external ear*

An external ear was treated with PVRP in a patient suffering from traumatic complete auricle amputation. Replanted auricle was postoperatively repetitively treated with local injections of PVRP, polydeoxyribonucleotide and hyperbaric oxygen therapy. Although this is a case report, authors report efficacy of PVRP [80].

### *3.5.2 The middle ear*

Multiple studies report the administration of PVRP for treatment of tympanic membrane perforations, acute or chronic. Acute traumatic perforations were treated efficiently with PVRP-gel only in a randomized controlled study [81]. The gel was also administered in chronic perforations during tympanoplasty due to better durability. It was administered as a plug lateral to the tympanic membrane after being reconstructed with tympanoplasty and perichondrium graft, due to central perforation. Success of graft taking was statistically significantly higher and infection rate lower in a group treated with PVRP compared to the control group [89]. In general, it appears that PVRP is more efficient in the treatment of tympanic membrane perforations if administered concomitantly with grafts or packing material. Namely, PVRP improves the fascial [82,88] or fat graft uptake [90] and improves healing if administered as PVRP-soaked atelocollagen sponge [91] or gelfoam [85].

Furthermore, PVRP was administered as the hourglass-shaped gel graft to the small tympanic membrane perforations after freshening perforation margins. A successful perforation repair was achieved in 84% in this study [87]. Despite the promising results, a randomized controlled study reported the low success rate of an hourglass-shaped PVRP for treatment of tympanic membrane perforations [95]. Therefore standardization of PVRP preparation protocols and reporting guidelines are required to perform adequate meta-analyses to evaluate the effectiveness of PVRP without grafts in tympanoplasty.

Osteogenic potential of PVRP has already been identified, therefore it was successfully applied in bone reconstruction in otorhinolaryngology. It was administered with bone pate to obliterate mastoid after canal wall down mastoidectomy. The case series study reported more than 85% total success rate with no reported complications. An advantage of PVRP and bone pate obliteration is that a disease recurrence can be easier radiologically identified compared to other obliteration materials [94]. Additionally, PVRP was used with bone pate and titanium mesh to reconstruct posterior ear canal wall after canal wall down mastoidectomy [92]. To conclude there are no randomized controlled clinical studies of PVRP application in mastoid surgery.

## 3.6 Platelet- and extracellular vesicle-rich plasma adverse effects

Adverse effects related to the application of PVRP are very rare. The majority of adverse effects are related to the surgical procedure during which the PVRP is administered. The risk to transmit infectious or malignant disease is minute when using autologous PVRP. Heterologous PVRP application can cause a rejection reaction also [96].

## 3.7 Future of platelet- and extracellular vesicle-rich plasma in otorhinolaryngology

PVRP offers potential for its use also in the treatment of inner ear, retrocochlear, vestibular, facial nerve, laryngeal, tracheal and voice disorders. According to published animal studies, PVRP could be combined with skin grafts for soft tissue reconstruction [97], as the treatment option for vocal cord paralysis [98], vocal cord injury [99], dysphagia [100], pharyngocutaneous fistula [101], anosmia [102], facial palsy [103] or hearing loss [104].

Due to a successful application and promising results of PVRP in the treatment of ear diseases, authors propose PVRP use in the treatment of chronic ear inflammation after canal wall down mastoidectomy required to treat cholesteatoma. For that reason, following subchapters describe cholesteatoma, its current treatment and new treatment options proposed by authors.

### *3.7.1 Treatment of chronic middle ear infection after cholesteatoma surgery with platelet- and extracellular vesicle-rich plasma*

#### 3.7.1.1 Cholesteatoma burden

Cholesteatoma is defined as the presence of squamous epithelial tissue in the middle ear and/or mastoid associated with chronic inflammation. The most common type of cholesteatoma arises from the tympanic membrane and is called an acquired cholesteatoma which is depicted in Fig. 4. The congenital type is very rare. The incidence of acquired cholesteatoma in the developed world is low, with an estimated annual incidence of 9.2 cases per 100 000 inhabitants [105]. However, the incidence may be higher in developing countries based on the higher incidence of chronic suppurative otitis media. Due to increased migration flows from the developing to developed countries, the incidence of cholesteatoma and chronic suppurative otitis media is rising in developed countries as well [106].

#### 3.7.1.2 Cholesteatoma treatment

Cholesteatoma is a disease treated surgically by removing squamous epithelial tissue trapped in middle ear space. Many surgical options and approaches exist based upon the nature and extent of the disease. They range from simple tympanoplasty to radical mastoidectomy. The latter is called open technique surgery which is depicted in Fig. 5. Cholesteatoma extending into the antrum and mastoid and those extending medial to the head of the malleus may require complete mastoidectomy. Many patients with cholesteatoma will undergo multiple surgical procedures due to recurrence of the disease. Each subsequent procedure is more ablative and less concerned with hearing status as opposed to the eradication of cholesteatoma. Despite the advance in surgical techniques, an open technique surgery is still used in approximately 40% of cholesteatoma surgery [107]. These patients end up with large

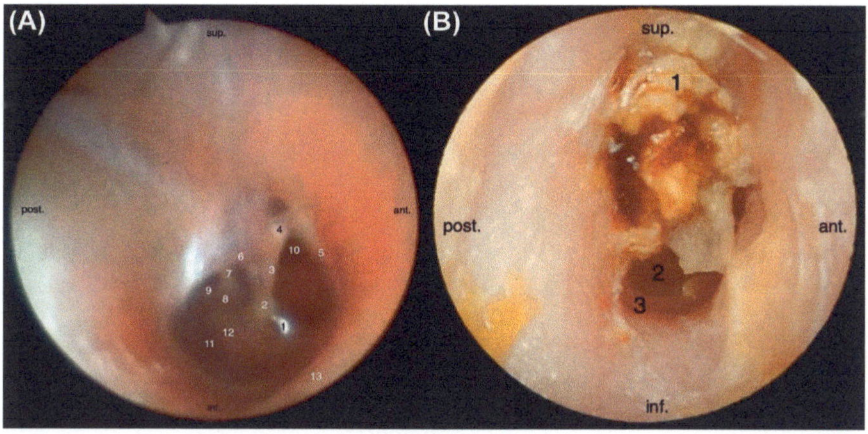


**Fig. 4** Endoscopic photographs of the normal tympanic membrane (A) and cholesteatoma (B) of the right ear. Comment: A shows normal tympanic membrane, which is pale, gray and in neutral position (i.e. not retracted toward middle ear or bulging outwards), light reflection (1) is cone-shaped located in an anteroinferior quadrant of the tympanic membrane. Umbo (2), handle of malleus (3), lateral processus of malleus (4), tympanic annulus (5), chorda tympani (6), incus (7), incudostapedial joint (8), stapes (9), shadow of eustachian tube (10), shadow of round window niche (11) and promontory (12) are visible through transparent tympanic membrane. Anterior external ear canal wall is bulging posteriorly (13). B shows typical keratinizing masses (i.e. squamous epithelial tissue) of cholesteatoma (1) in posterolateral aspects of the tympanic membrane, remnant of pars tensa (2) and tympanic annulus (3). Posterior external ear canal wall and attic are probably eroded, which is a sign of a destructive chronic ear inflammation requiring surgical treatment.

postsurgical cavity due to removal of the posterior external ear canal wall (i.e. radical cavity).

Due to multiple suppurative periods from the radical cavity and formation of granulation tissue, a large spectrum of surgical and nonsurgical treatment options have been applied to reduce patient discomfort [108,109]. Granulations in a once-healed radical cavity arise when self-cleaning ability and ventilation are hampered and epidermal debris is allowed to collect and become infected with bacteria [110]. This causes a persistent discharge from moist radical cavities which is depicted in Fig. 6 [111].

The most common bacteria isolated from a moist radical cavity is *Staphylococcus aureus* [110]. Consequently, frequent cleaning and conservative treatment measures that include removal of the granulation followed by chemocautery with substances as 2% aqueous gentian violet are needed to control the discharge from moist radical cavities. However, frequently an

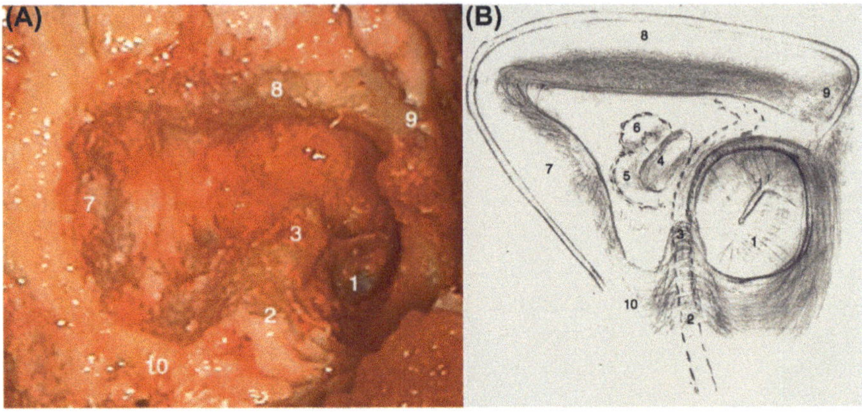


**Fig. 5** Intraoperative otomicroscopical photography of right radical cavity after removal of cholesteatoma (A) and an illustration (B) with anatomical landmarks. Comment: The mastoid and middle ear with air cells are drilled and deepithelialised during surgery to remove the complete cholesteatoma and make a single cavity (i.e. radical cavity) which enables good postoperative ventilation and drainage of the middle ear. Important structures are identified during surgery: pars tensa of tympanic membrane (1), base of former posterior external ear canal wall (2), facial ridge (3), facial nerve (dashed), lateral (4), posterior (5) and superior semicircular canal (6), sigmoid sinus (7), middle cranial fossa dura (8), zygomatic root (9), mastoid tip (10). The radical cavity reepithelialises postoperatively.

area of granulation resists epidermisation owing to a small underlying focus of osteitic bone or sequestrum [112]. Although there is little published evidence to indicate the incidence of moisture or discharge following open technique surgery, two 10-year studies show that improvement in surgical techniques has reduced this complication [113]. Discharge is reported in 3%–20% of radical cavities [114,115].

Another option to prevent moist radical cavity is to obliterate it with synthetic material as bioactive glass, hydroxyapatite, calcium phosphate ceramic granules, coral, demineralized bone matrix, titanium, silicone or with autologous tissue as musculoperiosteal flaps, bone chips, bone paste and cartilage [108,116].

#### 3.7.1.3 Proposed platelet- and extracellular vesicle-rich plasma application in moist radical cavities after cholesteatoma surgery

Not only the sequelae of cholesteatoma and its surgical treatment are debilitating, above-mentioned long-lasting conservative measures of chronically inflamed moist radical cavities are significantly worsening the patients' quality of life and put a significant burden on health care. Due to the knowledge

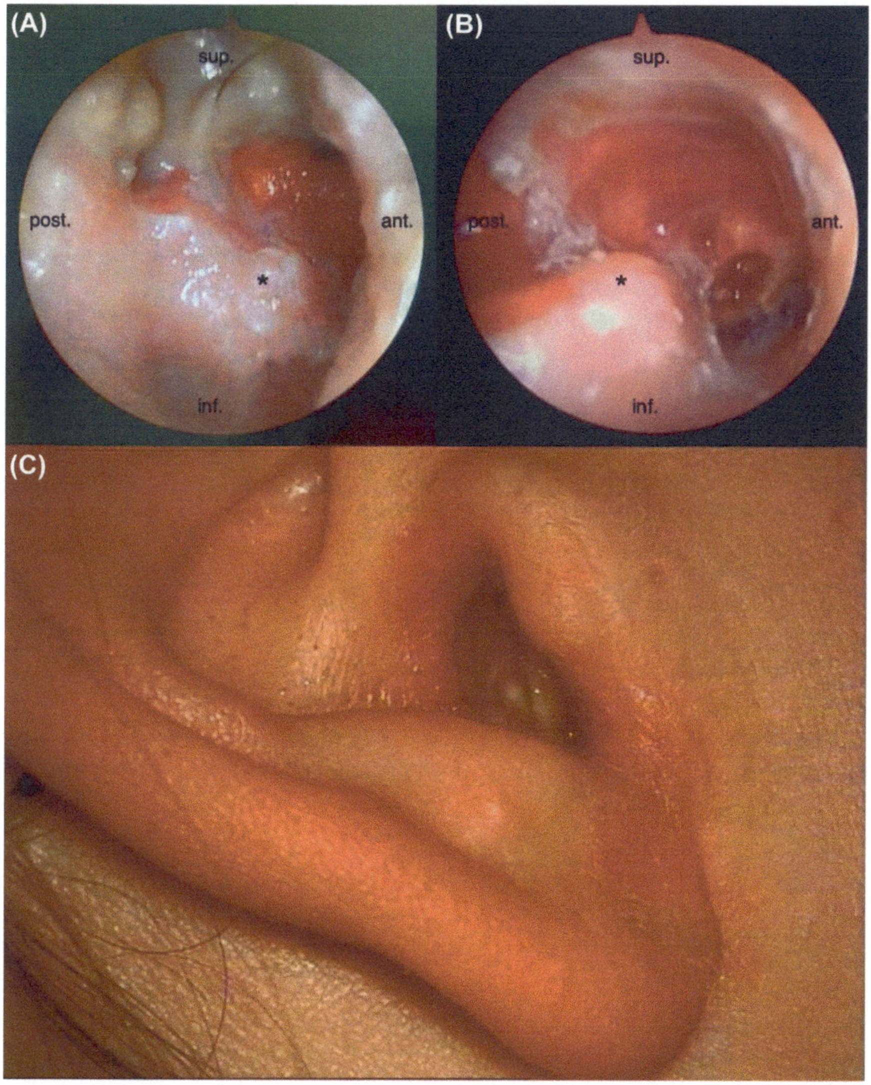


**Fig. 6** Endoscopic photographs of inflamed, i.e. moist radical cavities after cholesteatoma surgery and photography of inflammation extending to auricle. Comment: A and B depict diffuse inflammation of radical cavities. An abundant discharge is visible on A. Facial ridges (*) are relatively spared from the inflammatory process. Inflammation is extending to auricle (C).

of local immune response mechanisms in radical cavities and exhausted treatment options, new conservative treatment options are being sought. We have identified PVRP as promising treatment option for moist radical cavities based on published preclinical and clinical studies. PVRP could be prepared by 2-step plasma-based centrifugation protocol and administered to moist radical cavity via PVRP-soaked ear wick (Fig. 3).

## 4. Future of extracellular vesicle isolates in otorhinolaryngology

Since PVRP is considered as effective treatment option in otorhinolaryngology and PDEV are its main effectors, there is a need to design a PDEV-based isolate in the future. PDEV isolates have to our best knowledge not been used in clinical settings for wound healing disorders so far.

However, other cell-derived EV isolates have already been proposed to be used either as diagnostic or therapeutic tool [6]. Namely, EVs have been studied thoroughly to clarify the pathogenesis of disease.

Pathogenesis of vestibular schwannoma-associated hearing loss has not been clarified yet. For that reason, EVs derived from vestibular schwannoma cells have been studied and identified as important mediators in the pathogenesis of sensorineural hearing loss in vestibular schwannoma. These findings enable further research of vestibular schwannoma diagnosis and treatment [117]. Pathogenetic role of exosomes was also identified in otitis media [118].

For the evaluation of inner ear status, exosomes could be used as biomarkers, since animal studies show an important role of inner ear exosomes in drug ototoxicity [119]. EVs, especially exosomes possess therapeutic potential also. They could be used as drug vehicles [120]. Exosome-associated adeno-associated virus (Exo-AAV) could be used to insert the deafness-responsible missing genes to inner ear hair cells. In animal studies, Exo-AAV transduced all inner ear hair cells (i.e. inner and outer) compared to adeno-associated virus which transduced inner hair cells only [120]. EVs isolated from inner ear hair cells may reduce the noise- and drug-related cochlear damage as they may incorporate pro-resolving mediators, anti-inflammatory drugs and may carry drugs. Therefore, these EVs could be used intracochlearly via concomitant cochlear implantation to reduce post-implantation inflammation in scala tympani [121].

Due to known pathogenetic roles of EVs in carcinogenesis [4,7], they possess a screening, diagnostic and prognostic roles in head and neck cancer also. Salivary and serum exosomes with these roles were identified in oral squamous cell cancer [122–125] and laryngeal cancer [126]. Also, exosomes that negatively regulate radiosensitivity of laryngeal cancer cells were identified, therefore novel insights into laryngeal cancer treatment are under development [127]. However, to fulfill claimed expectations of EV isolates as promising diagnostic and therapeutic tools, more evidence on EV-based disease mechanisms should be gathered and methods of EV harvesting and assessment improved [128]. For further application of these preparations in the sterile areas of the human body, it is inevitable to test their sterility.

## Acknowledgments

Authors acknowledge support from EU Commission (Ves4us), and Slovenian Research Agency (ARRS), (grants P3-0388, J1-9162, J2-8166 and J2-8169).

## References

[1] C. Théry, K.W. Witwer, E. Aikawa, M.J. Alcaraz, J.D. Anderson, R. Andriantsitohaina, A. Antoniou, T. Arab, F. Archer, G.K. Atkin-Smith, D.C. Ayre, J.-M. Bach, D. Bachurski, H. Baharvand, L. Balaj, S. Baldacchino, N.N. Bauer, A.A. Baxter, M. Bebawy, C. Beckham, A.B. Zavec, A. Benmoussa, A.C. Berardi, P. Bergese, E. Bielska, C. Blenkiron, S. Bobis-Wozowicz, E. Boilard, W. Boireau, A. Bongiovanni, F.E. Borràs, S. Bosch, C.M. Boulanger, X. Breakefield, A.M. Breglio, M.Á. Brennan, D.R. Brigstock, A. Brisson, M.L. Broekman, J.F. Bromberg, P. Bryl-Górecka, S. Buch, A.H. Buck, D. Burger, S. Busatto, D. Buschmann, B. Bussolati, E.I. Buzás, J.B. Byrd, G. Camussi, D.R. Carter, S. Caruso, L.W. Chamley, Y.-T. Chang, C. Chen, S. Chen, L. Cheng, A.R. Chin, A. Clayton, S.P. Clerici, A. Cocks, E. Cocucci, R.J. Coffey, A. Cordeiro-da-Silva, Y. Couch, F.A. Coumans, B. Coyle, R. Crescitelli, M.F. Criado, C. D'Souza-Schorey, S. Das, A.D. Chaudhuri, P. de Candia, E.F.D.S. Junior, O.D. Wever, H.A. del Portillo, T. Demaret, S. Deville, A. Devitt, B. Dhondt, D.D. Vizio, L.C. Dieterich, V. Dolo, A.P.D. Rubio, M. Dominici, M.R. Dourado, T.A. Driedonks, F.V. Duarte, H.M. Duncan, R.M. Eichenberger, K. Ekström, S.E. Andaloussi, C. Elie-Caille, U. Erdbrügger, J.M. Falcón-Pérez, F. Fatima, J.E. Fish, M. Flores-Bellver, A. Försönits, A. Frelet-Barrand, F. Fricke, G. Fuhrmann, S. Gabrielsson, A. Gámez-Valero, C. Gardiner, K. Gärtner, R. Gaudin, Y.S. Gho, B. Giebel, C. Gilbert, M. Gimona, I. Giusti, D.C. Goberdhan, A. Görgens, S.M. Gorski, D.W. Greening, J.C. Gross, A. Gualerzi, G.N. Gupta, D. Gustafson, A. Handberg, R.A. Haraszti, P. Harrison, H. Hegyesi, A. Hendrix, A.F. Hill, F.H. Hochberg, K.F. Hoffmann, B. Holder, H. Holthofer, B. Hosseinkhani, G. Hu, Y. Huang, V. Huber, S. Hunt, A.G.-E. Ibrahim, T. Ikezu, J.M. Inal, M. Isin, A. Ivanova, H.K. Jackson, S. Jacobsen, S.M. Jay, M. Jayachandran, G. Jenster, L. Jiang, S.M. Johnson, J.C. Jones, A. Jong, T. Jovanovic-Talisman, S. Jung, R. Kalluri, S. Kano, S. Kaur, Y. Kawamura, E.T. Keller, D. Khamari, E. Khomyakova, A. Khvorova, P. Kierulf, K.P. Kim, T. Kislinger, M. Klingeborn, D.J.K. II, M. Kornek, M.M. Kosanović, Á.F. Kovács, E.-M. Krämer-Albers, S. Krasemann, M. Krause, I.V. Kurochkin, G.D. Kusuma, S. Kuypers, S. Laitinen, S.M. Langevin, L.R. Languino, J. Lannigan, C. Lässer, L.C. Laurent, G. Lavieu, E. Lázaro-Ibáñez, S.L. Lay, M.-S. Lee, Y.X.F. Lee, D.S. Lemos, M. Lenassi,

A. Leszczynska, I.T. Li, K. Liao, S.F. Libregts, E. Ligeti, R. Lim, S.K. Lim, A. Linē, K. Linnemannstöns, A. Llorente, C.A. Lombard, M.J. Lorenowicz, Á.M. Lörincz, J. Lötvall, J. Lovett, M.C. Lowry, X. Loyer, Q. Lu, B. Lukomska, T.R. Lunavat, S.L. Maas, H. Malhi, A. Marcilla, J. Mariani, J. Mariscal, E.S. Martens-Uzunova, L. Martin-Jaular, M.C. Martinez, V.R. Martins, M. Mathieu, S. Mathivanan, M. Maugeri, L.K. McGinnis, M.J. McVey, K.L. Meehan, I. Mertens, V.R. Minciacchi, A. Möller, M.M. Jørgensen, A. Morales-Kastresana, J. Morhayim, F. Mullier, M. Muraca, L. Musante, V. Mussack, D.C. Muth, K.H. Myburgh, T. Najrana, M. Nawaz, I. Nazarenko, P. Nejsum, C. Neri, T. Neri, R. Nieuwland, L. Nimrichter, J.P. Nolan, E.N.N. t Hoen, N.N. Hooten, L. O'Driscoll, T. O'Grady, A. O'Loghlen, T. Ochiya, M. Olivier, A. Ortiz, L.A. Ortiz, X. Osteikoetxea, O. Østergaard, M. Ostrowski, J. Park, D.M. Pegtel, H. Peinado, F. Perut, M.W. Pfaffl, D.G. Phinney, B.C. Pieters, R.C. Pink, D.S. Pisetsky, E.P. von Strandmann, I. Polakovicova, I.K. Poon, B.H. Powell, I. Prada, L. Pulliam, P. Quesenberry, A. Radeghieri, R.L. Raffai, S. Raimondo, J. Rak, M.I. Ramirez, G. Raposo, M.S. Rayyan, N. Regev-Rudzki, F.L. Ricklefs, P.D. Robbins, D.D. Roberts, S.C. Rodrigues, E. Rohde, S. Rome, K.M. Rouschop, A. Rughetti, A.E. Russell, P. Saá, S. Sahoo, E. Salas-Huenuleo, C. Sánchez, J.A. Saugstad, M.J. Saul, R.M. Schiffelers, R. Schneider, T.H. Schøyen, A. Scott, E. Shahaj, S. Sharma, O. Shatnyeva, F. Shekari, G.V. Shelke, A.K. Shetty, K. Shiba, P.R.-M. Siljander, A.M. Silva, A. Skowronek, R.P. Soares, B.W. Sódar, C. Soekmadji, J. Sotillo, P.D. Stahl, W. Stoorvogel, S.L. Stott, E.F. Strasser, S. Swift, H. Tahara, M. Tewari, K. Timms, S. Tiwari, R. Tixeira, M. Tkach, W.S. Toh, R. Tomasini, A.C. Torrecilhas, J.P. Tosar, V. Toxavidis, L. Urbanelli, P. Vader, B.W. van Balkom, S.G. van der Grein, J.V. Deun, M.J. van Herwijnen, K.V. Keuren-Jensen, G. van Niel, M.E. van Royen, A.J. van Wijnen, M.H. Vasconcelos, T.D. Veit, L.J. Vella, É. Velot, F.J. Verweij, B. Vestad, J.L. Viñas, T. Visnovitz, K.V. Vukman, J. Wahlgren, D.C. Watson, M.H. Wauben, A. Weaver, J.P. Webber, V. Weber, A.M. Wehman, D.J. Weiss, J.A. Welsh, S. Wendt, A.M. Wheelock, Z. Wiener, L. Witte, J. Wolfram, A. Xagorari, P. Xander, J. Xu, X. Yan, M. Yáñez-Mó, H. Yin, Y. Yuana, V. Zappulli, J. Zarubova, V. Žėkas, J. Zhang, Z. Zhao, L. Zheng, A.R. Zheutlin, A.M. Zickler, P. Zimmermann, A.M. Zivkovic, D. Zocco, E.K. Zuba-Surma, Minimal information for studies of extracellular vesicles 2018 (MISEV2018): a position statement of the International Society for Extracellular Vesicles and update of the MISEV2014 guidelines, J. Extracell. Vesicles 7 (2018) 1535750, https://doi.org/10.1080/20013078.2018.1535750.

[2] G. van Niel, G. D'Angelo, G. Raposo, Shedding light on the cell biology of extracellular vesicles, Nat. Rev. Mol. Cell Biol. 19 (2018) 213–228, https://doi.org/10.1038/nrm.2017.125.

[3] M. Yáñez-Mó, P.R.-M. Siljander, Z. Andreu, A.B. Zavec, F.E. Borràs, E.I. Buzas, K. Buzas, E. Casal, F. Cappello, J. Carvalho, E. Colás, A.C. Silva, S. Fais, J.M. Falcon-Perez, I.M. Ghobrial, B. Giebel, M. Gimona, M. Graner, I. Gursel, M. Gursel, N.H.H. Heegaard, A. Hendrix, P. Kierulf, K. Kokubun, M. Kosanovic, V. Kralj-Iglic, E.-M. Krämer-Albers, S. Laitinen, C. Lässer, T. Lener, E. Ligeti, A. Linē, G. Lipps, A. Llorente, J. Lötvall, M. Manček-Keber, A. Marcilla, M. Mittelbrunn, I. Nazarenko, E.N.M.N. t Hoen, T.A. Nyman, L. O'Driscoll, M. Olivan, C. Oliveira, É. Pállinger, H.A. del Portillo, J. Reventós, M. Rigau, E. Rohde, M. Sammar, F. Sánchez-Madrid, N. Santarém, K. Schallmoser, M.S. Ostenfeld, W. Stoorvogel, R. Stukelj, S.G. Van der Grein, M.H. Vasconcelos, M.H.M. Wauben, O. De Wever, Biological properties of extracellular vesicles and their physiological functions, J. Extracell. Vesicles 4 (2015), https://doi.org/10.3402/jev.v4.27066.

[4] E. Ogorevc, V. Kralj-Iglič, P. Veranič, The role of extracellular vesicles in phenotypic cancer transformation, Radiol. Oncol. 47 (2013) 197–205, https://doi.org/10.2478/raon-2013-0037.
[5] V. Kralj-Iglič, Stability of membranous nanostructures: a possible key mechanism in cancer progression, Int. J. Nanomed. 7 (2012) 3579–3596, https://doi.org/10.2147/IJN.S29076.
[6] S. Fais, L. O'Driscoll, F.E. Borras, E. Buzas, G. Camussi, F. Cappello, J. Carvalho, A. Cordeiro da Silva, H. Del Portillo, S. El Andaloussi, T. Ficko Trček, R. Furlan, A. Hendrix, I. Gursel, V. Kralj-Iglic, B. Kaeffer, M. Kosanovic, M.E. Lekka, G. Lipps, M. Logozzi, A. Marcilla, M. Sammar, A. Llorente, I. Nazarenko, C. Oliveira, G. Pocsfalvi, L. Rajendran, G. Raposo, E. Rohde, P. Siljander, G. van Niel, M.H. Vasconcelos, M. Yáñez-Mó, M.L. Yliperttula, N. Zarovni, A.B. Zavec, B. Giebel, Evidence-based clinical use of nanoscale extracellular vesicles in nanomedicine, ACS Nano 10 (2016) 3886–3899, https://doi.org/10.1021/acsnano.5b08015.
[7] E. Ogorevc, S. Hudoklin, P. Veranič, V. Kralj-Iglič, Extracellular vesicle-mediated transfer of membranous components from the highly malignant T24 urinary carcinoma cell line to the non-malignant RT4 urinary papilloma cell line, Protoplasma 251 (2014) 699–702, https://doi.org/10.1007/s00709-013-0544-5.
[8] V. Šuštar, M. Frank, V. Janša, P. Šušanj, H. Hägerstrand, P. Veranič, V. Kralj-Iglič, Microvesicles from blood plasma observed by electron microscopy, Med. Razgl. 49 (2010) 137–144.
[9] A. Mrvar-Brečko, V. Šuštar, V. Janša, R. Štukelj, R. Janša, E. Mujagić, P. Kruljc, A. Iglič, H. Hägerstrand, V. Kralj-Iglič, Isolated microvesicles from peripheral blood and body fluids as observed by scanning electron microscope, Blood Cells, Mol. Dis. 44 (2010) 307–312, https://doi.org/10.1016/j.bcmd.2010.02.003.
[10] I. Junkar, V. Šuštar, M. Frank, V. Janša, A.B. Zavec, B. Rozman, M. Mozetic, H. Hagerstrand, V. Kralj-Iglic, Blood and synovial microparticles as revealed by atomic force and scanning electron microscope, Open Autoimmun. J. 1 (2009) 50–58, https://doi.org/10.2174/1876894600901010050.
[11] V. Šuštar, A. Bedina-Zavec, R. Štukelj, M. Frank, G. Bobojević, R. Janša, E. Ogorevc, P. Kruljc, K. Mam, B. Šimunič, M. Manček-Keber, R. Jerala, B. Rozman, P. Veranič, H. Hägerstrand, V. Kralj-Iglič, Nanoparticles isolated from blood: a reflection of vesiculability of blood cells during the isolation process, Int. J. Nanomed. 6 (2011) 2737–2748, https://doi.org/10.2147/IJN.S24537.
[12] S.-C. Tao, S.-C. Guo, C.-Q. Zhang, Platelet-derived extracellular vesicles: an emerging therapeutic approach, Int. J. Biol. Sci. 13 (2017) 828–834, https://doi.org/10.7150/ijbs.19776.
[13] A.R. Brisson, S. Tan, R. Linares, C. Gounou, N. Arraud, Extracellular vesicles from activated platelets: a semiquantitative cryo-electron microscopy and immuno-gold labeling study, Platelets 28 (2017) 263–271, https://doi.org/10.1080/09537104.2016.1268255.
[14] B. Uršič, D. Vozel, V. Šuštar, B. Kocjančič, D. Dolinar, V. Kralj-Iglič, Extracellular vesicles from platelet-rich plasma as conveyors of regeneration potential in orthopedics, J. Hematol. Thromboembolic Dis. 2 (2014), https://doi.org/10.4172/2329-8790.1000163.
[15] R. Alves, R. Grimalt, A review of platelet-rich plasma: history, biology, mechanism of action, and classification, Skin Appendage Disord. 4 (2018) 18–24, https://doi.org/10.1159/000477353.
[16] S.-C. Guo, S.-C. Tao, W.-J. Yin, X. Qi, T. Yuan, C.-Q. Zhang, Exosomes derived from platelet-rich plasma promote the re-epithelization of chronic cutaneous wounds

via activation of YAP in a diabetic rat model, Theranostics 7 (2017) 81–96, https://doi.org/10.7150/thno.16803.

[17] J.M. DeLong, R.P. Russell, A.D. Mazzocca, Platelet-rich plasma: the PAW classification system, Arthrosc. J. Arthrosc. Relat. Surg. Off. Publ. Arthrosc. Assoc. N. Am. Int. Arthrosc. Assoc. 28 (2012) 998–1009, https://doi.org/10.1016/j.arthro.2012.04.148.

[18] M. Koupenova, L. Clancy, H.A. Corkrey, J.E. Freedman, Circulating platelets as mediators of immunity, inflammation, and thrombosis, circ, Res. 122 (2018) 337–351, https://doi.org/10.1161/CIRCRESAHA.117.310795.

[19] R.A. Ali, L.M. Wuescher, R.G. Worth, Platelets: essential components of the immune system, Curr. Trends Immunol. 16 (2015) 65–78.

[20] M. Holinstat, Normal platelet function, Canc. Metastasis Rev. 36 (2017) 195–198, https://doi.org/10.1007/s10555-017-9677-x.

[21] C.N. Morrell, A.A. Aggrey, L.M. Chapman, K.L. Modjeski, Emerging roles for platelets as immune and inflammatory cells, Blood 123 (2014) 2759–2767, https://doi.org/10.1182/blood-2013-11-462432.

[22] M.D. Fabbro, M. Bortolin, S. Taschieri, C. Ceci, R.L. Weinstein, Antimicrobial properties of platelet-rich preparations. A systematic review of the current pre-clinical evidence, Platelets 27 (2016) 276–285, https://doi.org/10.3109/09537104.2015.1116686.

[23] E. Anitua, I. Andia, B. Ardanza, P. Nurden, A. Nurden, Autologous platelets as a source of proteins for healing and tissue regeneration, Thromb. Haemostasis 91 (2004) 4–15, https://doi.org/10.1160/TH03-07-0440.

[24] J. Etulain, Platelets in wound healing and regenerative medicine, Platelets 29 (2018) 556–568, https://doi.org/10.1080/09537104.2018.1430357.

[25] J. Chahla, M.E. Cinque, N.S. Piuzzi, S. Mannava, A.G. Geeslin, I.R. Murray, G.J. Dornan, G.F. Muschler, R.F. LaPrade, A call for standardization in platelet-rich plasma preparation protocols and composition reporting: a systematic review of the clinical orthopaedic literature, J. Bone Joint Surg. Am. 99 (2017) 1769–1779, https://doi.org/10.2106/JBJS.16.01374.

[26] R. Štukelj, I.H. Ignaščenko, S. Peternelj, M. Peruško, T. Blažič, M. Pajnič, Š.B. Vnuk, V. Šuštar, A.B. Zavec, K. Schara, R. Janša, V. Kralj-Iglič, Role of blood sampling in assessment of concentration of extracellular nanovesicles in isolates from peripheral blood, in: Adv. Planar Lipid Bilayers Liposomes, Elsevier, 2014, pp. 175–189, https://doi.org/10.1016/B978-0-12-418699-6.00007-2.

[27] R. Štukelj, K. Schara, A. Bedina - Zavec, V. Šuštar, M. Pajnič, L. Pađen, J.L. Krek, V. Kralj-Iglič, A. Mrvar-Brečko, R. Janša, Effect of shear stress in the flow through the sampling needle on concentration of nanovesicles isolated from blood, Eur. J. Pharm. Sci. Off. J. Eur. Fed. Pharm. Sci. 98 (2017) 17–29, https://doi.org/10.1016/j.ejps.2016.10.007.

[28] R.J.F.C. do Amaral, N.P. da Silva, N.F. Haddad, L.S. Lopes, F.D. Ferreira, R.B. Filho, P.A. Cappelletti, W. de Mello, E. Cordeiro-Spinetti, A. Balduino, Platelet-rich plasma obtained with different anticoagulants and their effect on platelet numbers and mesenchymal stromal cells behavior in vitro, Stem Cell. Int. 2016 (2016) 7414036, https://doi.org/10.1155/2016/7414036.

[29] K. Huang, G. Giddins, L. Wu, Platelet-rich plasma versus corticosteroid injections in the management of elbow epicondylitis and plantar fasciitis: an updated systematic review and meta-analysis, Am. J. Sports Med. (2019), https://doi.org/10.1177/0363546519888450.

[30] Q. Wu, X. Luo, Y. Xiong, G. Liu, J. Wang, X. Chen, B. Mi, Platelet-rich plasma versus hyaluronic acid in knee osteoarthritis: a meta-analysis with the consistent ratio of injection, J. Orthop. Surg. 28 (2020), https://doi.org/10.1177/2309499019887660.

[31] P. Kossev, T. Sokolov, Platelet-rich plasma (PRP) in orthopedics and traumatology — review, in: K. Metodiev (Ed.), Immunopathol. Immunomodulation, InTech, 2015, pp. 173—195, https://doi.org/10.5772/61326.
[32] L. Di Sante, C. Villani, V. Santilli, M. Valeo, E. Bologna, L. Imparato, M. Paoloni, A. Iagnocco, Intra-articular hyaluronic acid vs platelet-rich plasma in the treatment of hip osteoarthritis, Med. Ultrason. 18 (2016) 463—468, https://doi.org/10.11152/mu-874.
[33] D. Man, H. Plosker, J.E. Winland-brown, The use of autologous platelet-rich plasma (platelet gel) and autologous platelet-poor plasma (fibrin glue) in cosmetic surgery, Plast. Reconstr. Surg. 107 (2001) 229—236.
[34] Y.-T. Wu, T.-Y. Ho, Y.-C. Chou, M.-J. Ke, T.-Y. Li, G.-S. Huang, L.-C. Chen, Six-month efficacy of platelet-rich plasma for carpal tunnel syndrome: a prospective randomized, single-blind controlled trial, Sci. Rep. 7 (2017) 94, https://doi.org/10.1038/s41598-017-00224-6.
[35] M.-A. Gkini, A.-E. Kouskoukis, G. Tripsianis, D. Rigopoulos, K. Kouskoukis, Study of platelet-rich plasma injections in the treatment of androgenetic alopecia through an one-year period, J. Cutan. Aesthetic Surg. 7 (2014) 213, https://doi.org/10.4103/0974-2077.150743.
[36] E.-E. Betsi, E. Germain, D.F. Kalbermatten, M. Tremp, V. Emmenegger, Platelet-rich plasma injection is effective and safe for the treatment of alopecia, Eur. J. Plast. Surg. 36 (2013) 407—412, https://doi.org/10.1007/s00238-013-0816-5.
[37] A.S. Dawood, H.A. Salem, Current clinical applications of platelet-rich plasma in various gynecological disorders: an appraisal of theory and practice, Clin. Exp. Reprod. Med. 45 (2018) 67—74, https://doi.org/10.5653/cerm.2018.45.2.67.
[38] Z. Hu, S. Qu, J. Zhang, X. Cao, P. Wang, S. Huang, F. Shi, Y. Dong, J. Wu, B. Tang, J. Zhu, Efficacy and safety of platelet-rich plasma for patients with diabetic ulcers: a systematic review and meta-analysis, Adv. Wound Care 8 (2019) 298—308, https://doi.org/10.1089/wound.2018.0842.
[39] L.M. Rappl, Effect of platelet rich plasma gel in a physiologically relevant platelet concentration on wounds in persons with spinal cord injury, Int. Wound J. 8 (2011) 187—195, https://doi.org/10.1111/j.1742-481X.2011.00770.x.
[40] D. Semenič, Vloga Alogenskega Trombocitnega Gela Pri Celjenju Kroničnih Ran Spodnjih Okončin: Doktorsko Delo, 2018. https://repozitorij.uni-lj.si/IzpisGradiva.php?lang=slv&id=102421. (Accessed 9 July 2019).
[41] Y. Xia, J. Zhao, J. Xie, Y. Lv, D. Cao, The efficacy of platelet-rich plasma dressing for chronic nonhealing ulcers: a meta-analysis of 15 randomized controlled trials, Plast. Reconstr. Surg. 144 (2019) 1463—1474, https://doi.org/10.1097/PRS.0000000000006281.
[42] X. Hou, J. Yuan, A. Aisaiti, Y. Liu, J. Zhao, The effect of platelet—rich plasma on clinical outcomes of the surgical treatment of periodontal intrabony defects: a systematic review and meta—analysis, BMC Oral Health 16 (2016) 71, https://doi.org/10.1186/s12903-016-0261-5.
[43] A. Albanese, M.E. Licata, B. Polizzi, G. Campisi, Platelet-rich plasma (PRP) in dental and oral surgery: from the wound healing to bone regeneration, Immun. Ageing A. 10 (2013) 23, https://doi.org/10.1186/1742-4933-10-23.
[44] M. Franchini, M. Cruciani, C. Mengoli, F. Masiello, G. Marano, E. D'Aloja, C. Dell'Aringa, I. Pati, E. Veropalumbo, S. Pupella, S. Vaglio, G.M. Liumbruno, The use of platelet-rich plasma in oral surgery: a systematic review and meta-analysis, Blood Transfus. Trasfus. Sangue. 17 (2019) 357—367, https://doi.org/10.2450/2019.0177-19.
[45] P.-Y. Chung, M.-T. Lin, H.-P. Chang, Effectiveness of platelet-rich plasma injection in patients with temporomandibular joint osteoarthritis: a systematic review and meta-

analysis of randomized controlled trials, Oral Surg. Oral Med. Oral Pathol. Oral Radiol. 127 (2019) 106–116, https://doi.org/10.1016/j.oooo.2018.09.003.
[46] A. Cieślik-Bielecka, J. Glik, R. Skowroński, T. Bielecki, Benefit of leukocyte- and platelet-rich plasma in operative wound closure in oral and maxillofacial surgery, BioMed Res. Int. 2016 (2016) 1–5, https://doi.org/10.1155/2016/7649206.
[47] E.T. Daif, Effect of autologous platelet-rich plasma on bone regeneration in mandibular fractures, Dent. Traumatol. 29 (2013) 399–403, https://doi.org/10.1111/edt.12021.
[48] R. Tabrizi, F. Pourdanesh, S. Jafari, P. Behnia, Can platelet-rich fibrin accelerate neurosensory recovery following sagittal split osteotomy? A double-blind, split-mouth, randomized clinical trial, Int. J. Oral Maxillofac. Surg. 47 (2018) 1011–1014, https://doi.org/10.1016/j.ijom.2018.04.010.
[49] E.L. Matz, A.M. Pearlman, R.P. Terlecki, Safety and feasibility of platelet rich fibrin matrix injections for treatment of common urologic conditions, Investig. Clin. Urol. 59 (2018) 61–65, https://doi.org/10.4111/icu.2018.59.1.61.
[50] V. García-Conca, M. Abad-Collado, J.R. Hueso-Abancens, E. Mengual-Verdú, D.P. Piñero, F. Aguirre-Balsalobre, J.C. Molina, Efficacy and safety of treatment of hyposecretory dry eye with platelet-rich plasma, Acta Ophthalmol. 97 (2019) e170–e178, https://doi.org/10.1111/aos.13907.
[51] J.L. Alio, A.E. Rodriguez, A.A. Abdelghany, R.F. Oliveira, Autologous platelet-rich plasma eye drops for the treatment of post-LASIK chronic ocular surface syndrome, J. Ophthalmol. 2017 (2017) 1–6, https://doi.org/10.1155/2017/2457620.
[52] M.Y. Avila, A.M. Igua, A.M. Mora, Randomised, prospective clinical trial of platelet-rich plasma injection in the management of severe dry eye, Br. J. Ophthalmol. 103 (5) (2018) 648–653, https://doi.org/10.1136/bjophthalmol-2018-312072.
[53] J.L. Alio, M. Abad, A. Artola, J.L. Rodriguez-Prats, S. Pastor, J. Ruiz-Colecha, Use of autologous platelet-rich plasma in the treatment of dormant corneal ulcers, Ophthalmology 114 (2007) 1286–1293, https://doi.org/10.1016/j.ophtha.2006.10.044, e1.
[54] A.N. Patel, C.H. Selzman, G.S. Kumpati, S.H. McKellar, D.A. Bull, Evaluation of autologous platelet rich plasma for cardiac surgery: outcome analysis of 2000 patients, J. Cardiothorac. Surg. 11 (2016) 62, https://doi.org/10.1186/s13019-016-0452-9.
[55] B.H. Kirmani, S.G. Jones, S. Datta, E.K. McLaughlin, A.J. Hoschtitzky, A meta-analysis of platelet gel for prevention of sternal wound infections following cardiac surgery, Blood Transfus 15 (2017) 57–65, https://doi.org/10.2450/2016.0231-15.
[56] D.H. Alamdari, M. Asadi, A.N. Rahim, G. Maddah, S. Azizi, S. Shahidsales, M. Mehrabibahar, Efficacy and safety of pleurodesis using platelet-rich plasma and fibrin glue in management of postoperative chylothorax after esophagectomy, World J. Surg. 42 (2018) 1046–1055, https://doi.org/10.1007/s00268-017-4242-x.
[57] B. Gunaydin, M. Acar, G. Emmez, D. Akcali, N. Tokgoz, Epidural patch with autologous platelet rich plasma: a novel approach, J. Anesth. 31 (2017) 907–910, https://doi.org/10.1007/s00540-017-2400-9.
[58] R. Tarantino, P. Donnarumma, C. Mancarella, M. Rullo, G. Ferrazza, G. Barrella, S. Martini, R. Delfini, Posterolateral arthrodesis in lumbar spine surgery using autologous platelet-rich plasma and cancellous bone substitute: an osteoinductive and osteoconductive effect, Global Spine J. 4 (2014) 137–142, https://doi.org/10.1055/s-0034-1376157.
[59] T. Theys, A. Van Hoylandt, C.-E. Broeckx, L. Van Gerven, J. Jonkergouw, M. Quirynen, J. van Loon, Plasma-rich fibrin in neurosurgery: a feasibility study, Acta Neurochir. 160 (2018) 1497–1503, https://doi.org/10.1007/s00701-018-3579-8.

[60] D.V. Pachito, C. de O.C. Latorraca, R. Riera, Efficacy of platelet-rich plasma for non-transfusion use: overview of systematic reviews, Int. J. Clin. Pract. 73 (2019) e13402, https://doi.org/10.1111/ijcp.13402.

[61] A.L. Maisel-Campbell, A. Ismail, K.A. Reynolds, E. Poon, L. Serrano, S. Grushchak, C. Farid, D.P. West, M. Alam, A systematic review of the safety and effectiveness of platelet-rich plasma (PRP) for skin aging, Arch. Dermatol. Res. (2019), https://doi.org/10.1007/s00403-019-01999-6.

[62] H.I. Gawdat, A.M. Tawdy, R.A. Hegazy, M.M. Zakaria, R.S. Allam, Autologous platelet-rich plasma versus readymade growth factors in skin rejuvenation: a split face study, J. Cosmet. Dermatol. 16 (2017) 258–264, https://doi.org/10.1111/jocd.12341.

[63] T.-S. Hsieh, W.-K. Chiu, T.-F. Yang, H.-J. Wang, C. Chen, A meta-analysis of the evidence for assisted therapy with platelet-rich plasma for atrophic acne scars, Aesthetic Plast. Surg. 43 (2019) 1615–1623, https://doi.org/10.1007/s00266-019-01471-w.

[64] S. Gode, A. Ozturk, E. Kısmalı, V. Berber, G. Turhal, The effect of platelet-rich fibrin on nasal skin thickness in rhinoplasty, Facial Plast. Surg. FPS. 35 (2019) 400–403, https://doi.org/10.1055/s-0039-1693436.

[65] S. Gode, A. Ozturk, V. Berber, E. Kısmalı, Effect of injectable platelet-rich fibrin on diced cartilage's viability in rhinoplasty, Facial Plast. Surg. FPS. 35 (2019) 393–396, https://doi.org/10.1055/s-0039-1693035.

[66] I. Kuzucu, G.K. Beriat, H. Ezerarslan, S. Ozdemir, S. Kocaturk, Effects of the autologous platelet-rich plasma in nasal pack on postoperative quality of life, J. Craniofac. Surg. 28 (2017) e299–e302, https://doi.org/10.1097/SCS.0000000000003664.

[67] J.J. Mendonça-Caridad, P. Juiz-Lopez, J.P. Rubio-Rodriguez, Frontal sinus obliteration and craniofacial reconstruction with platelet rich plasma in a patient with fibrous dysplasia, Int. J. Oral Maxillofac. Surg. 35 (2006) 88–91, https://doi.org/10.1016/j.ijom.2005.06.017.

[68] J. Mendonça-Caridad, P.J. Lopez, F.V. Fayos, G. Miery, A novel approach to human cranial tissue regeneration and frontal sinus obliteration with an autogenous platelet-rich/fibrin-rich composite matrix: 10 patients with a 6-10 year follow-up, J. Tissue Eng. Regen. Med. 7 (2013) 491–500, https://doi.org/10.1002/term.534.

[69] C. Taglialatela Scafati, S. Taglialatela Scafati, A. Aveta, M. Cassese, C. Vitale, Chronic frontal sinus disease: combined use of platelet-rich plasma and calvarial bone grafts for sinus obliteration in aggressive and secondary cases, Rev. Stomatol. Chir. Maxillofac. 111 (2010) 216–220, https://doi.org/10.1016/j.stomax.2010.07.004.

[70] Y.W. Khafagy, A.M. Abd Elfattah, W. Moneir, E.H. Salem, Leukocyte- and platelet-rich fibrin: a new graft material in endoscopic repair of spontaneous CSF leaks, Eur. Arch. Oto-Rhino-Laryngol. 275 (2018) 2245–2252, https://doi.org/10.1007/s00405-018-5048-7.

[71] S.M. Refahee, M.A. Aboulhassan, O. Abdel Aziz, D. Emara, H.M. Seif El Dein, B.G. Moussa, M. Abu Sneineh, Is PRP effective in reducing the scar width of primary cleft lip repair? A randomized controlled clinical study, Cleft Palate-Craniofacial J. Off. Publ. Am. Cleft Palate-Craniofacial Assoc. 57 (5) (2019) 581–588, https://doi.org/10.1177/1055665619884455.

[72] H. Pathak, S. Mohanty, A.B. Urs, J. Dabas, Treatment of oral mucosal lesions by scalpel excision and platelet-rich fibrin membrane grafting: a review of 26 sites, J. Oral Maxillofac. Surg. 73 (2015) 1865–1874, https://doi.org/10.1016/j.joms.2015.03.041.

[73] M. Mahajan, M.K. Gupta, C. Bande, V. Meshram, Comparative evaluation of healing pattern after surgical excision of oral mucosal lesions by using platelet-rich fibrin (PRF) membrane and collagen membrane as grafting materials—a randomized clinical

trial, J. Oral Maxillofac. Surg. 76 (2018) 1469.e1–1469.e9, https://doi.org/10.1016/j.joms.2018.02.031.
[74] M.H.M. EL-Komy, N.A. Saleh, M.A. Saleh, Autologous platelet-rich plasma and triamcinolone acetonide intralesional injection in the treatment of oral erosions of pemphigus vulgaris: a pilot study, Arch. Dermatol. Res. 310 (2018) 375–381, https://doi.org/10.1007/s00403-018-1824-x.
[75] M. Scala, M. Gipponi, P. Mereu, P. Strada, R. Corvò, A. Muraglia, M. Massa, S. Bertoglio, P. Santi, F. Cafiero, Regeneration of mandibular osteoradionecrosis defect with platelet rich plasma gel, Vivo Athens Greece 24 (2010) 889–893.
[76] M. Elkahwagi, M. Elokda, D. Elghannam, A. Elsobki, Role of autologous platelet-rich fibrin in relocation pharyngoplasty for obstructive sleep apnoea, Int. J. Oral Maxillofac. Surg. 49 (2020) 200–206, https://doi.org/10.1016/j.ijom.2019.05.008.
[77] M. Joudi, D. Hamidi Alamdari, M. Hyradfar, H.R. Rahimi, E. Saremi, M. Fathi, R. Shojaeian, G. Koliakos, Lateral traumatic esophago-cutaneous fistula in a child; platelet-rich fibrin glue challenge, Iran, Red Crescent Med. J. 15 (2013) 256–259, https://doi.org/10.5812/ircmj.7975.
[78] E. Ricci, G. Riva, F. Dagna, A.L. Cavalot, The use of platelet-rich plasma gel in superficial parotidectomy, Acta Otorhinolaryngol. Ital. Organo Uff. Della Soc. Ital. Otorinolaringol. E Chir. Cerv.-Facc. 39 (6) (2019) 363–366, https://doi.org/10.14639/0392-100X-2093.
[79] M. Scala, P. Mereu, F. Spagnolo, M. Massa, A. Barla, S. Mosci, G. Forno, A. Ingenito, P. Strada, The use of platelet-rich plasma gel in patients with mixed tumour undergoing superficial parotidectomy: a randomized study, Vivo Athens Greece 28 (2014) 121–124.
[80] S.K. Lee, Y.M. Lim, D.H. Lew, S.Y. Song, Salvage of unilateral complete ear amputation with continuous local hyperbaric oxygen, platelet-rich plasma and polydeoxyribonucleotide without micro-revascularization, Arch. Plast. Surg. 44 (2017) 554–558, https://doi.org/10.5999/aps.2017.00451.
[81] M. Habesoglu, C. Oysu, S. Sahin, A. Sahin-Yilmaz, D. Korkmaz, A. Tosun, A. Karaaslan, Platelet-rich fibrin plays a role on healing of acute-traumatic ear drum perforation, J. Craniofac. Surg. 25 (2014) 2056–2058, https://doi.org/10.1097/SCS.0000000000001140.
[82] S.P.S. Yadav, J.S. Malik, P. Malik, P.K. Sehgal, J.S. Gulia, R.K. Ranga, Studying the result of underlay myringoplasty using platelet-rich plasma, J. Laryngol. Otol. 132 (2018) 990–994, https://doi.org/10.1017/S0022215118001846.
[83] D. Sharma, S. Mohindroo, R.K. Azad, Efficacy of platelet rich fibrin in myringoplasty, Int. J. Otorhinolaryngol. Head Neck Surg. 4 (2018) 677–681, https://doi.org/10.18203/issn.2454-5929.ijohns20181670.
[84] G. Sankaranarayanan, V. Prithiviraj, R.V. Kumar, A study on efficacy of autologous platelet rich plasma in myringoplasty, Otolaryngol. Online J. 3 (2013) 1–15.
[85] M. Saeedi, M. Ajalloueian, E. Zare, A. Taheri, J. Yousefi, S.M.J. Mirlohi, N.M. Aref, M.J. Saeedi, M.H. Khosravi, The effect of PRP-enriched gelfoam on ChronicTympanic membrane perforation: a double-blind randomized clinical trial, Int. Tinnitus J. 21 (2017) 108–111, https://doi.org/10.5935/0946-5448.20170021.
[86] T. Fawzy, M. Hussein, S. Eid, S. Guindi, Effect of adding platelet-rich plasma to fat grafts in myringoplasty, Egypt, J. Otolaryngol. 34 (2018) 224–228, https://doi.org/10.4103/ejo.ejo_53_18.
[87] M.W. El-Anwar, I. Elnashar, Y.A. Foad, Platelet-rich plasma myringoplasty: a new office procedure for the repair of small tympanic membrane perforations, Ear Nose Throat J. 96 (2017) 312–326, https://doi.org/10.1177/014556131709600818.
[88] N.P. Nair, A. Alexander, B. Abhishekh, J.S. Hegde, S. Ganesan, S.K. Saxena, Safety and efficacy of autologous platelet-rich fibrin on graft uptake in myringoplasty: a

randomized controlled trial, Int. Arch. Otorhinolaryngol. 23 (2019) 77–82, https://doi.org/10.1055/s-0038-1649495.
[89] M.W. El-Anwar, M.A.S. El-Ahl, A.A. Zidan, M.A.-R.A.-S. Yacoup, Topical use of autologous platelet rich plasma in myringoplasty, Auris Nasus Larynx 42 (2015) 365–368, https://doi.org/10.1016/j.anl.2015.02.016.
[90] T. Ersözlü, E. Gultekin, A comparison of the autologous platelet-rich plasma gel fat graft myringoplasty and the fat graft myringoplasty for the closure of different sizes of tympanic membrane perforations, Ear Nose Throat J. 99 (5) (2020) 331–336, https://doi.org/10.1177/0145561319900388.
[91] Y. Shiomi, Y. Shiomi, Surgical outcomes of myringoplasty using platelet-rich plasma and evaluation of the outcome-associated factors, Auris Nasus Larynx 47 (2) (2019) 191–197, https://doi.org/10.1016/j.anl.2019.06.005.
[92] M. Elbary, W. Nasr, S. Sorour, Platelet-rich plasma in reconstruction of posterior meatal wall after canal wall down mastoidectomy, Int. Arch. Otorhinolaryngol. 22 (2018) 103–107, https://doi.org/10.1055/s-0037-1602694.
[93] C.H. Jang, C.H. Choi, Y.B. Cho, Effect of BMP2–platelet-rich plasma–biphasic calcium phosphate scaffold on accelerated osteogenesis in mastoid obliteration, In Vivo 30 (2016) 835–840, https://doi.org/10.21873/invivo.11002.
[94] S.M. Askar, I.M. Saber, M. Omar, Mastoid reconstruction with platelet-rich plasma and bone pate after canal wall down mastoidectomy: a preliminary report, Ear Nose Throat J. (2019), https://doi.org/10.1177/0145561319879789.
[95] S.F. Alhabib, I. Saliba, Hyaluronic acid fat graft myringoplasty versus autologous platelet rich plasma, J. Clin. Med. Res. 9 (2017) 30–34, https://doi.org/10.14740/jocmr2800w.
[96] R.S. Dhillon, E.M. Schwarz, M.D. Maloney, Platelet-rich plasma therapy - future or trend? Arthritis Res. Ther. 14 (2012) 219, https://doi.org/10.1186/ar3914.
[97] Y.R. Jeon, E.H. Kang, C.E. Yang, I.S. Yun, W.J. Lee, D.H. Lew, The effect of platelet-rich plasma on composite graft survival: Plast. Reconstr. Surg. 134 (2014) 239–246, https://doi.org/10.1097/PRS.0000000000000392.
[98] S.K. Özgürsoy, F. Tunçkaşık, M.E. Tunçkaşık, E. Akıncıoğlu, H. Doğan, G.K. Beriat, Histopathologic evaluation of hyaluronic acid and plasma-rich platelet injection into rabbit vocal cords: an experimental study, Turk. Arch. Otolaryngol. 56 (2018) 30–35, https://doi.org/10.5152/tao.2018.2942.
[99] S.B. Cobden, K. Oztürk, S. Duman, H. Esen, T.M. Aktan, M.C. Avunduk, C. Elsurer, Treatment of acute vocal fold injury with platelet-rich plasma, J. Voice 30 (2016) 731–735, https://doi.org/10.1016/j.jvoice.2015.07.012.
[100] A. Vidovic, U. Delling, Aryepiglottic fold augmentation as treatment for late-onset dysphagia following surgical treatment of recurrent laryngeal neuropathy, Tierarztl. Prax. Ausg. G Grosstiere Nutztiere. 45 (2017) 219–225, https://doi.org/10.15653/TPG-160712.
[101] A. Eryılmaz, B. Demirci, C. Gunel, F.K. Doger, O. Yukselen, I.K. Omurlu, Y. Basal, F. Agdas, S. Basak, Can tissue adhesives and platelet-rich plasma prevent pharyngocutaneous fistula formation? Auris Nasus Larynx 43 (2016) 62–67, https://doi.org/10.1016/j.anl.2015.06.012.
[102] A.G. Yasak, O. Yigit, E. Araz Server, S. Durna Dastan, M. Gul, The effectiveness of platelet-rich plasma in an anosmia-induced mice model, Laryngoscope 128 (2018) E157–E162, https://doi.org/10.1002/lary.27029.
[103] M. Sánchez, A. Garate, A.M. Bilbao, J. Oraa, F. Yangüela, P. Sánchez, J. Guadilla, B. Aizpurua, J. Azofra, N. Fiz, D. Delgado, Platelet-rich plasma for injured peripheral nerves: biological repair process and clinical application guidelines, in: Demystifying Polyneuropathy - Recent Adv. New Dir, 2018, https://doi.org/10.5772/intechopen.81104.

[104] M. Stolle, J. Schulze, A. Roemer, T. Lenarz, M. Durisin, A. Warnecke, Human plasma rich in growth factors improves survival and neurite outgrowth of spiral ganglion neurons in vitro, Tissue Eng. 24 (2018) 493–501, https://doi.org/10.1089/ten.TEA.2017.0120.

[105] H.O. Kemppainen, H.J. Puhakka, P.J. Laippala, M.M. Sipilä, M.P. Manninen, P.H. Karma, Epidemiology and aetiology of middle ear cholesteatoma, Acta Otolaryngol. 119 (1999) 568–572, https://doi.org/10.1080/00016489950180801.

[106] J. Benson, L. Mwanri, Chronic suppurative otitis media and cholesteatoma in Australia's refugee population, Aust. Fam. Physician 41 (2012) 978–980.

[107] D. Lucidi, E. De Corso, G. Paludetti, B. Sergi, Quality of life and functional results in canal wall down vs canal wall up mastoidectomy, Acta Otorhinolaryngol. Ital. 39 (2019) 53–60, https://doi.org/10.14639/0392-100X-2005.

[108] C. Skoulakis, P. Koltsidopoulos, A. Iyer, G. Kontorinis, Mastoid obliteration with synthetic materials: a review of the literature, J. Int. Adv. Otol. 15 (2019) 400–404, https://doi.org/10.5152/iao.2019.7038.

[109] A.T. Harris, B. Mettias, T.H.J. Lesser, Pooled analysis of the evidence for open cavity, combined approach and reconstruction of the mastoid cavity in primary cholesteatoma surgery, J. Laryngol. Otol. 130 (2016) 235–241, https://doi.org/10.1017/S0022215116000013.

[110] M.B. Gluth, B.Y.B. Tan, P.L. Santa Maria, M.D. Atlas, Unique microbiology of chronically unstable canal wall down tympanomastoid cavities: considerations for surgical revision, J. Laryngol. Otol. 127 (2013) 458–462, https://doi.org/10.1017/S0022215113000583.

[111] D. Henatsch, F. Wesseling, J.J. Briedé, R.J. Stokroos, Treatment of chronically infected open mastoid cavities with medical honey: a randomized controlled trial, Otol. Neurotol. 36 (2015) 782–787, https://doi.org/10.1097/MAO.0000000000000728.

[112] G.E. Shambaugh, Part two: surgery of infections of the ear, in: Surg. Ear, second ed., W. B. Saunders Company, Philadelphia & London, 1967, pp. 286–287.

[113] J.G. Toner, G.D. Smyth, Surgical treatment of cholesteatoma: a comparison of three techniques, Am. J. Otol. 11 (1990) 247–249.

[114] D. Henatsch, S. Alsulami, A.M. Duijvestijn, J.P. Cleutjens, C.J. Peutz-Kootstra, R.J. Stokroos, Histopathological and inflammatory features of chronically discharging open mastoid cavities secondary analysis of a randomized clinical trial, JAMA Otolaryngol. Neck Surg. 144 (2018) 211–217, https://doi.org/10.1001/jamaoto.2017.2801.

[115] Mastoidectomy: canal wall down techniques, in: D.E. Brackmann, C. Shelton, M.A. Arriaga (Eds.), Otol. Surg., 3th ed., W.B. Saunders Company, Philadelphia, 1994, 239–239.

[116] Biomaterials in tympanoplasty: mastoid obliteration and ear canal reconstruction, in: S.E. Kountakis (Ed.), Encycl. Otolaryngol. Head Neck Surg., Springer-Verlag, Berlin Heidelberg, 2013, pp. 333–334. https://www.springer.com/gp/book/9783642234989. (Accessed 27 January 2020).

[117] V.Y.R. Soares, N.A. Atai, T. Fujita, S. Dilwali, S. Sivaraman, L.D. Landegger, F.H. Hochberg, C.A.P.C. Oliveira, F. Bahmad, X.O. Breakefield, K.M. Stankovic, Extracellular vesicles derived from human vestibular schwannomas associated with poor hearing damage cochlear cells, Neuro Oncol. 18 (2016) 1498–1507, https://doi.org/10.1093/neuonc/now099.

[118] S. Val, S. Jeong, M. Poley, A. Krueger, G. Nino, K. Brown, D. Preciado, Purification and characterization of microRNAs within middle ear fluid exosomes: implication in otitis media pathophysiology, Pediatr. Res. 81 (2017) 911–918, https://doi.org/10.1038/pr.2017.25.

[119] E.H.C. Wong, Y.Y. Dong, M. Coray, M. Cortada, S. Levano, A. Schmidt, Y. Brand, D. Bodmer, L. Muller, Inner ear exosomes and their potential use as biomarkers, PloS One 13 (2018), https://doi.org/10.1371/journal.pone.0198029.
[120] B. György, C. Sage, A.A. Indzhykulian, D.I. Scheffer, A.R. Brisson, S. Tan, X. Wu, A. Volak, D. Mu, P.I. Tamvakologos, Y. Li, Z. Fitzpatrick, M. Ericsson, X.O. Breakefield, D.P. Corey, C.A. Maguire, Rescue of hearing by gene delivery to inner-ear hair cells using exosome-associated AAV, Mol. Ther. 25 (2017) 379–391, https://doi.org/10.1016/j.ymthe.2016.12.010.
[121] G.M. Kalinec, L. Gao, W. Cohn, J.P. Whitelegge, K.F. Faull, F. Kalinec, Extracellular vesicles from auditory cells as nanocarriers for anti-inflammatory drugs and pro-resolving mediators, Front. Cell. Neurosci. 13 (2019), https://doi.org/10.3389/fncel.2019.00530.
[122] W.-Q. Zhong, J.-G. Ren, X.-P. Xiong, Q.-W. Man, W. Zhang, L. Gao, C. Li, B. Liu, Z.-J. Sun, J. Jia, W.-F. Zhang, Y.-F. Zhao, G. Chen, Increased salivary microvesicles are associated with the prognosis of patients with oral squamous cell carcinoma, J. Cell Mol. Med. 23 (2019) 4054–4062, https://doi.org/10.1111/jcmm.14291.
[123] L. He, F. Ping, Z. Fan, C. Zhang, M. Deng, B. Cheng, J. Xia, Salivary exosomal miR-24-3p serves as a potential detective biomarker for oral squamous cell carcinoma screening, Biomed. Pharmacother. Biomedecine Pharmacother. 121 (2020) 109553, https://doi.org/10.1016/j.biopha.2019.109553.
[124] C. Li, Y. Zhou, J. Liu, X. Su, H. Qin, S. Huang, X. Huang, N. Zhou, Potential markers from serum-purified exosomes for detecting oral squamous cell carcinoma metastasis, Cancer Epidemiol. Biomark. Prev. Publ. Am. Assoc. Cancer Res. Cosponsored Am. Soc. Prev. Oncol. 28 (2019) 1668–1681, https://doi.org/10.1158/1055-9965.EPI-18-1122.
[125] S. Rodríguez Zorrilla, M. Pérez-Sayans, S. Fais, M. Logozzi, M. Gallas Torreira, A. García García, A pilot clinical study on the prognostic relevance of plasmatic exosomes levels in oral squamous cell carcinoma patients, Cancers 11 (2019), https://doi.org/10.3390/cancers11030429.
[126] J. Wang, Y. Zhou, J. Lu, Y. Sun, H. Xiao, M. Liu, L. Tian, Combined detection of serum exosomal miR-21 and HOTAIR as diagnostic and prognostic biomarkers for laryngeal squamous cell carcinoma, Med. Oncol. Northwood Lond. Engl. 31 (2014) 148, https://doi.org/10.1007/s12032-014-0148-8.
[127] X. Cui, D. Xiao, Y. Cui, X. Wang, Exosomes-derived long non-coding RNA HOTAIR reduces laryngeal cancer radiosensitivity by regulating microRNA-454-3p/E2F2 Axis, OncoTargets Ther. 12 (2019) 10827–10839, https://doi.org/10.2147/OTT.S224881.
[128] D. Vozel, B. Uršič, J.L. Krek, R. Štukelj, V. Kralj-Iglič, Applicability of extracellular vesicles in clinical studies, Eur. J. Clin. Invest. 47 (2017) 305–313, https://doi.org/10.1111/eci.12733.

CHAPTER FOUR

# Plant extracellular vesicles: the safe for bioactive compounds

**Marzia Pucci[a] and Stefania Raimondo[a,b,*]**
[a]Department of Biomedicine, Neuroscience and Advanced Diagnostics (Bi.N.D.), Section of Biology and Genetics, University of Palermo, Palermo, Italy
[b]Navhetec s.r.l, Palermo, Italy
[*]Corresponding author: E-mail: stefania.raimondo@unipa.it

## Contents



## Abstract

If researchers within the field of extracellular vesicles are fascinated by the discovery of what was previously defined as cell rubbish bags and are today considered as important vehicles of biological information, it is even more fascinating to understand that each day, through the diet, we come into contact with vesicles from the plant kingdom, a safe for various compounds with different properties. This chapter will, therefore, focus on the description of studies carried out to date on vesicles originating from the plant kingdom. Above all we will describe the content of the vesicles, looking at the proteomic, lipidomic and nucleic acid composition; we will further discuss the properties they possess and therefore the potential application in human health.

*Advances in Biomembranes and Lipid Self-Assembly*, Volume 33
ISSN 2451-9634
https://doi.org/10.1016/bs.abl.2020.04.002

## 1. Introduction

For a long time, plants were the main or even the only source of natural solutions available to physicians. This led to a deepening of knowledge about plant species that exhibit curative properties in order to increase their use.

The first doctors were passionate botanists who, with their knowledge, discovered the therapeutic value of numerous plant species. Researchers such as Hippocrates, Theophrastus, Dioscorides, and Galen provided the basis for the development of medicine, both in the field of what we today call diagnostics and in that of pharmacology, anatomy and preventive medicine. Hippocrates (460-377 BC) was called the "father of medicine" because he practiced the art of healing with intelligence and a deep critical spirit. He was able to put into practice the therapeutic activity of many plants such as opium poppy, scilla, black hellebore, belladonna and castor seeds.

Galen is, surely, the most cited and most consulted author of late ancient and medieval medicine. He established a distinction between the use of single medicinal plants, that he defined as "simple", and the use of plants with synergistic action. He also highlighted the effectiveness of the individual active components of plant origin obtained through the use of common solvents such as vinegar, wine, and oil. Galen, who in view of his considerable contributions in the field of pharmaceutical science, was called the "father of pharmacy"; was the first to make a systematic list of medicinal species and describe the therapeutic indications in the treatment of individual diseases. He was the founder of galenic recipes and his work will maintain a dogmatic role for many centuries.

From the ancient times to the present, scientific progress has not only highlighted and enabled characterization of the organic molecules responsible for the beneficial effects of plants but has also provided explanation of the mechanisms of their origin and synthesis, as well as their localization in the various organs of plants and cellular organelles. Vacuoles were considered to be the main site of the accumulation of waste products of cellular metabolism. They are associated with so-called secondary metabolites, many of which are bioactive. These include phenols, tannins, alkaloids, glucosides, balsams, essential oils, and resins, some of which can be accumulated inside the cell or secreted outside. The plant's internal and external secretory tissues are also important in this respect.

The Golgi apparatus plays a decisive role since it is responsible for the synthesis of polysaccharides and glycolipids and the sorting of proteins coming from the endoplasmic reticulum [1].

In recent decades, studies that aimed at investigating plant endosomal trafficking have led to the discovery of the existence of multivesicular bodies. In particular, it was shown in 1967 [2] that the fusion of multivesicular bodies with the plasma membrane leads to the extracellular release of small vesicles; this concept was subsequently brought to attention [3].

Nowadays it is widely described that plants release "exosome-like" extracellular vesicles (EVs). Similarly to those of animal origin, plant EVs are lipoprotein structures of heterogeneous dimensions, which carry numerous bioactive substances including miRNAs (Fig. 1). Although their origin in different plant species has not been fully clarified, their role in cross-kingdom interactions has recently attracted the attention of researchers. The purpose of this chapter is to describe the rich and unique composition of vesicles of plant origin, highlight their known properties to date, and evaluate their possible applications for human health.

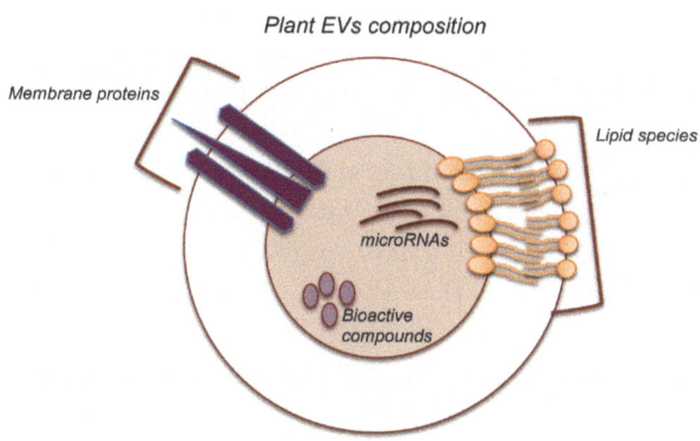


**Fig. 1** Schematic representation of plant-derived EVs.

## 2. Proteomic composition of plant EVs

The protein composition of plant vesicles has been determined in several plant species using proteomic approaches to both, characterization of these structures and to comparison of their profile with that of mammalian vesicles. Although specific markers of vesicles of plant origin are not yet known, some protein families have been found in vesicles isolated from different species; further studies will be needed to assess whether these can be considered specific markers.

The protein family, widely described as a constituent of vesicles of animal origin, and also found in plant vesicles is that of Heat Shock Proteins (HSPs); HSPs are molecular chaperones with a critical role in protein homeostasis that are found in the membrane of vesicles secreted by various cell types [4]. In mammalian EVs, HSPs play a prominent role in immune system regulation [4–6]. In plant species, HSPs were initially recognized as proteins induced in cells undergoing stress [7], usually in response to high temperature. It is now acknowledged that these proteins play a role in physiologic conditions; HSPs are responsible for plant growth and development, and for the nutrient internalization [8].

In the first study of EVs isolated from C. limon, the presence of HSP70 and HSP90 was found through the combination of Gel-enhanced Liquid Chromatography-Mass Spectrometry (GeLC-MS/MS) and Liquid Chromatography tandem Mass Spectrometry (LC-MS/MS analyses) [9]. In accordance with this study, further proteomic analysis of the vesicles isolated from four different citrus species (*C. sinensis, C. limon, C. paradise,* and *C. aurantium*) identified HSP70 and HSP80 as highly expressed proteins. Hsp60 and Hsp70 were found in EVs from sunflower seedlings [10], together with HSP90 that were also present in grapefruit EVs [11]. Finally, HSC70 was detected in grape-EVs [12].

Another protein family that is found in plant vesicles and also in mammalian vesicles is Annexin. In particular, annexin A1 and annexin A2 are critical in EV biogenesis, in the budding of the intraluminal vesicles and in the multivesicular bodies generation [13,14]. In the plant kingdom, annexins were found in EVs from Citrus species [9,15], sunflower seedlings [10], and EV isolates from the leaf apoplast [16]. An interesting group of proteins identified in EVs from different plant species is the Aquaporins (AQPs) [17]. Aquaporins have been identified in both animal and plant kingdoms; they are integral membrane proteins that constitute selective channels for water molecules transport [18]; considering their function, AQPs are critical in kidney pathophysiology [19]. Several studies have identified AQPs as constituents of EVs; AQP1 and AQP2 have been detected in urinary EVs [20–22] and may represent a biomarker of kidney function. In the plant kingdom, AQPs are involved in the regulation of cell turgor. AQPs were detected in EVs from citrus fruits [9,15,23], grape [12], and in broccoli-derived EVs [24], playing a role in maintaining vesicle stability.

In addition to the proteins described above, as is the case in the animal kingdom, the content of plant vesicles reflects the composition of the producing cells. Therefore the content of the vesicles reflects the composition of the plant species as well as the conditions in which the plants live. The proteomic analysis of vesicles purified from the apoplastic fluid of *Arabidopsis thaliana* revealed enrichment in proteins associated with stress and defense as well as in proteins involved in reactive oxygen species pathways [16]. The presence of these protein families confirms the key role of vesicles in mediating the response to external stimuli. Similarly, proteins related to stress response were also found in the proteomic profile of pollensomes, specific vesicles isolated from olive (Oleaeuropaea) pollen grains during pollen germination [25]. These vesicle populations also contained proteins involved in metabolism and signaling, cytoskeleton remodeling and membrane transport as well as all proteins necessary for pollen tube growth.

## 3. Lipidomic composition of plant EVs and other bioactive compounds

The lipid composition of EVs has been extensively investigated both in animal and plant kingdom. In addition to being related to the stability of the vesicles, the lipid component plays a fundamental role in the biological effect that these mediate. In fact, most of the lipids found in plant vesicles have demonstrated biological activity, thus explaining part of the biological effect that vegetal vesicles have on mammalian cells. This issue will be discussed in the present chapter. Besides having an important role in the formation of cell membranes, phosphatidic acid (PA) of both, exogenous (dietary) and endogenous origin, is an intermediate in the synthesis of triacylglycerols and other phospholipids. PA is involved in numerous cellular biochemical and physiological processes [26]. In particular, it is known to have a role in the fusion and fission processes of biological membranes and therefore in the dynamics of membranes; moreover, it is a known activator of cellular signaling pathways: mTOR is a serine/threonine-protein kinase recognized for the regulation of cell proliferation, growth, and survival. Phosphatidic acid is present in many of the plant vesicles today characterized at the lipidomic level. The presence of this lipid was first found in 2009 in the extracellular fluid of sunflower seeds, suggesting the existence of exosome-like vesicles in sunflower fluids [27]. Grape nanoparticles are enriched in PA (53.2%) as well. Interestingly, the authors observed that compared to the whole fruit, vesicles are more enriched in PA, suggesting selective sorting in vesicles [12]. Phosphatidic acid was also found in ginger vesicles (43%); in this study, the authors also demonstrated how vesicles, which can influence gut microbiota, use phosphatidic acid to be internalized by intestinal Lactobacillus rhamnosus [28]. Another lipid found in plant vesicles is phosphatidylethanolamine (PE), found in grape and grapefruit vesicles [11,12]. This lipid plays a role in the process of cytokinesis in mammalian cells [29]. Grapefruit vesicles [11] and ginger vesicles [28] are also enriched in phosphatidylcholine (PC); its presence in ginger vesicles is crucial for their internalization by intestinal Ruminococcaceae. Interestingly, PC has been demonstrated as an anti-inflammatory lipid, showing protective effects in the treatment of inflammatory bowel disease [30,31]. In the study, authors investigated the effect of PC on lipopolysaccharide (LPS)-stimulated intestinal epithelial cells; the pre-treatment with PC inhibited TNF-$\alpha$ secretion and reduced the nuclear translocation and phosphorylation of p65. Similarly, it is conceivable that the presence of PC in plant vesicles could be the mediator of the observed anti-inflammatory effects. In addition to the other lipid species, ginger vesicles are enriched in monogalactosylmonoacylglycerol and digalactosyldiacylglycerol [32,33]. Galactolipids are a class of compounds largely found in edible plants; interestingly, these molecules were previously described for their promising biological properties [34]. The administration of spinach monogalactosyldiacylglycerol blocks colon tumor growth in mice [35] while several pieces of evidence highlighted the anti-inflammatory effects of these lipid species on-ear edema [36] and in human articular cartilage [37]. MGDG is able to reduce some critical inflammatory mediators as COX-2, inducible nitric oxide synthase (iNOS), nitrogen oxide (NO) and prostaglandin E2 (PGE2) [38].

In addition to the lipid species discussed above, edible plant EVs are a source of other compounds with attributed beneficial properties to human health. The analysis of Grapefruit-EVs indicated the presence of naringenin, the functional component of the flavonone naringin [11]. To date, several bioactive characteristics have been attributed to naringenin; it was described for its antioxidant properties [39,40], as an anti-inflammatory compound [41,42] and as a modulator of the immune system [43,44].

Ginger EVs are enriched in shogaol [45], a phytochemical with recognized biological properties [46], as an anticancer [47,48] and an anti-inflammatory [49,50] agent. Finally, broccoli-derived EVs contain sulforaphane

[51], a compound from cruciferous vegetables, with anti-inflammatory [52,53] and anticancer effects [54]. In particular, its activity is correlated to the activation of the Nrf2 transcription factor that, by regulating the expression of antioxidant genes, protects cells from oxidative events [55].

## 4. Plant miRNAs

MicroRNAs (miRNAs) are a class of small non-coding RNAs, that play a key role in post-transcriptional regulation by targeting specific mRNAs for their degradation or translational repression. The involvement of human miRNAs in different biological processes is well known [56–58]; furthermore, over the last ten years, the role of plant miRNAs as a new functional component of food with therapeutic effects (e.g., antitumor immune function enhancing and preventing disease) has been hypothesized. In the past, the beneficial properties of some plant species were exclusively attributed to some functional components such as polyphenols, alkaloids, saponins, tannins, vitamins, and minerals [59]. Recent data in the literature showed that, in addition to those mentioned above, plant miRNAs are absorbed by the mammalian digestive tract and are able to regulate mammalian genes [60,61], representing a novel class of cross-kingdom modulators that mediates animal-plant interactions at the molecular level [62]. Plant miRNAs were first discovered in 2002 in Arabidopsis [63] and are characterized for their methylation at their 3′ ends by HEN1 (a plant mRNA methyltransferase), that distinguishes plant miRNAs from those of the animals [64]. Moreover, while in animals, the complementary sites related to target recognition are located at the 3′UTR, in plants, these sites can occur anywhere within the mRNA [65].

### 4.1 Plant miRNAs interact with mammalian cells

Studies aimed at understanding the mechanisms by which plant miRNAs can regulate mammalian gene expression have shown that, unexpectedly, these molecules are stable; they have in fact been found in different body fluids such as serum, plasma, urine, saliva, and breast milk. Interestingly, plant-miRNAs have also been found in mammalian EVs, suggesting that, once internalized by mammalian cells they can be shuttled within the organism through vesicle packaging. In the following paragraphs, we will discuss the recent evidence of the interaction of plant-miRNAs, free or vesicle-associated, with mammalian cells (Fig. 2).

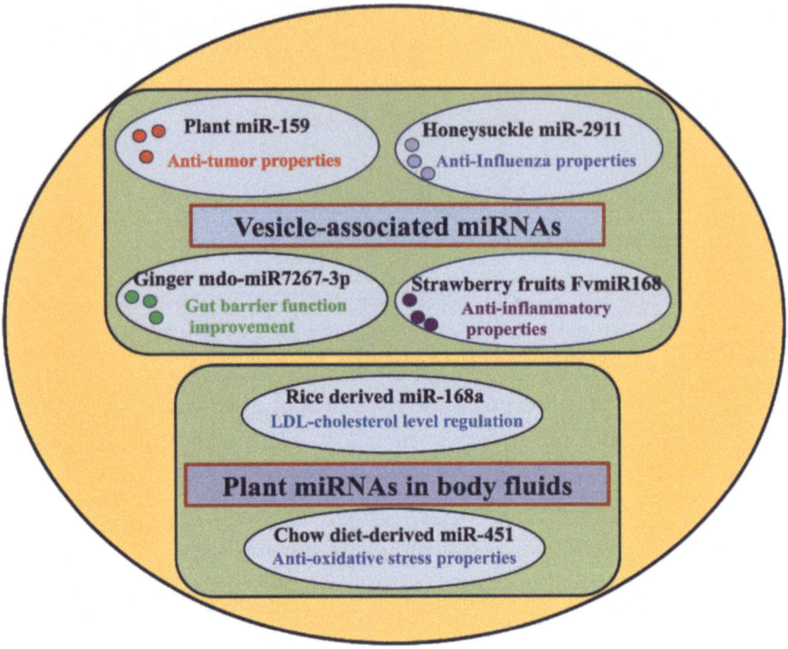


**Fig. 2** Summary of plant-miRNAs found in mammalians, free or vesicle-associated (in mammalian EVs or in plant EVs).

### *4.1.1 Plant miRNAs in mammalian body fluid*

Kai Wang et al. detected exogenous RNA sequences from foods in human plasma. In particular, they identified RNA sequences from cereal (e.g., corn Zea mays, rice Oryza sativa Japonica group and Sorghum bicolor), soybeans (Glycine max), tomato (Solanum Lycopersicum) and grape (Vitis vinifera) [66]. Zhang et al., in 2012, demonstrated that plant miRNAs influence gene expression in mammalian organs through food intake [61]. Specifically, the authors identified plant miR168a, normally contained in rice, in the Chinese human serum. miR168a reduces the expression of its target, low-density lipoprotein receptor adapter protein 1 (LDLRAP1) mRNA in mouse liver, decreasing LDL clearance from mouse plasma and thus increasing plasma LDL-cholesterol level. Plant miR168a level was elevated already after 6 h of rice feeding and a decreased LDLRAP1 in mouse liver was detected at the same time; these results indicate that plant miR168a, derived from food intake, rapidly modulate mouse liver target gene expression. The downregulation of LDLRAP1 level by miR168a, and an increase of plasma LDL cholesterol levels, was also reported after 3 days. Moreover, the physiological role of miR168a in the reduction of LDLRAP1 target and in the increase of plasma LDL-cholesterol level was inhibited by *anti*-miR168a [61]. Wang et al. (2017) showed that miR144/451-null mice can orally uptake miR-451 from a daily chow diet, enhancing the level of miR-451 in the circulating blood. miR-451 in food can pass through the digestive system to enter the bloodstream and protect the erythroid cells from oxidative stress by suppressing 14-3-3ζ and allowing the transcription of two anti-oxidant genes cat and gpx1 [67].

### 4.1.2 Plant miRNAs in mammalian EVs

Zhou Z et al. (2015) described for the first time that miR2911, extracted from honeysuckle (HS), targets various Influenza A viruses (IAVs), including H1N1, H5N1, and H7N9, inhibiting viral replication "in vitro" and "in vivo". Moreover, miR2911 or HS decoction administration strongly reduced mouse mortality caused by H5N1 infection. High level of miR2911 in mouse peripheral blood was detected mainly in microvesicles (MVs), suggesting that miR2911 in HS decoction might be initially absorbed through the gastrointestinal tract, then packaged into MVs, by intestinal epithelial cells, and finally delivered through the circulatory system [68]. Intriguingly, other miRNAs contained in HS were destroyed during the boiling process, but miR2911 was found to be intact in the final HS decoction suggesting atypical miRNA stability [68]. Supporting the above data on miR2911 stability, Yang et al., (2015) showed that the detection of miR2911 in murine sera and urine strongly correlated with dietary intake abundance; this result confirmed the miRNA digestive stability and bioavailability compared to other plant miRNAs, and this is probably attributed to modifications conferred by the host [69]. Exogenous plant miR2911 was carried from a variety of common vegetables such as broccoli, cabbage, cauliflower, lettuce and spinach [70]; using a chow diet containing honeysuckle (*Lonicera japonica*), plant-miR2911 was found in the sera and urine of mice already after 3—5 days demonstrating that this atypical miRNA is able to survive circulation and is expelled also in urine. Gavage-feeding with a concentrated dose of honeysuckle (honeysuckle decoction) supplemented with a synthetic plant miR168a caused miR2911 and miR168a detection in the sera of mice, suggesting that further work is necessary to optimize dietary uptake of some plant-based small RNAs [71]. As first, Li et al. (2015) revealed the presence of plant miRNAs in fetal umbilical cord blood and in amniotic fluid of healthy Chinese pregnant women, suggesting that exogenous miRNAs could be transferred through the placenta to the fetal side. To further investigate the trans-placental transmission of exogenous plant miRNAs in mice, an exogenous mature miR2911 was used as a tool. In particular, following the gavage feeding of Honeysuckle, rich in miR2911, the authors detected an increasing circulating exogenous plant-miR2911 mainly present in microvesicles (MVs) suggesting that the trans-placental transmission of miRNAs could be mediated by MVs pathway [72].

The antitumor capacity of plant-derived miRNAs was also reported in the literature. Chin AR et al., demonstrated for the first time the role of a plant miR159 in reducing breast cancer cell growth. In particular, they isolated plant miR159 extracellular vesicles from human serum and showed that their level was inversely correlated with breast cancer (BC) incidence and progression. In MDA-MB-231 breast cancer cells, treatment with synthetic miR159 mimics inhibited cancer cell proliferation by targeting the transcription factor TCF7, finally leading to a decrease in MYC protein levels. Moreover, oral administration of miR159 mimics strongly inhibited xenograft breast tumors in mice [73]. Mlotshwa et al., (2015) reported that the oral administration of a cocktail containing miR-34a, miR-143, and miR-145, whose expression is reduced in colon cancer [74—76], decreases the tumor burden in the ApcMin/+ mouse model [76]. In particular, the three tumor suppressor miRNAs were synthesized, mimicking plant-miRNAs, with the nucleotide sequence of the mouse miRNAs, but with methylation at their 3′ ends, which is a characteristic of plant miRNAs [77].

## 4.2 MiRNAs in plant EVs

It has been reported that some miRNAs are resistant to acidic conditions, high temperature and RNAse cleavage through their packaging in various lipid vesicles such as EVs or in association with components of the RNA-induced silencing complex (RISC), such as argonaute 2 (AGO2) [78—83]. In particular, it has been proposed that ingested plant miRNAs or in general "food-derived miRNAs" are absorbed by the gastrointestinal (GI) epithelial cells of consuming animals [61] which subsequently package them into EVs. EVs protect miRNAs from degradation and deliver them to neighboring or distant cells, where they are capable of interaction with endogenous RNAs and regulation of their expression [61,68,84,85]. Edible plant-derived exosome-like nanoparticles (EPDELNs) are novel plant ultrastructure that has been characterized and showed to display structures similar to mammalian exosomes. Moreover, data in the literature demonstrated that EPDELNs are absorbed in the mammalian gastrointestinal (GI) tract and have the potential to regulate human mRNA by mediating plant-animal intercellular communication [11,12,86,87]. Xiao et al. (2018) identified miRNA profiles of EPDELNs of 11 different edible fruits and vegetables [88]. The authors identified 418 miRNAs and divided them into three classes according to their EPDELN distribution: 26 frequent miRNAs, 39 moderately present miRNAs, and 353 rare miRNAs. The most abundant miRNAs were related to inflammatory and tumor response pathways. Grape, grapefruit, ginger and carrot nanoparticles possess anti-inflammatory properties and sustain intestine homeostasis [89]; ginger-derived nanoparticles showed a key role in counteracting liver-related diseases such as alcohol-induced damage [45,88].

Zhao et al. (2018) isolated exosome-like nanoparticles from coconut water. They demonstrated that the diameter of these particles was smaller than animal-derived exosomes and a difference in miRNA content between immature and mature coconut water was reported [90]. Moreover, data in the literature revealed that ELNs could alleviate colitis disease. Teng et al. (2018) demonstrated that plant-derived exosome-like nanoparticles (ELNs) are taken up by the gut bacteria, improving antimicrobial immunity of the intestinal barrier and thereby alleviating colitis. In particular, the authors demonstrated that in Lactobacillus rhamnosus (LGG), a gut probiotic indole-3-carboxaldehyde (I3A) induced by vesicles from grape juice (GELNs) mdo-miR7267-3p promotes the expression of IL-22 through activation of the AHR (aryl hydrocarbon receptor) signaling pathway [28]. Grapefruit vesicles were considered as modulators of the inflammatory response in a murine model of dextran sulfate sodium (DSS) - induced colitis; vesicles are, in fact, internalized by intestinal macrophages, promoting the expression of hemeoxygenase-1 (HO-1) and reducing the levels of the pro-inflammatory cytokine IL-1β and TNF- α [11]. Cavalieri D et al. (2016) showed that plant miRNA modulates dendritic cells' reaction to inflammatory stimuli by limiting T cell proliferation and reducing inflammation. In particular, the authors showed that exposure of dendritic cells to synthetic strawberry fruit FvmiR168, delivered in exosome-like structures, does not directly affect their capacity to induce the proliferation and differentiation of T cells; nevertheless, the authors observed a significant reduction in inflammatory response, following exposure to LPS, when dendritic cells were treated with FvmiR168; this response was probably related to decreased levels of TRIF transcript, downstream transducer of TLR4 and TLR3 [91].

# 5. Biological functions of plant EVs and future opportunities

The growing interest in the characterization and comprehension of the biological activity of plant vesicles originates both from the desire to understand how plants use this mechanism as a means of cellular communication and also from the wish to explore the opportunities they offer. In fact, as mentioned in the introduction, it was fascinating to discover how plant vesicles represent a safe for numerous bioactive compounds, enclosed in a single structure. The evidence showing how plant vesicles are able to interact with mammalian cells and change their phenotype, leads today to consider them important tools to be used in nutraceuticals. Finally, since their vegetable origin allows their purification from large volumes, their industrial application is feasible.

After having described the composition of plant vesicles from a proteomic, lipidomic and miRNAs point of view in detail in the previous paragraphs, below we will discuss the results of the biological activities of plant vesicles and finally the opportunities that could arise from their use as vehicles carrying other therapeutic compounds (Fig. 3).

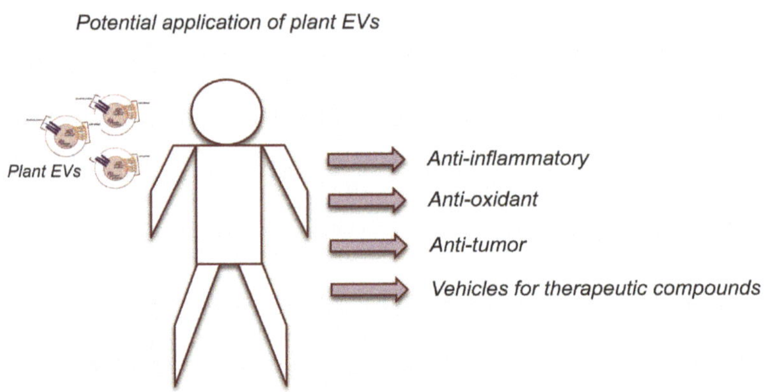


**Fig. 3** Schematic representation of the potential applications of plant-derived EVs.

## 5.1 Plant EVs intrinsic properties

Since plant vesicles are introduced into the organism daily through the diet, many studies have focused on the effects of vesicles on the gastrointestinal tract and on their anti-inflammatory properties.

One of the first studies was carried out in 2013; the authors isolated GELNs by ultracentrifugation on a sucrose gradient and characterized them at morphological and dimensional levels, and by analyzing the lipid, protein and miRNAs content. The administration of the vesicles labeled with the lipophilic dye DiR to mice by oral gavage revealed that these were already in 6 h internalized in the gut, in particular in intestinal stem cells, and that the vesicles promoted their proliferation. In addition, further extensive analyses showed that GELNs activate the Wnt/β-catenin pathway and thus increase the levels of genes related to stem cell growth [12]. In the same study, the authors highlighted that seven-day treatment with GELNs protects animals from damage caused by colitis induced with DSS (Dextran Sulfate Sodium) [12]. In a following study, the authors isolated vesicles also from other plant species such as grapefruit, ginger, and carrot (EPDENs). One of the first significant results shown in this study emphasizes that these vesicles are resistant and stable in solutions that mimic the environment of the stomach and intestine, although their diameter and charge changes depending on the species. Similar to what was reported in the previous study, the vesicles, administered orally in vivo, are internalized by stem cells and intestinal macrophages, suggesting a functional role also on the immune component. In fact, in vitro treatment of the murine macrophage line, RAW 264.7 with EPDENs resulted in increased levels of the anti-inflammatory factors IL10 and heme oxygenase (HO-1), while the levels of the pro-inflammatory cytokines IL6 and TNF-a were reduced. Consistent with these findings, authors also found an increase of the nuclear erythroid 2-related factor 2 (Nrf2), a transcription factor involved in the anti-inflammatory and anti-oxidant response [89]. Altogether, these results suggest a key role of plant vesicles in modulating the inflammatory response; as discussed above this effect can be related to their rich content. Also of considerable interest is a recently published paper that highlights how plant vesicles are internalized by the gut microbiota, thus affecting the intestinal microbiome. In this study the authors observed that ginger vesicles are preferentially internalized by Lactobacillaceae [28] and that they are able, through their miRNAs content, to directly regulate the gene expression of bacteria. The authors conclude that the interaction between plant vesicles and intestinal bacteria mediates the anti-inflammatory response by increasing IL22 [28].

In addition to their anti-inflammatory properties, some studies have highlighted the role of plant vesicles in inhibiting tumor cell proliferation. Vesicles isolated from lemon juice inhibit the growth of different types of tumor cells by inducing tumor necrosis factor-alpha-related apoptosis-inducing ligand (TRAIL) - mediated apoptosis [9]. At the same time, these vesicles cause, in colon cancer cells, the downregulation of certain lipid species such as acetyl-CoA carboxylase alpha, largely associated with cancer progression [92]. Recently, Cao et al., (2019) purified EVs from ginseng roots and demonstrated that these vesicles induce apoptosis of murine melanoma cell lines by inducing reactive oxygen species. In particular, the authors found that EVs reduced CD206 levels in M2-like macrophages and concomitantly increased the levels of CD80, CD86, MHC-II, and TLR2/4. These results suggest that ginseng EVs inhibit macrophages M2 polarization. In addition, the treatment of B16F10 melanoma cells with the conditioned medium of M2 macrophages treated with EVs lead to tumor cell apoptosis [93]. The results provided in this study suggest that plant EVs may act as immunomodulators, contributing to the immune response to tumors.

Finally, the effect of plant vesicles has recently been evaluated on in vitro skin regeneration models, suggesting new application possibilities. For example, vesicles isolated from wheat-grass juice could be employed in the wound-healing, as they enhance the proliferation and migration of different cell types (such as dermal fibroblasts) and stimulate collagen type I expression [94]. Vesicles purified from the leaves and stems of Dendropanax morbifera decrease the protein expression of MITF, the microphthalmia-associated transcription factor involved in melanocyte differentiation, and of other proteins involved in melanogenesis as tyrosinase (TYR), the human tyrosinase related proteins -1 and -2 (TRP-1 and TRP-2) in Bl6BL6 melanoma cells [95]; in addition, the treatment reduces the intracellular synthesis of melanin in a 3-dimensional human epidermis model [95].

## 5.2 Plant EVs as vehicles for therapeutic compounds

Continuous investigation of vehicles for the drug delivery of therapeutic substances which are unstable if administered alone, has revealed the necessity for development of safe systems, that are not rejected by the organism, and that interact with the target cell. For several years, extracellular vesicles, especially those of mesenchymal stem cell origin, have been considered as emerging promising controlled delivery systems [96,97].

Numerous studies have shown that vesicles of mammalian origin can be vehicles for therapeutic compounds and siRNAs. Hydrophobic substances can easily be incorporated into vesicles by incubation with them [98], while other drugs or siRNAs are added by electroporation techniques [99] and new approaches are being developed.

The possibility of isolating vesicles from the plant kingdom, the awareness that they are assimilated daily through diet, and that their purification is feasible on a large scale, makes the use of plant vesicles attractive.

However, to date, few studies have evaluated the possibility of using plant vesicles for drug delivery. Early research in 2014 showed that nanovesicles isolated from grapefruit represent a system for the delivery of anti-inflammatory substances to intestinal macrophages. In particular, the authors have shown that 2 h of incubation of grapefruit nanovesicles with the anti-inflammatory compound Methotrexate allows the compound to be incorporated and to maintain its functionality; moreover, administration of the vesicles loaded with the drug to a mouse model of colitis significantly reduced tissue damage and inflammatory infiltration [11].

In a later study, Zhuang et al. (2016) developed plant nanovectors consisting of lipids isolated from grapefruit vesicles. The authors showed that these nanovectors when administered intranasally to mice, were quickly located in the olfactory bulb, hippocampus, thalamus, and cerebellum, thus representing a possible delivery system to the brain. In fact, in the same study, researchers demonstrated that it was possible to encapsulate miRNAs, such as miR17, in plant nanovectors and that the administration led to the reduction of GL-26 brain tumor [100].

The implementation of the large-scale purification systems for plant vesicles as well as an in-depth analysis of regulatory aspects are necessary for their application as drug delivery systems.

## 6. Conclusions

In conclusion, in this chapter we have reported the studies on the composition of plant vesicles, focusing on the bioactive compounds they carry and their interactions with animal species. We have therefore highlighted their beneficial properties and potential area of application for human health.

## Acknowledgements

Stefania Raimondo is supported by PON "Ricerca e Innovazione" 2014—2020 - Azione 1.2 "Mobilità dei Ricercatori" - AIM "Attraction and International Mobility". Marzia Pucci is supported by an AIRC fellowship (project n°22542). We kindly thank Prof. Riccardo Alessandro and Prof. Francesco M. Raimondo for critical suggestions on the manuscript.

**Conflict of interest:** Stefania Raimondo is the co-founder of the Innovative Start-up Navhetec s.r.l. focused on the isolation and application of Citrus juice-derive extracellular vesicles. The authors declare no conflict of interest.

## References

[1] U. Neumann, F. Brandizzi, C. Hawes, Protein transport in plant cells: in and out of the Golgi, Ann. Bot. 92 (2) (2003) 167–180.
[2] W. Halperin, W.A. Jensen, Ultrastructural changes during growth and embryogenesis in carrot cell cultures, J. Ultrastruct. Res. 18 (3) (1967) 428–443.
[3] Q. An, A.J. van Bel, R. Huckelhoven, Do plant cells secrete exosomes derived from multivesicular bodies? Plant Signal. Behav. 2 (1) (2007) 4–7.
[4] V.S. Reddy, et al., Extracellular small heat shock proteins: exosomal biogenesis and function, Cell Stress Chaperones 23 (3) (2018) 441–454.
[5] L.H. Lv, et al., Anticancer drugs cause release of exosomes with heat shock proteins from human hepatocellular carcinoma cells that elicit effective natural killer cell antitumor responses in vitro, J. Biol. Chem. 287 (19) (2012) 15874–15885.
[6] R. Aneja, et al., Extracellular heat shock protein-70 induces endotoxin tolerance in THP-1 cells, J. Immunol. 177 (10) (2006) 7184–7192.
[7] Z.S. Xu, et al., Heat shock protein 90 in plants: molecular mechanisms and roles in stress responses, Int. J. Mol. Sci. 13 (12) (2012) 15706–15723.
[8] S. Ul Haq, et al., Heat shock proteins: dynamic biomolecules to counter plant biotic and abiotic stresses, Int. J. Mol. Sci. 20 (21) (2019).
[9] S. Raimondo, et al., Citrus limon-derived nanovesicles inhibit cancer cell proliferation and suppress CML xenograft growth by inducing TRAIL-mediated cell death, Oncotarget 6 (23) (2015) 19514–19527.
[10] M. Regente, et al., Plant extracellular vesicles are incorporated by a fungal pathogen and inhibit its growth, J. Exp. Bot. 68 (20) (2017) 5485–5495.
[11] B. Wang, et al., Targeted drug delivery to intestinal macrophages by bioactive nanovesicles released from grapefruit, Mol. Ther. 22 (3) (2014) 522–534.
[12] S. Ju, et al., Grape exosome-like nanoparticles induce intestinal stem cells and protect mice from DSS-induced colitis, Mol. Ther. 21 (7) (2013) 1345–1357.
[13] I.J. White, et al., EGF stimulates annexin 1-dependent inward vesiculation in a multivesicular endosome subpopulation, EMBO J. 25 (1) (2006) 1–12.
[14] N. Mayran, R.G. Parton, J. Gruenberg, Annexin II regulates multivesicular endosome biogenesis in the degradation pathway of animal cells, EMBO J. 22 (13) (2003) 3242–3253.
[15] G. Pocsfalvi, et al., Protein biocargo of citrus fruit-derived vesicles reveals heterogeneous transport and extracellular vesicle populations, J. Plant Physiol. 229 (2018) 111–121.
[16] B.D. Rutter, R.W. Innes, Extracellular vesicles isolated from the leaf apoplast carry stress-response proteins, Plant Physiol. 173 (1) (2017) 728–741.
[17] M.C. Martinez-Ballesta, et al., The expanding role of vesicles containing aquaporins, Cells 7 (10) (2018).
[18] M. Ozu, et al., Aquaporins: more than functional monomers in a tetrameric arrangement, Cells 7 (11) (2018).
[19] Y. Noda, et al., Aquaporins in kidney pathophysiology, Nat. Rev. Nephrol. 6 (3) (2010) 168–178.
[20] T. Pisitkun, R.F. Shen, M.A. Knepper, Identification and proteomic profiling of exosomes in human urine, Proc. Natl. Acad. Sci. U. S. A. 101 (36) (2004) 13368–13373.
[21] S. Oshikawa, H. Sonoda, M. Ikeda, Aquaporins in urinary extracellular vesicles (exosomes), Int. J. Mol. Sci. 17 (6) (2016).

[22] S. Corciulo, et al., AQP1-containing exosomes in peritoneal dialysis effluent as biomarker of dialysis efficiency, Cells 8 (4) (2019).
[23] C. Stanly, et al., Membrane transporters in citrus clementina fruit juice-derived nanovesicles, Int. J. Mol. Sci. 20 (24) (2019).
[24] M.D.C. Martinez-Ballesta, et al., Plasma membrane aquaporins mediates vesicle stability in broccoli, PLoS One 13 (2) (2018) e0192422.
[25] N. Prado, et al., Nanovesicles are secreted during pollen germination and pollen tube growth: a possible role in fertilization, Mol. Plant 7 (3) (2014) 573–577.
[26] J. Zegarlinska, et al., Phosphatidic acid - a simple phospholipid with multiple faces, Acta Biochim. Pol. 65 (2) (2018) 163–171.
[27] M. Regente, et al., Vesicular fractions of sunflower apoplastic fluids are associated with potential exosome marker proteins, FEBS Lett. 583 (20) (2009) 3363–3366.
[28] Y. Teng, et al., Plant-derived exosomal MicroRNAs shape the gut microbiota, Cell Host Microbe 24 (5) (2018) 637–652 e8.
[29] K. Emoto, M. Umeda, An essential role for a membrane lipid in cytokinesis. Regulation of contractile ring disassembly by redistribution of phosphatidylethanolamine, J. Cell Biol. 149 (6) (2000) 1215–1224.
[30] I. Treede, et al., Anti-inflammatory effects of phosphatidylcholine, J. Biol. Chem. 282 (37) (2007) 27155–27164.
[31] M. Chen, et al., Phosphatidylcholine regulates NF-κB activation in attenuation of LPS-induced inflammation: evidence from in vitro study, Anim. Cell Syst. 22 (1) (2018) 7–14.
[32] M. Zhang, et al., Edible ginger-derived nanoparticles: a novel therapeutic approach for the prevention and treatment of inflammatory bowel disease and colitis-associated cancer, Biomaterials 101 (2016) 321–340.
[33] M. Zhang, et al., Oral administration of ginger-derived nanolipids loaded with siRNA as a novel approach for efficient siRNA drug delivery to treat ulcerative colitis, Nanomedicine 12 (16) (2017) 1927–1943.
[34] L.P. Christensen, Galactolipids as potential health promoting compounds in vegetable foods, Recent Pat. Food Nutr. Agric. 1 (1) (2009) 50–58.
[35] N. Maeda, et al., Oral administration of monogalactosyl diacylglycerol from spinach inhibits colon tumor growth in mice, Exp. Ther. Med. 5 (1) (2013) 17–22.
[36] A. Bruno, et al., Selective in vivo anti-inflammatory action of the galactolipid monogalactosyldiacylglycerol, Eur. J. Pharmacol. 524 (1–3) (2005) 159–168.
[37] V. Ulivi, et al., Anti-inflammatory activity of monogalactosyldiacylglycerol in human articular cartilage in vitro: activation of an anti-inflammatory cyclooxygenase-2 (COX-2) pathway, Arthritis Res. Ther. 13 (3) (2011) R92.
[38] C.C. Hou, et al., A galactolipid possesses novel cancer chemopreventive effects by suppressing inflammatory mediators and mouse B16 melanoma, Cancer Res. 67 (14) (2007) 6907–6915.
[39] L. Mu, et al., Protective effects of naringenin in a rat model of sepsis-triggered acute kidney injury via activation of antioxidant enzymes and reduction in urinary angiotensinogen, Med. Sci. Monit. 25 (2019) 5986–5991.
[40] N.H. Zaidun, Z.C. Thent, A.A. Latiff, Combating oxidative stress disorders with citrus flavonoid: Naringenin, Life Sci. 208 (2018) 111–122.
[41] C. Nguyen-Ngo, J.C. Willcox, M. Lappas, Anti-diabetic, anti-inflammatory, and anti-oxidant effects of naringenin in an in vitro human model and an in vivo murine model of gestational diabetes mellitus, Mol. Nutr. Food Res. 63 (19) (2019) e1900224.
[42] Z. Zhao, et al., Naringenin inhibits migration of breast cancer cells via inflammatory and apoptosis cell signaling pathways, Inflammopharmacology 27 (5) (2019) 1021–1036.

[43] L.F. Mendes, et al., Flavonoid-mediated immunomodulation of human macrophages involves key metabolites and metabolic pathways, Sci. Rep. 9 (1) (2019) 14906.
[44] J. Wang, et al., Dietary naringenin supplementation attenuates experimental autoimmune encephalomyelitis by modulating autoimmune inflammatory responses in mice, J. Nutr. Biochem. 54 (2018) 130–139.
[45] X. Zhuang, et al., Ginger-derived nanoparticles protect against alcohol-induced liver damage, J. Extracell. Vesicles 4 (2015) 28713.
[46] X. Kou, et al., Occurrence, biological activity and metabolism of 6-shogaol, Food Funct. 9 (3) (2018) 1310–1327.
[47] T.Y. Li, B.H. Chiang, 6-shogaol induces autophagic cell death then triggered apoptosis in colorectal adenocarcinoma HT-29 cells, Biomed. Pharmacother. 93 (2017) 208–217.
[48] U.M. Nazim, S.Y. Park, Attenuation of autophagy flux by 6-shogaol sensitizes human liver cancer cells to TRAIL-induced apoptosis via p53 and ROS, Int. J. Mol. Med. 43 (2) (2019) 701–708.
[49] J. Luettig, et al., The ginger component 6-shogaol prevents TNF-alpha-induced barrier loss via inhibition of PI3K/Akt and NF-kappaB signaling, Mol. Nutr. Food Res. 60 (12) (2016) 2576–2586.
[50] R.B. Semwal, et al., Gingerols and shogaols: important nutraceutical principles from ginger, Phytochemistry 117 (2015) 554–568.
[51] Z. Deng, et al., Broccoli-derived nanoparticle inhibits mouse Colitis by activating dendritic cell AMP-activated protein kinase, Mol. Ther. 25 (7) (2017) 1641–1654.
[52] R.T. Ruhee, S. Ma, K. Suzuki, Sulforaphane protects cells against lipopolysaccharide-stimulated inflammation in murine macrophages, Antioxidants 8 (12) (2019).
[53] L. Subedi, et al., Anti-inflammatory effect of sulforaphane on LPS-activated microglia potentially through JNK/AP-1/NF-kappaB inhibition and Nrf2/HO-1 activation, Cells 8 (2) (2019).
[54] X. Su, et al., Anticancer activity of sulforaphane: the epigenetic mechanisms and the Nrf2 signaling pathway, Oxid. Med. Cell Longev. 2018 (2018) 5438179.
[55] S.M. Ahmed, et al., Nrf2 signaling pathway: pivotal roles in inflammation, Biochim. Biophys. Acta (BBA) - Mol. Basis Dis. 1863 (2) (2017) 585–597.
[56] K.J. Kwon-Chung, et al., Recent advances in biology and immunology of *Cryptococcus neoformans*, J. Med. Vet. Mycol. 30 (Suppl. 1) (1992) 133–142.
[57] V. Ambros, The functions of animal microRNAs, Nature 431 (7006) (2004) 350–355.
[58] N. Bushati, S.M. Cohen, microRNA functions, Annu. Rev. Cell Dev. Biol. 23 (2007) 175–205.
[59] D. Tungmunnithum, et al., Flavonoids and other phenolic compounds from medicinal plants for pharmaceutical and medical aspects: an overview, Medicines (Basel) 5 (3) (2018).
[60] S.I. Ivashuta, et al., Endogenous small RNAs in grain: semi-quantification and sequence homology to human and animal genes, Food Chem. Toxicol. 47 (2) (2009) 353–360.
[61] L. Zhang, et al., Exogenous plant MIR168a specifically targets mammalian LDLRAP1: evidence of cross-kingdom regulation by microRNA, Cell Res. 22 (1) (2012) 107–126.
[62] R.A. Rosenbloom, J. Chaudhary, D. Castro-Eschenbach, Traditional botanical medicine: an introduction, Am. J. Therapeut. 18 (2) (2011) 158–161.
[63] C. Llave, et al., Endogenous and silencing-associated small RNAs in plants, Plant Cell 14 (7) (2002) 1605–1619.
[64] B. Zhang, et al., Plant microRNA: a small regulatory molecule with big impact, Dev. Biol. 289 (1) (2006) 3–16.

[65] R.S. Pillai, C.G. Artus, W. Filipowicz, Tethering of human Ago proteins to mRNA mimics the miRNA-mediated repression of protein synthesis, RNA 10 (10) (2004) 1518−1525.
[66] K. Wang, et al., The complex exogenous RNA spectra in human plasma: an interface with human gut biota? PLoS One 7 (12) (2012) e51009.
[67] W. Wang, et al., Dietary miR-451 protects erythroid cells from oxidative stress via increasing the activity of Foxo3 pathway, Oncotarget 8 (63) (2017) 107109−107124.
[68] Z. Zhou, et al., Honeysuckle-encoded atypical microRNA2911 directly targets influenza A viruses, Cell Res. 25 (1) (2015) 39−49.
[69] J. Yang, et al., Detection of an abundant plant-based small RNA in healthy consumers, PLoS One 10 (9) (2015) e0137516.
[70] J. Yang, et al., Anomalous uptake and circulatory characteristics of the plant-based small RNA MIR2911, Sci. Rep. 6 (2016) 26834.
[71] J. Yang, et al., Detection of dietary plant-based small RNAs in animals, Cell Res. 25 (4) (2015) 517−520.
[72] J. Li, et al., Small non-coding RNAs transfer through mammalian placenta and directly regulate fetal gene expression, Protein Cell 6 (6) (2015) 391−396.
[73] A.R. Chin, et al., Cross-kingdom inhibition of breast cancer growth by plant miR159, Cell Res. 26 (2) (2016) 217−228.
[74] C. Faber, T. Kirchner, F. Hlubek, The impact of microRNAs on colorectal cancer, Virchows Arch. 454 (4) (2009) 359−367.
[75] Y. Takaoka, et al., Forced expression of miR-143 represses ERK5/c-Myc and p68/p72 signaling in concert with miR-145 in gut tumors of Apc(Min) mice, PLoS One 7 (8) (2012) e42137.
[76] S. Mlotshwa, et al., A novel chemopreventive strategy based on therapeutic microRNAs produced in plants, Cell Res. 25 (4) (2015) 521−524.
[77] H.A. Ebhardt, et al., Extensive 3' modification of plant small RNAs is modulated by helper component-proteinase expression, Proc. Natl. Acad. Sci. U. S. A. 102 (38) (2005) 13398−13403.
[78] J.D. Arroyo, et al., Argonaute2 complexes carry a population of circulating microRNAs independent of vesicles in human plasma, Proc. Natl. Acad. Sci. U. S. A. 108 (12) (2011) 5003−5008.
[79] P.S. Mitchell, et al., Circulating microRNAs as stable blood-based markers for cancer detection, Proc. Natl. Acad. Sci. U. S. A. 105 (30) (2008) 10513−10518.
[80] Y. Zhang, et al., Secreted monocytic miR-150 enhances targeted endothelial cell migration, Mol. Cell 39 (1) (2010) 133−144.
[81] B. Zhu, X. Wang, L. Li, Human gut microbiome: the second genome of human body, Protein Cell 1 (8) (2010) 718−725.
[82] T. Kogure, et al., Intercellular nanovesicle-mediated microRNA transfer: a mechanism of environmental modulation of hepatocellular cancer cell growth, Hepatology 54 (4) (2011) 1237−1248.
[83] K.C. Vickers, et al., MicroRNAs are transported in plasma and delivered to recipient cells by high-density lipoproteins, Nat. Cell Biol. 13 (4) (2011) 423−433.
[84] P. Ballo, et al., Impact of obesity on left ventricular systolic function in hypertensive subjects with normal ejection fraction, Int. J. Cardiol. 141 (3) (2010) 316−320.
[85] H. Valadi, et al., Exosome-mediated transfer of mRNAs and microRNAs is a novel mechanism of genetic exchange between cells, Nat. Cell Biol. 9 (6) (2007) 654−659.
[86] T. Kimura, et al., Lymphatic dysfunction attenuates tumor immunity through impaired antigen presentation, Oncotarget 6 (20) (2015) 18081−18093.
[87] Q. Wang, et al., Grapefruit-derived nanovectors use an activated leukocyte trafficking pathway to deliver therapeutic agents to inflammatory tumor sites, Cancer Res. 75 (12) (2015) 2520−2529.

[88] J. Xiao, et al., Identification of exosome-like nanoparticle-derived microRNAs from 11 edible fruits and vegetables, PeerJ 6 (2018) e5186.
[89] J. Mu, et al., Interspecies communication between plant and mouse gut host cells through edible plant derived exosome-like nanoparticles, Mol. Nutr. Food Res. 58 (7) (2014) 1561–1573.
[90] Z. Zhao, et al., Isolation of exosome-like nanoparticles and analysis of microRNAs derived from coconut water based on small RNA high-throughput sequencing, J. Agric. Food Chem. 66 (11) (2018) 2749–2757.
[91] D. Cavalieri, et al., Plant microRNAs as novel immunomodulatory agents, Sci. Rep. 6 (2016) 25761.
[92] S. Raimondo, et al., Label-free quantitative proteomic profiling of colon cancer cells identifies acetyl-CoA carboxylase alpha as antitumor target of Citrus limon-derived nanovesicles, J. Proteomics 173 (2018) 1–11.
[93] M. Cao, et al., Ginseng-derived nanoparticles alter macrophage polarization to inhibit melanoma growth, J. Immunother. Cancer 7 (1) (2019) 326.
[94] F. Sahin, et al., In vitro wound healing activity of wheat-derived nanovesicles, Appl. Biochem. Biotechnol. 188 (2) (2019) 381–394.
[95] R. Lee, et al., Anti-melanogenic effects of extracellular vesicles derived from plant leaves and stems in mouse melanoma cells and human healthy skin, J. Extracell. Vesicles 9 (1) (2020) 1703480.
[96] S. Raimondo, et al., Extracellular vesicles as biological shuttles for targeted therapies, Int. J. Mol. Sci. 20 (8) (2019).
[97] P. Akuma, O.D. Okagu, C.C. Udenigwe, Naturally occurring exosome vesicles as potential delivery vehicle for bioactive compounds, Front. Sustain. Food Syst. 3 (23) (2019).
[98] D. Sun, et al., A novel nanoparticle drug delivery system: the anti-inflammatory activity of curcumin is enhanced when encapsulated in exosomes, Mol. Ther. 18 (9) (2010) 1606–1614.
[99] M.S. Kim, et al., Development of exosome-encapsulated paclitaxel to overcome MDR in cancer cells, Nanomedicine 12 (3) (2016) 655–664.
[100] X. Zhuang, et al., Grapefruit-derived nanovectors delivering therapeutic miR17 through an intranasal route inhibit brain tumor progression, Mol. Ther. 24 (1) (2016) 96–105.

CHAPTER FIVE

# Standardization and reproducibility in EV research: the support of a Quality Management System

**Giovanna L. Liguori[a,*] and Annamaria Kisslinger[b,*]**
[a]Institute of Genetics and Biophysics (IGB), National Research Council (CNR) of Italy, Naples, Italy
[b]Institute of Experimental Endocrinology and Oncology (IEOS), National Research Council (CNR) of Italy, Naples, Italy
[*]Corresponding author: E-mails: giovanna.liguori@igb.cnr.it, a.kisslinger@ieos.cnr.it

## Contents



## Abstract

Scientific research is becoming increasingly competitive, requiring researchers to continuously improve and enhance their performance, and technologically complex, raising issues as reliability and reproducibility of the data as well as standardization of the methodologies. Moreover, globalization and emergence of new markets had a great impact on funding agencies, that encouraged scientists to develop visionary ideas and outstanding technologies to respond to the needs of society. The study of extracellular vesicles (EVs) is emerging as an extremely promising field for biomedical, nutraceutical and cosmetic applications, and the scientific community working on EVs is facing key challenges to realize the EV promise. The implementation of Quality

*Advances in Biomembranes and Lipid Self-Assembly*, Volume 33
ISSN 2451-9634
https://doi.org/10.1016/bs.abl.2020.05.005

Management (QM) tools can be extremely valid for scientists to successfully overcome these challenges. It is fundamental, however, that QM is customized on the requirements, needs, expectations and resources of investigators, to be really helpful in research project and activities. In this chapter, we introduce the history, principles and standards of QM and show some examples of its implementation in the context of public research institutions. Then, we focus on key features and main challenges of the EV research, including the standardization of procedures and technologies to guarantee reliability and reproducibility of the data. Finally, we present the QM approach developed for the VES4US research project, whose objective is to realize a radical new platform for production, isolation and functionalization of EVs from a natural source. The tools utilized and the models developed can be incorporated in a roadmap for the effective and efficient scientific management of innovative and multidisciplinary projects, specially focusing on EVs.

## 1. Introduction

A structured approach to Quality Management (QM) in research activities and projects has a great potential to improve the rigor, reproducibility, reliability and ultimately value of both basic and applied biomedical research, without curtailing the freedom of researchers. Doing research in a competitive and dynamic environment requires researchers to continuously improve and enhance their scientific performance and capabilities to perform reproducible experiments using high Quality standards, especially if it is earmarked for technology transfer. As an example, standardization in scientific research and development is emerging as an essential tool to ensure valuable and robust outcomes, within a framework of best practice [1–6]. Strategies for integrating best practices into basic research environments are not new, and recent examples of the application of International Organization for Standardization—ISO 9001 standards in research institutions have indicated many advantages in terms of governance, control, efficiency, and results [7–11]. In line with the need of integrating Quality standards within biological research, recent Horizon 2020 program of the European Commission calls either requires or strongly suggests good research data management as a prerequisite leading to knowledge discovery and innovation, and to subsequent data and knowledge integration and reuse. Indeed, Data Management Plans (DMPs) are now a key element of H2020 proposals. As part of making research data findable, accessible, interoperable and re-useable (FAIR) [12,13], DMPs have to describe the data management life cycle for the data that will be collected, processed and generated by a Horizon 2020 project. In addition, the implementation of a QM approach may support the development and execution of research projects, especially the ones that are more interdisciplinary, multi-site and high-risk, providing a roadmap toward better quality of results and processes.

The last decade has seen a flourishing development of research in extracellular vesicles (EVs) with an increase in the number of scientific publications describing the impact of EVs in several physiological and pathological processes. EV characterization could provide new biomarkers for the diagnosis and prognosis of the different pathologies in which they are involved. Moreover, due to their ability to interact with specific types of cells and to mediate numerous complex cellular responses, EVs are being actively explored as therapeutic agents, as themselves or as nanocarriers for drug delivery. However, specific issues arise when working with vesicles, whose size and amount often make their isolation and characterization very difficult to obtain. These issues have to be addressed to assure reliability and reproducibility of scientific data as well as the possibility of manufacturing at large scale so to exploit EVs for diagnostic and therapeutic applications. In the last years, the scientific community working on EVs has increasingly recognized the need to standardize and share methodologies and technologies for the collection, isolation/purification, characterization and functional analysis of EVs [14—16]. Therefore, the EV field is very sensible to standardization and reproducibility issues and could take great advantages on implementation of one or more QM tools to support research activities as well as to set-up and validate common scientific procedures.

In this chapter, we, first, describe the general concepts on which QM is based and how it has been implemented in several research institutions. Then, we introduce a specific case study in which a QM approach has been developed in EV research. The study refers to the project funded by the Horizon 2020 — Future and Emerging Technology (FET) Open program, entitled "Extracellular vesicles from a natural source for tailor-made nanomaterials" "VES4US" (https://ves4us.eu). We describe the general QM approach we followed inside VES4US and the main methodologies and tools applied. The VES4US QM approach could be an operating model for EV research project networks and more generally for multisite research projects, based on the development of radically new technologies, in which interdisciplinarity, innovation and interconnection are key features.

## 2. A travel into Quality Management: history, culture, standards and principles

### 2.1 An historical excursus through Quality

Quality culture evolved from the medieval master-apprentice crafting level to actual standardized Quality systems, where all the processes and outcomes are measured, documented and analyzed (Fig. 1). The concept of Quality happened with the first Industrial Revolution, which arose in the United Kingdom before spreading throughout Europe and the rest of the world, and the scaling up of the production at the end of the 18th century [17,18]. In the 1920s, Quality began to emerge with the advent of the first big companies with complex organizational systems. In this context, the term Quality was referred to final products which fulfilled foreseen requirements, whereas the idea that work could be studied and the work process formalized and improved did not even exist. There was no management theory before the thoughts of Frederick Winslow Taylor, which in the early 1900s formulated a new approach to factory management, called scientific management. Taylor divided the planning function from the production, focusing on the efficiency and productivity, and introduced pioneer ideas, still valid nowadays, such as employee training and implementation of standardized best practices to improve productivity [19]. As time went by, the

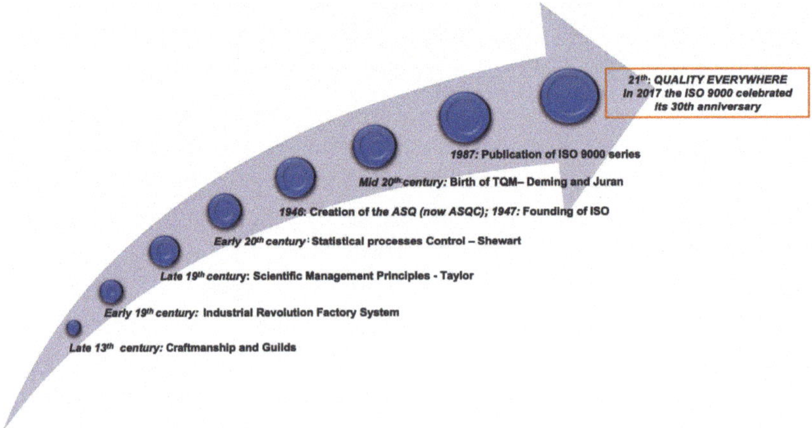


**Fig. 1** ***History and evolution of Quality.*** A brief excursus trough the milestones of QM from the Craftmanship and Guilds of the late thirteenth century to the actual ISO:9000 series of international standards. *ASQ*, American Society of Quality; *ASQC*, American Society of Quality Control; *ISO*, International Organization for Standardization; *TQM*, Total Quality Management.

complexity of several manufactured products increased and jobs became more specialized, thus introducing the need to inspect products for quality after manufacture. Therefore, Quality was removed from the manufacturer's auspices and assigned to the control or Quality department. In the 1930s, Walter Shewhart of the Bell Telephone Laboratories ushered the era of Statistical Quality Control by starting to use statistical charts for the control of product variables, that shortly after became widely known and gradually adopted in manufacturing industries. Shewhart demonstrated that variation in the production process led to the variation in the product, thus by eliminating the variation of the process a good standard of end product could be achieved [20]. In 1944 the first professional journal on Quality, named "Quality Control" was published and in 1946 the American Society of Quality, now called American Society of Quality Control, was founded with the aim to develop, promote and apply Quality concepts. At the end of the second World War, A. Feigenbaum recognized the importance of a comprehensive organization approach to Quality and, thanks to the work of Deming and Juran, "modern Quality" was born, based on the prevention of accidents through the design and implementation of a formal Quality system. Top managers supported integration of Quality throughout organizations and developed the culture of continuous improvement, Kaizen in Japanese, for identifying opportunities for streamlining work and reducing waste.

Among the most widely used tools for continuous improvement, we can find the Plan-Do-Check-Act (PDCA) or Deming cycle, developed by Deming in Japan in the 1950s, consisting in a logical sequence of four key stages [21–23]. The first phase, Plan, defines a problem and makes an hypothesis about possible causes and solutions, the second one, Do, implements a solution, the third one, Check, evaluates the results, and the last one, Act, corresponds to standardization if the results are satisfactory or to make corrective actions if not and formulate a new hypothesis. Then the cycle goes back to Plan and restarts again in the optic of a continuous improvement. This Quality tool was deeply used in QM: the continuous PDCA loop allowed analysis, measurement and identification of sources of variation from customer requirements and enabled taking corrective actions. Nowadays, PDCA cycle helps project managers in maintaining greater control over a given project, and is a powerful tool in performance management of individual employees as well as project teams on a daily routine basis. PDCA tool was the basis for implementation of a Total Quality Management (TQM) approach. The term "Total Quality" was coined for the first time by Japanese Quality specialists and a TQM System was adopted in most of the Japanese factories. TQM is defined as a customer-oriented process aiming at long-term success through customer satisfaction and benefits to all members of the organization and to society. TQM engages all divisions, departments and levels of the organization focusing on the systematic management of data in all processes and practices to eliminate waste and pursue continuous improvement, while changing organizational culture [24–26]. Later on, also in the United States more business started to experience the benefits of adopting the TQM and nowadays TQM is definitely a very well-accepted system of management in different organizations.

## 2.2 Quality Management standards and principles: the International Organization for Standardization (ISO)

While it is a common saying that standardization is "as old as the hills", the birth of "modern standardization" has been identified in the 19th century with the coming of industrialization and, at the end of the same century, with the very beginning of globalization, the need for standards became internationally recognized. The first international organization fully devoted to standardization was the International Electrotechnical Commission born in 1906, whereas the first international standardizing body with general competence, the International Standardization Association (ISA) was created in 1930. After the Second World War, ISA disappeared and was substituted by the International Organization for Standardization (ISO) founded in 1947 [27]. ISO grew rapidly in the mid 1960s and in its activity shifted from the harmonization of existing national standards to direct development of standards at international level. In 1971 the result of the technical work of ISO, then called recommendations, took the official name of international standard [27]. In 1987, ISO published the series of Quality standards which is now known as ISO 9000, widely accepted as "gold standards" in QM, and successively revised in 1994, 2000, 2008 and 2015 [28–30]. ISO 9000 family comprises different standards: ISO 9000 – Fundamentals and Vocabulary; ISO 9001 – Requirements for Certification; ISO 9002 – Guidelines for Applying ISO 9001 and ISO 9004 – Guidelines for Maintaining the Success of an Organization. ISO 9001 Quality standards can be applied to any process or service organization and not only to the products. It sets out all the criteria for the implementation of a Quality Managment System (QMS) and is the international standard most widely used by organizations of different type, from energy management to social responsibility. In general, a QMS is defined as a formalized system that documents processes, procedures and responsibilities for achieving Quality policies and objectives. A QMS helps in coordinating and directing the organization activities to meet customer and regulatory requirements and improve its effectiveness and efficiency on a continuous basis [28–30]. In the context of research laboratories, the accuracy, reliability, and timeliness of the reported results define its quality. All aspects of scientific activities and procedures should be controlled as well as every variable involved in (e.g. staff, equipment, and methods). The application of ISO 9001 is the first step in starting to be confident with a QMS. It is the only standard in the family that can be certified, where certification represents a written assurance by a third party of the conformity of a product, process or service to defined requirements and of the company as whole.

The actual ISO series is based on the 7 principles developed by ISO Technical Committee (ISO/TC176), responsible for maintaining ISO QM standards (Fig. 2). These principles are the basis for the ISO 9000

**Fig. 2** ***Principles of QM.*** ISO 9001:2015 is based on the 7 QM principles here reported, that represent a foundation to guide organizations toward improvement of their processes and performance.

certification and guide in the right direction the organizations that seek to continuously improve their performance and ensure the high quality of their products, focusing on the needs and expectations of all interested parties [31]. Because Quality is defined by the customer, the first principle is the *Customer focus* and highlights the attitude that organizations should have toward their customers, understanding (and anticipating) their specific needs and requirements, and even striving to exceed customer expectations. For this reason, *Leadership* (second principle) has a fundamental role in establishing the unity of purpose and direction, and creating and maintaining an environment in which everyone is encouraged and fully committed to participate in achieving the organization objectives, as foreseen by *Engagement of people* (third principle). *Process approach* (fourth principle) pushes the management to recognize that an objective is achieved more efficiently when activities and associated resources are managed together as a process. At the same time, identifying and understanding interrelated processes, and managing them as a system, is more efficient and effective in achieving the goals of the organization. The fifth principle underlines that *Improvement* of the overall performance of the organization should be a permanent objective of the management. The constant analysis of how the activities are carried out identifies more effective and efficient ways of making improvements and achieving excellence, counteracting internal and external risks and creating new opportunities. Obtaining the sufficient and adequate data and information represents the basis for taking right and effective decisions, as is stressed in the sixth principle, *Evidence-based decision making*, in order to improve operational effectiveness and efficiency, assessment of process performance and ability to achieve objectives. The last principle, *Relationship management*, suggests that the management should enhance the interdependent relationships with all its interested parties, such as suppliers, partners, customers, investors, members of the organization, and society as a whole. This assures mutual benefits and creation of real value by sharing resources and competence and managing Quality-related risks [31].

In the ISO 9001:2015 risk management is more explicit than in the previous version of 2008 and it is built into the whole management focusing on risk-based thinking [30]. In order to meet the QMS objectives, organizations must consider risk as part of their management plan, which requires a higher level of commitment of top management. This systematic approach to management toward risk benefits organizations, reduces risk and highlights opportunities, adding value across organizations, from internal operations to market exposure and stakeholders. The implementation of research projects inherently carry some risks, because there is always the risk that objectives are not or only partially achieved. Risk management is especially important for ambitious, highly innovative and interdisciplinary projects, that go beyond the state of the art to develop radically new ideas and/or breakthrough technologies. In order to guarantee progress and successful implementation, risk should be assessed at the beginning of the project and constantly controlled, reassessed and reviewed through project life cycle [32,33].

An additional level of quality above certification, is the accreditation, that is the formal recognition by an authoritative body of the competence to work with specified standards, and to perform specific types of testing, measurement and calibration. In many cases accreditation is already a requirement for suppliers and regulatory authorities will not accept test or calibration results from a lab that is not accredited. ISO/IEC 17025:2005 is the standard for which most laboratories, owned and operated by government, industry or, in fact, any other organization, must hold accreditation in order to be deemed technically competent. The ISO/IEC 17025:2005 standard is also useful to universities, research centers, governments, regulators, inspection bodies, product certification organizations and other conformity assessment bodies with the need to do testing, sampling or calibration [34].

## 2.3 Good Practice in Laboratory, Clinics, Manufacturing and Research

The principles of the ISO portfolio of QM standards serve as basis for the establishment of Good Laboratory Practice (GLP), Good Clinical Practice (GCP) and Good Manufacturing Practice (GMP), required by most government regulatory bodies and referring to specific stages in the development and life cycle of a drug product (Fig. 3). Following issues regarding the quality and integrity of data records in support of marketing applications for drugs and pesticides, Environmental Protection Agency (EPA) and Food and Drug Administration (FDA) issued GLP regulations in 1983 and in 1997, respectively. GLP identifies a set of principles and requirements for planning, executing, monitoring, recording, reporting, and archiving nonclinical laboratory studies. These studies are performed to generate data for determining the safety of medical products. The application of GLP assures regulatory authorities that the data submitted in research or marketing applications accurately reflect the results obtained during studies, and that it can be relied upon in making risk/safety assessments. In many countries the studies for human medical products must be performed in

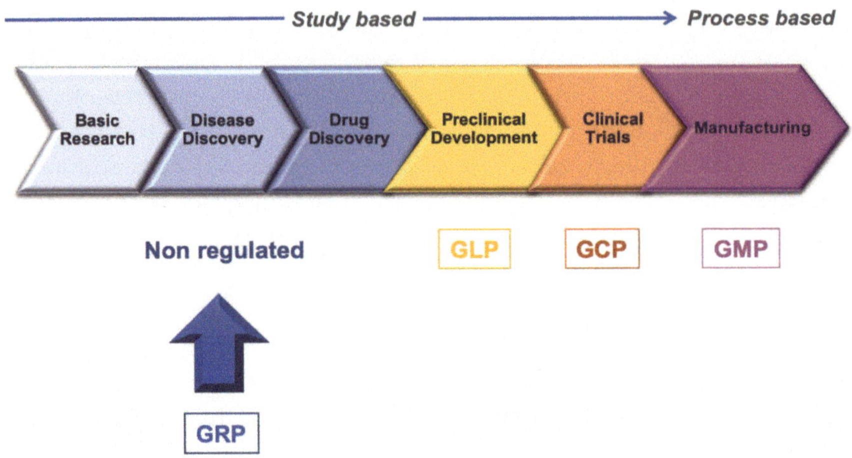


**Fig. 3** ***Good Practice regulations along the life cycle of a drug product.*** The development and life cycle of a drug product is characterized by different stages along a continuum from study-based to process-based research. The first stages, including basic research, disease discovery and drug discovery, are conducted in a non-regulated environment, that is not intended to be submitted to regulatory Quality systems. The successive stages, preclinical development, clinical trials and drug manufacturing are regulated by Good Laboratory Practice (GLP), Good Clinical Practice (GCP) and Good Manufacturing Practice (GMP), respectively. However, also in non-regulated research several Good Research Practices (GRPs) were issued from international bodies and universities. Differently from the other GXPs, GRPs are not formalized standards, but represent guidelines and recommendations for research activities.

accordance with Organization for Economic Development (OECD) GLP guidelines [35].

In some cases, clinical studies in humans are required to further demonstrate product safety and/or efficacy. If clinical trials are necessary, they must be performed under GCP, a set of ethical and scientific Quality standards for designing, conducting, recording, and reporting trials that involve the participation of human subjects. Finally, after serious safety problems involving the quality and safety of medical products, the FDA introduced also GMP. The current drug GMP was first codified in 1963, then expanded in 1978, and minimally revised in 2008 [36]. GMP regulation identifies the minimum practices, processes and standards required for manufacturing, packaging, testing, storage and distribution of drugs, biologics, human cells, tissues, cellular and tissue-based products (HCT/P), certain Class I medical devices, and all Class II and III medical products. The World Health Organization (WHO) published its own GLP, GCP and GMP requirements or regulations, similar to those put in place by FDA [37–39].

In the case of biological substances composed of complex macromolecular structures that may induce pleiotropic activities, as it is the case for EVs, the development of complex biological EV-based therapeutics in compliance with GMP is reported in different studies. Many efforts have been made to develop methods and processes suitable for EV manufacturing scheme covering the upstream processing phase and the downstream technologies until filling and storage established in a GMP-compliant strategy [40,41]. Nowadays, EVs derived from human and plants specimens are registered in clinical trials. The main sources of human-derived EVs, in compliance with GMP, include bone marrow mesenchymal stem cells (BM-MSCs), adipose tissue-derived stem cells (ADSCs), monocyte-derived dendritic cells (MDDCs), cardiac progenitor cells and also HEK293 human embryonic kidney cells. Human-derived EVs were applied in clinical trials having complete clinical reports; their application is major in cancer indication and also in inflammatory or chronic disease (chronic kidney disease). In contrast, the development in clinical trials of plant-derived EVs, namely grape, ginger and aloe is in the early stage and patients have not yet been recruited [42,43]. However, EV-based products are considered as biological medicines, falling into the category of biopharmaceuticals as defined in EU, United States, and Australia [44]. Due to the complexity of EVs and the possibility that they could have much in common with their parental cells source, development of EV-based therapeutics should take into account both as the regulations for biologicals and as those for cell-based products and advanced therapy medicinal products (ATMPs) [45,46]. In conclusion, in order to promote the translation from experimental research to stringently executed and documented therapeutic application of EVs, many efforts are necessary for the standardization of GMP-compliant manufacturing processes and Quality Control (QC) since a very early stage of manufacturing and during early clinical safety and proof of concepts testing.

In biomedical science, many research investigations (basic and discovery) are conducted in non-regulated research environments because the studies are exploratory and data are not intended to be submitted to regulatory Quality systems. However, data quality is a key requirement for research integrity, data reliability and reproducibility and therefore should be maintained under controlled and verifiable conditions, irrespective of scientific discipline or field, along the research spectrum (non-regulated vs. regulated or discovery vs. translational; Fig. 3). Strategies for integrating best practices into basic research environments are not new [1,7,47–50]. Even though Good Research Practice (GRP), as a QM standard dedicated to the research environment, is not yet organized in an international guidance document, different prescriptive documents were drafted and/or personalized by several research institutions. International bodies, universities and councils produced specific guidelines and recommendations to guide scientists in guaranteeing research data quality and integrity. Among others, the Joint Code of Practice (for quality) in Research—JCoPR, written in 2012 and updated in 2015, represents the first formal statement of Quality Assurance (QA) requirements. It provides a set of recommendations strongly suggested as standards for carrying out scientific research for the main funding agencies in particular on behalf of the United Kingdom Department for Environment, Food and Rural Affairs (UK DEFRA) [51]. The Research Quality Association (RQA) published in 2013 a handbook to provide clear references for the application of Quality and in 2014 a guideline for QM in non-regulated research [52,53], but the most comprehensive reference for Quality in research is still the Quality Practices in Basic Biomedical Research (QPBR) Handbook drafted in 2011 by WHO and then revised in 2013 [54].

## 3. Quality Management in research institutions

The implementation of a structured QMS according to international standards for continuous improvement and increasing of performances is a recent acquisition of academic and research institutions. Different models and standards for the quality of science and the quality of research processes have been proposed in different research institutions and a Quality approach is becoming an important requirement for project proposals submitted for funding. For example, the National Institute of Health (NIH) has expanded guidelines to enhance research rigor and reproducibility to ensure the quality of the data of the studies they fund, publish and support [55,56]. However, a complex organization like a multidisciplinary public research institute is extremely difficult to manage due to the intrinsic factors characterizing research activities. Obstacles to implementation of a QMS derive from the public sector peculiarity, the presence of multiple stakeholders with different needs, the difficulty in the identification of performance indicators, the shortage of the resources, and, last but not least, the lack of specific skills in Quality together to the need of flexibility typical in research [57]. On the other hand, most QMS have been developed for companies or service providers and have therefore limited applicability to research institutions, which makes it hard to motivate scientists to work with the QMS on a daily basis. A complete agreement and rapid adoption of Quality standards in the scientific community seem hard to achieve, but the use of general Quality principles and processes that are available is not only a viable option but rather the cornerstone in promoting data quality and reproducibility [9]. Despite these obstacles, QMS implementation is increasing in research institutions, showing that this cultural approach has enormous potential to improve the quality of research. We will present here some of the most relevant examples of successful QMS implementation. In 2012 the Medical Research Council (MRC) published Guidelines for Good Research Practice, a document that describes how MRC-funded research, within universities, medical schools, hospitals and other research organizations, is planned and conducted. The principles and standards regards all those involved in MRC-funded research, including researchers, research support staff, students, research managers and administrators [58].

Through a voluntary approach, the French National Institute for Agronomic Research INRA, implemented a QMS in order to improve the management of risks specific to research programs and to increase the reliability of results [9]. The investment in an effective QMS provides the considerable benefits and return of cost multifold expanding the opportunities for the scientific community as demonstrates the implementation of a QMS (based on ISO 9001 and GLP) in preclinical experimental (animal) research environments in both Europe, at the German Mouse Clinic, and in the United States, at the University of Kentucky [59]. In Germany, in the Department of Experimental Neurology at the Charité Universitätsmedizin Berlin, the project "Productiveness and Robustness through Modular Improvement of Experimental Research" (PREMIER) was started with the help of the Volkswagen Foundation and an expert advisory board [2]. The aim of the project was to design and implement a research- and science-driven QMS, providing a biomedically oriented, structured but flexible QA system that is acceptable for academic researchers. Using the language and nomenclature of the laboratory, the key processes of the QMS reflect the life cycle of the scientific research: experimental design, execution, evaluation, and reporting while novel form of auditing shall assess compliance with its Quality standards. PREMIER has been preregistered at the Open Science Framework and its QMS, including tools and training modules, will be freely available.

In Italy, different research institutions of the National Research Council (CNR) successfully developed and applied a QMS based on ISO standards. This approach improved both the management and the experimental activities of the laboratories, enhancing reliability and robustness of scientific results, and at the same time introduced new skills in research environment. It is the case of the implementation of a QA system according to GMP and ISO 9001:2008 standard for radiopharmaceutical production [60], and the application of a QMS based on the scheme of regional accreditation of industrial research laboratories which was inspired by ISO 9001:2008 and ISO 17025:2005 and set up by the regional authority [10]. In 2011 a project, named "Quality and Project Management OpenLab" (qPMO), was developed in five institutes of the CNR, belonging to two different Departments (Biomedicine and Agri-Food), with the aim of realizing a TQM model for biomedical research laboratories, applying Quality, Project and team-working methodologies [7]. The project focused on different aspects of research management: (i) management of knowledge, based on the definition and the subsequent applications of standards for the drafting of guidelines in Life Sciences [8], as well as the development of a web platform for the collection, cataloging and dissemination of scientific data (https://quality4-lab.cnr.it/en); (ii) management of experimental procedures centered on the application of the risk management method Failure Mode and Effect Analysis (FMEA) to scientific protocols; (iii) QM of a research laboratory, realizing a QMS certified according to the ISO 9001:2008 standards; (iv) management of multivariable assays based on the application of Design of Experiments (DoE) models to different kinds of scientific assays [61]. The QM models realized by the CNR qPMO research network, made by scientists for scientists, are powerful tools to identify best practice, to ensure the reliability and reproducibility of the results an enhance data and protocols sharing among research laboratories.

All these examples showed that a proper distribution of Quality culture from areas of high management development to intellectual and scientific production might facilitate and speed up research. The application of QM to scientific research positively impacts the research groups and laboratories by implementing interoperability and coordination, fostering a common language, increasing efficiency and effectiveness, facilitating technology transfers, helping in data management, augmenting the satisfaction of the interested parties and ultimately increasing the prestige of laboratories and public research institutions.

## 4. Challenges and perspectives in EV research

In the last years, the scientific community working on EVs has increasingly recognized the need to set-up and standardize methodological approaches for the collection, isolation, characterization and functional studies of EVs [14–16]. Therefore, the EV field is very prone to address standardization and reproducibility issues. The implementation of a QMS could have a great impact in supporting research networks to set-up, standardize and validate common scientific procedures as well as in building and performing highly competitive and innovative scientific projects.

EVs are cell-derived membranous particles able to carry inside proteins, lipids, nucleic acids, metabolites, and even entire organelles, and to mediate cell-cell communication at close or distant sites. All cells, prokaryotes and eukaryotes, are able to release EVs as part of their normal physiology and/or after different types of stimulation. This process is conserved throughout evolution from bacteria to humans and plants and can represent a key means of communication between species [62]. In addition to their function in normal physiology, EVs have also been implicated in several pathological processes, such as metabolic and cardiovascular diseases, pathologies of the nervous system and cancer [63–66]. Depending on their content, EVs can act both promoting or counteracting the disease [64,67–69]. EVs have been detected in a variety of biological fluids, including blood, urine, semen and cerebrospinal fluid. The possibility to collect, purify and characterize EVs by liquid biopsies has a great impact for the diagnosis and prognosis of the different pathologies in which they are involved. Disease progression and response to therapy could also be followed through an EV multicomponent analysis [66].

EVs are very heterogeneous in size (from 30 nm to even 10 μm), biophysical and biochemical properties as well as cargo content. The EVs cargo (both soluble and membrane components) depends on the cell of origin as well as the physiological state of the cell and specifies both the type of recipient cells and the intracellular effect elicited [70,71]. Due to their ability to target specific types of cells and to mediate a lot of complex cellular responses, EVs are being actively explored as therapeutic agents. Moreover, being naturally present in every living organism, endogenous EVs at physiological concentrations are also intrinsically characterized by high stability, biocompatibility, low toxicity, and permeability to biological barriers. Hence, increasing research efforts have been devoted to evaluate the outstanding potential of EVs purified by different sources as natural nanocarriers for drug delivery in medicine, nutraceutics and cosmetics [72–75].

However, despite the great interest in this field, various misconceptions and methodological pitfalls have hampered progress in understanding the biogenesis and biological function of these vesicles and consequently their use in diagnosis, therapy and industrial applications. Key aspects still to solve are connected to low EV yield, purity and heterogeneity of the samples, that complicate further characterization and functional studies [65,75]. Moreover, a confusing and misleading nomenclature has been widely used in literature to refer to EVs [76,77]. In 2014, the board members of the International Society on Extracellular Vesicles (ISEV), created in 2012, provided researchers with a minimal set of biochemical, biophysical and functional standards that should be used to attribute any specific biological cargo or functions to EVs [14]. Successively, the minimal information for studies of EVs (MISEV) 2014 was revised, collecting and integrating feedbacks from a larger number of investigators working in the field [15,78]. Moreover, the MISEV 2018 strongly endorses the use of the general term EVs, instead of terms such as exosomes and ectosomes that are historically burdened by contradictory definitions and associated to expectations of unique biogenesis. Particular care requires the isolation from body fluids: standardization is a prerequisite to validate EV-associated biomarkers. All the issues regarding the isolation and characterization of EVs have to be considered with a special care before making strong conclusions on their functional role in specific pathologies or their use as providers of biological markers [15]. Both ISEV and the International Society on Thrombosis and Hemostasis have provided guidelines for this [14–16,79]. Nowadays, several national societies for scientists studying EVs are also being formed, among them the Italian Society, named EVIta (www.evitasociety.org), was founded and had its first meeting in 2019. In summary, all the scientific community working in EV field is strongly committed to optimize, standardize and share new methodological approaches for isolation and analysis of EVs to increase robustness, reproducibility and reliability of the huge amount of data that is being produced in the field. A great effort has also been devoted to collect the data produced by the generation and continuous update of specific EV databases, such as Vesiclepedia [80], EVpedia [81] and exRNA Atlas [82] that represent a crucial tool to improve comprehension of EV complexity. The first two include data not only on proteins but also on nucleic acids and lipids found associated to EVs, as well as on the purification procedures used; the last one is specifically devoted to RNA data. More recently, the

EV-TRACK knowledgebase has been developed for recording the experimental parameters of EV-related studies. EV-TRACK invites submission of details on EV isolation and characterization via a multi-step online template, then associates each with a particular study or publication and also returns an "EV-METRIC" as an aggregate measure of the level of detail provided [83]. It is a unique resource that aimed to implement a bottom-up community consensus approach about relevant experimental parameters and improve the rigor and interpretation of studies on EVs, in addition to record the evolution of EV research. ISEV has strongly encouraged researchers to consult these databases as well as to submit their experimental protocols and data to EV-TRACK to facilitate standardization of EV research through increased systematic reporting on EV biology and methodology [14,15].

To support the development of successful, high-growth innovative enterprises, able to shape and capture the new emergent markets, the European Innovation Council (EIC) in the context of Horizon 2020, launched new pilot initiatives, aiming to fund high-risk cutting edge projects focusing on developing radical and innovative technologies. Among these, the Future and Emerging Technologies (FET) Open program is devoted to support interdisciplinary consortia exploring visionary ideas for radically new future technologies that challenge current scientific and technological paradigms. Of note, in the last three years, the FET Open program has funded five projects focusing on the implementation of new technologies in the EV field, that are INDEX (www.indexproject.eu), VES4US (https://ves4us.eu), evFOUNDRY (www.evfoundry.eu), GLADIATOR (www.fet-gladiator.eu) and MINDGAP (www.biomark.isep.ipp.pt./mindgap). Recently the five networks organized a clustering event and concomitantly wrote a document, the EVmanifesto, to highlight the importance of introducing and securing more firmly the EV-related theme in the European research agenda.

## 5. A Quality Management System supporting EV research: the VES4US approach

The VES4US project was financed by the Horizon 2020 FET Open Program in 2018 to develop a radically new platform for the efficient production and functionalization of EVs from a natural source and their consequent exploitation as tailor-made nanodelivery systems. The Consortium is formed by a highly interdisciplinary and interconnected network made up by ten scientific teams belonging to five different research centers and universities and one consultancy firm from six different European countries. The group is led by the National Research Council of Italy, that participates through five different institutes, belonging to three different CNR Scientific Departments that are Biomedicine, Agri-food and Physics. The other partners are the Institute of Technology Sligo (Ireland), the Swiss Federal Institute of Technology (Switzerland), the University of Ljubljana (Slovenia), the Max Planck Institute for Polymer Research (Germany) and the ZABALA Innovation Consulting (Spain). The network covers different disciplines, such as biophysics, biochemistry, cell biology, chemistry, ecology and engineering. The project has been conceived with a strong QM approach to efficiently deal with the targeted innovation level of the scientific project, the interdisciplinarity and the multicentre structure of the Consortium, and the high grade of complexity related to EV isolation and characterization. In fact, one of the seven work packages of VES4US project, is transversal to all the others and is devoted to implement Quality and Project Management Systems. QMS aims to support the scientific activities, to increase their performance and fulfill the standards set-up at the beginning of the project itself. The QMS of VES4US was released following an intense Quality training, and is constantly updated and remodeled based on the feedbacks of partners. VES4US QMS was planned and developed with a slim and flexible structure and customized for multicentre research project's needs. These needs do not necessarily target certification but rather the efficient organization of the available resources (time, money and people). Therefore, VES4US QMS has applied standardized and agreed procedures to increase the performance of the activities, to reach the objectives and to contribute to the reproducibility and reliability of the measurements/analysis/data. The VES4US QMS has been developed to be compatible with UNI EN ISO 9001:2015 standard, in order to favor subsequent technology transfer and exploitation of the results. Here we will present some of the QM tools we took advantage of in setting up and implementing our QMS, together to some examples of applications.

Our QMS follows the documentation foreseen by ISO 9001 standard. According to it, there are four levels of documentation that an organization should observe to document its QMS, that are schematized as a hierarchy in Fig. 4. At the top of the pyramid there is the Quality Policy, that defines WHY, represents a declarative statement by the organization including its goals, mission and vision and states the commitment of the entire organization to quality and improvement. Then, at the lower level we can find the Procedures that address the points WHO, WHEN and WHERE and define

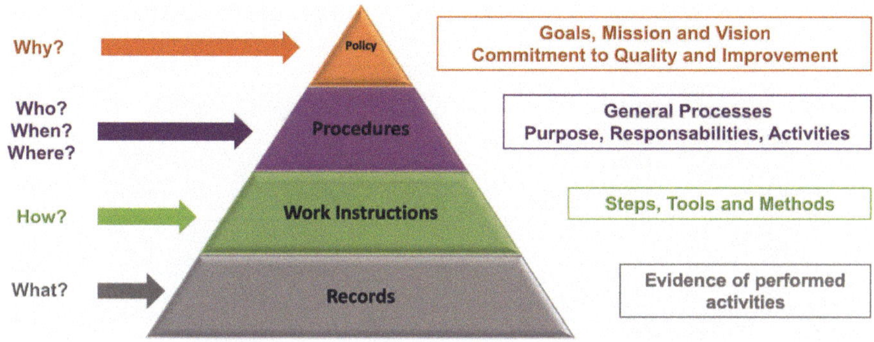


**Fig. 4** ***Documented Information required by the ISO 9001:2015 standard.*** The 9001 international standard specifies four levels of information that an organization must observe to document its Quality Management System. They are schematized as a pyramid at the center of the Figure. On the left are reported the questions that each level addresses and on the right the type of information that each level contains.

the General Processes in terms of purpose, responsibilities and description of the activities. Going down along the pyramid, there are the Work Instructions, that mainly address the question HOW, with a structure similar to the Procedures, they cover the same elements, but at a higher level of details. Work Instructions focus on the sequencing of the steps, tools, and methods to be used, they can be also part of a procedure, or can be referenced in there. Finally, at the bottom, the Records of the activities identify WHAT, and refer to all the activities performed, the data collected and also the records from the audits and the surveys performed to monitor the activities. To organize the documentation, we used a QM approach, that involves planning, doing, checking, and acting to reach the foreseen objectives, and specifically we took advantage of the PDCA cycle described above in the Chapter. The activities performed in each phase of the PDCA are shown in Fig. 5. In the first round of the PDCA (Fig. 5, Left), Quality documentation to implement in the QMS was selected, designed (Plan) and distributed to partners (Do) to collect feedbacks and suggestions (Check). Finally, the documentation was revised, approved and distributed to all partners for effective implementation (Act). In the successive rounds of PDCA cycle, activities of support and monitoring of QMS implementation were planned (Plan) and, in agreement, all staff members received specific training and implemented QMS according to their role and responsibilities (Do). Then, feedbacks after QMS implementation were collected through both Audits and Checklists (Check) and the QMS was update according to them and approved (Act). The repeating of different PDCA cycles

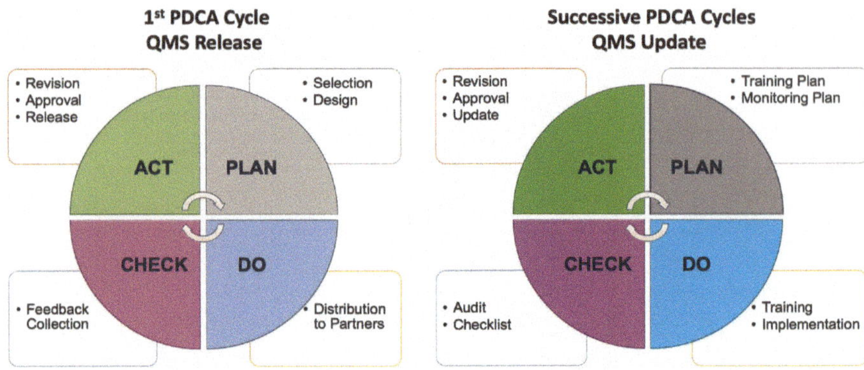


**Fig. 5** ***Application of the PDCA cycles to the implementation of VES4US Quality Management System.*** PDCA cycles showing the Planning, Doing, Checking and Acting activities related to VES4US QM System (QMS) implementation. On the left is reported the first PDCA cycle showing the activities related to the release of QMS, whereas on the right are reported the successive PDCA cycles, showing the activities necessary for the continuous monitoring and update of the QMS.

continuously monitors and improves the system, in relation to the upcoming issues and specific needs arising during project activities. Using this methodology, following the 9001 international standards, VES4US consortium identified its Policy, the main general processes to take care inside the consortium and Quality models for the formalization and standardization of scientific procedures as well as for data recording. The outcome is an agreed and flexible QMS, developed by regularly collecting suggestions and comments by partners, focusing on real impact on the ongoing activities and open to any adjustments necessary to achieve the project objectives.

Inside QM activities, is fundamental to identify and structure three main process groups, that are in order Quality Planning (QP), QA and QC. QP identifies the goals of the project in term of both the results to obtain and the methodology to use, the specific project deliverables and milestones and the Quality standards to meet. Moreover, QP develops all the guidelines, tools and timeline for the activities to perform as well the monitoring plan to control that the activities are conducted as foreseen.

QA monitors and verifies that the processes for the management and production of the deliverables are executed and are effective and that project outcomes fulfill the expected requirements. In other words, the aim of QA is to create the ideal conditions for any activity to be well performed and once only, so to minimize waste of time, money and resources. Finally, QC monitors and verifies that project deliverables meet defined Quality standards.

QC refers to the operational techniques, procedures and methods that are used to determine that final products comply with the defined Quality standards. This may be done in regularly scheduled project status meetings, at the end of any task activity, or as necessary throughout the project life cycle. To address all these points, the Consortium developed: (i) Quality templates for standardization of scientific procedures as well as for data registration (QP); (ii) templates for internal Audits and Checklists for specific items to collect feedbacks on the use of the Quality templates (QA); (iii) Checklists for the validation and delivery of the final outcomes (QC). The design and revision of the Quality templates, feedback collection and Audit activity has been performed by Quality Managers, who definitely are in charge of the set-up, implementation and monitoring of the QMS during project execution. The final approval and the QC of the deliverables and outcomes is the responsibility of the Project Coordinator together with the Steering Committee. Leaders of each task and work package are also strongly committed to promote and assure Quality culture and the correct implementation of the QMS. Finally, all staff members are key pieces of the QMS. It is fundamental that each one involved in the project is aware of the Quality policy and tools implemented by the Consortium, sufficiently trained to work in agreement with Quality standards, to use correctly Quality documents developed and eventually suggest ameliorations.

Among the different Quality templates designed, we will focus on the ones closer to the day-by-day activity in the laboratory, that are the Standard Operating Procedures or SOPs (Work Instructions). The advancement and use of SOPs are a basic part of a successful QMS [84]. SOPs may be written for any repetitive technical activity, as well as for any administrative or functional programmatic procedure, that is being followed within an organization. They are composed of detailed written instructions for the users to perform a specific activity to reduce the possibility of missed steps or other errors that impact the quality of the results and to achieve uniformity of the performance of a specific function [85]. In summary, SOPs are very powerful in increasing results reliability and experiment reproducibility and very precious for training of new personnel and transfer of competence, abilities and know-how. SOPs are also very useful for mitigating both health and safety risk as well as instrument damage or erroneous utilization [86]. It is advisable to have SOPs for any activity, but first of all for the ones that are especially relevant and strategic for the project and/or more complex or that have to be performed in several distinct laboratories. It is also advisable to define the SOPs as soon as possible to start the work with the right step. For sure, new SOPs need to be monitored, revised and updated to continuously improve performance, safety and reproducibility [87]. In defining our specific model for SOPs, we applied QM tools, some more specific as the Ishikawa diagram, others very well known, such as the flowchart, that concur to help in analyzing all the aspects related to a scientific procedure and also to organize it in a logical and simple form, that can be easily transferred. The Ishikawa diagram, also known as the fishbone diagram is the approach most frequently used in Quality to identify the possible causes of a problem and organize them in a causal chain [88]. The main effect to reach is represented at the end of a line called the causal line, whereas each primary branch extending from this line constitutes a class of potential primary causes (Fig. 6A). From each primary branch, secondary branches extend representing secondary causes related to the primary one in which they are inserted. Primary causes are classified in four standard categories (the 4 Ms), that are Manpower, Materials, Methods, and Machines. An optional fifth branch, named "Milieu," (environment) can also be added (Fig. 6A). This Ishikawa diagram has been previously used for the

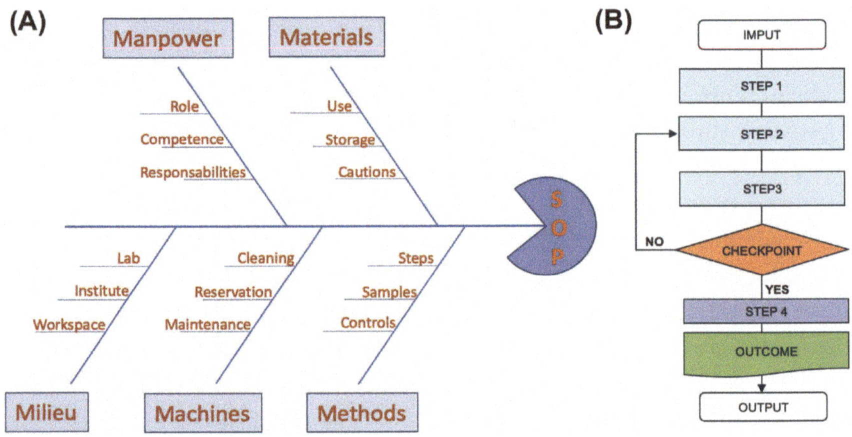


**Fig. 6** ***Application of Quality Management tools for the definition of the VES4US model for Standard Operating Procedures.*** (A) The Ishikawa diagram used to identify and organize the core topics of the VES4US Standard Operating Procedures (SOPs). Each primary ramification corresponds to a principal category (Manpower, Materials, Milieu, Machines and Methods), whereas each secondary ramification corresponds to a specific issue of the main category. All these issues have to be considered in a SOP to report all the information necessary for the correct and complete execution of the procedure. (B) Example of flow chart that can be used to describe the Method step by step. A key step is the checkpoint (rhombus); if the check is positive (YES), it is possible to go to the next step of the procedure, if not (NO) it could be necessary to go back to previous steps.

development of a Quality-based model for Life Sciences research guidelines [8]. This model has been followed for the designing of several guidelines covering main aspects of Life Sciences from basic guidelines for laboratories, research activities to instruments management, facilities management and the application of Quality methodology to research [8]. Guidelines were applied in different laboratories of CNR, causing both a reduction in time and cost of the different experiments and a more efficient management of facilities and procedures. Some guidelines were also included and used to provide the SOPs of a QMS for a research laboratory of CNR that was certified according to the UNI EN ISO 9001:2008 standard in 2014 [7]. Due to these impactful results, in a similar way we used the Ishikawa diagram in the definition of a model for SOPs (Fig. 6A). The aim of each SOP is the effective and efficient management of a procedure, minimizing the effects of potential influencing factors and sources of variation. All the Ishikawa categories, the personnel, the materials, the instruments, the methods and the milieu that concur to this goal, are inserted and must be explored in all their implications in developing a SOP. As schematized in Fig. 6A, it is necessary to indicate the role, the competence and the responsibilities of the personnel involved (Manpower), the required materials, with indications on storage, use, cautions (Materials), the necessary instruments, with indications on use, cautions, cleaning, maintenance and advance reservation (Machines), and finally the organization of the workspace, the laboratory and eventually the Institute (Milieu). However, the core of a SOP is represented by the experimental procedure (Method). The procedure has to be written step by step, very simply and clear, containing information on the samples in input, the external and internal controls and key steps which require particular care and/or precision. The adoption of flowchart can be helpful in describing the steps of the procedure, as shown in Fig. 6B. It is useful also to include in a SOP a QA and QC section, defining how to check that the procedure has been performed correctly and that the result reaches the foreseen standards. Using this model, several SOPs have been designed, regarding EV production, isolation, biophysical and biochemical characterization, as well as functionalization. These SOPs have been shared among partners and are being used in the different laboratories, and their validation is ongoing. We also designed a Lab Notebook, in both electronic and paper versions, (subject of a patent application), for the registration of the different activities without un-necessary and time-wasting details, to help data recording and sharing among partners. Staff members were trained on the use and the redaction of both SOPs and Lab Notebook, invited to consider the main parameters and controls of their own experimental procedures as well as of the ones developed by the other partners. They were also encouraged to give feedbacks and suggestions, so to be fully engaged in the QM of the scientific activities. In summary, a great effort has been performed to define QM tools specifically suitable for such interdisciplinary and innovative research activities and to continuously improve them collecting feedbacks by scientists performing research. The activity of QA and QC have been thought to be very simple and interactive to really monitor ongoing processes. The outcome is an agreed QMS with a slim, dynamic, customized and research-oriented structure, that at the end of the project we will validate and disseminate for successive application and implementation. Our ultimate goal is to disclosure a validated QMS that might represent an operating model for QM of innovation research projects characterized by high risks, strong interdisciplinary and multicentre character, and focusing on the development and standardization of new potentially breakthrough technologies. Research in EV field might take great advantage from this QMS model to face and win its challenges and go safer and faster toward deeper understanding and full exploitation of EV biology.

## 6. Concluding remarks

QM has been developing at a rapid pace over the last century, first in industrial and manufacturing sectors, from which moved to impact other areas, such as services, healthcare, education, government and also research. Dissemination in the other sectors demonstrates the great impact that QMS had in industries to potentiate effectiveness and efficiency of industrial processes as well as to increase the number and the quality level of the products. In other words, QM impacted on the economic strength and the prestige of an organization, increasing its competitiveness on the public markets. The possibility to increase the level of quality of both work processes and final outcomes attracted the attention of other sectors, among them education and research, whose objectives are to produce and divulgate knowledge and also to do it in the best way possible, managing efficiently available resources. However, it is not possible to efficiently translate a management system from a sector to another without seriously taking into accounts the specific needs, objectives, expectations and challenges of the new sector. It is not feasible to simply impose rules, standards and models that have been selected to function in a context even very far from the new one. It is intuitive to consider public research institutes very different from industries, but it has not been so intuitive to think that successful implementation of a QMS for research institutions needs its own standards, methodologies, tools and culture. For this reason, many researchers refuse or at least are very skeptic toward QM, even though most of their thinking is due to prejudice and not to knowledge or personal experience. It is undeniable that the propulsive strength of research is the intrinsic ability to discover, to always go beyond common paradigms, to conceive and realize visionary scenarios, to innovate and then positively impact on society health, wellness and development. But research is also rigor, accuracy of measures and tests, use of internal and external controls, validation and reproducibility of the experiments, and reliability of the results. Scientific rigor represents the foundation on which creativity and innovation can be erected. Research activities and QM should meet, mix and contaminate more and more, so to produce new tools and methodologies that are really tailored on research and researcher identity and can really support and not cage knowledge. To address this point, Research in Quality is needed, in other words a critical mass of researchers that deal with, develop and enhance Quality culture.

The scientific community working on EVs has internationally recognized the need of rigor and standardization in this promising field. Studying EV-mediated cell communication is strongly contributing to unravel still unknown mechanisms in basic cell biology and physiology as well as in several pathological processes, involved in cancer, neural, metabolic and cardiovascular diseases. Therefore, EVs have a great potential to be exploited in diagnostic and therapeutic approaches. However, all the issues regarding standardization of EV isolation and characterization procedures, as well as reliability and reproducibility of the increasing amount of data produced with different procedures and technologies, have to be fully addressed before exploitation. EV community is strongly committed to solve these issues and to clearly identify common validated practice, as also shown by the increasing number of position papers aiming to collect, define, update and share minimal information for studying EVs. In this scenario, the approach we have described in VES4US project is extremely suitable to ensure robust and valuable outcomes, within a framework of best practice. VES4US QMS was planned and constructed to be slim, focused and flexible as well as customized for the needs of a multicentre and multidisciplinary research project. The result is the implementation of a personalized QMS, compatible with UNI EN ISO 9001:2015 standard that represents an innovative approach with cross application in research environments, specially focusing on EVs. We auspicate that our QMS can inspire other colleagues to familiarize with Quality and eventually be modeled in research projects based on the development of radically new technologies, in which rigor, interdisciplinarity, innovation and interconnection are key features.

## Acknowledgments

The authors acknowledge all the partners of the VES4US project for their enthusiasm and engagement in QMS implementation, Mrs. Anna Aliperti for English language editing and Mrs. Mariarosaria Aletta for bibliographic support. The authors also acknowledge support by European Union's Horizon 2020 research and innovation program (Grant no. 801338) to GLL and AK and by PO FESR 2014–2020 "SATIN" from Regione Campania to GLL.

## References

[1] M.M. Robins, Quality Assurance can be applied to fundamental research, Biochemist 26 (2004) 49–53.

[2] U. Dirnagl, C. Kurreck, E. Castaños-Vélez, R. Bernard, Quality management for academic laboratories: burden or boon? EMBO Rep. 19 (2018) 1–5.

[3] A.M. Trotter, R. Calabrese, U. Palm, A. Krumenaker, ASQ TRI-2012 Best Quality Practices for Biomedical Research in Drug Development, ASQ, 2012.

[4] E. Giesen, Quality management for robust and reliable research, Int. J. Metrol. Qual. Eng. 6 (2015), https://doi.org/10.1051/ijmqe/2015029.

[5] A. Lanati, Quality Management in Scientific Research, Springer, 2018.

[6] J.M. Payne, K.E. France, N. Henley, H.A. D'Antoine, A.E. Bartu, E.J. Elliott, C. Bower, Researchers' experience with project management in health and medical research: results from a post-project review, BMC Publ. Health 11 (2011), https://doi.org/10.1186/1471-2458-11-424.

[7] A. Bongiovanni, G. Colotti, G.L. Liguori, M. Di Carlo, F.A. Digilio, G. Lacerra, A. Mascia, A.M. Cirafici, A. Barra, A. Lanati, A. Kisslinger, Applying Quality and Project Management methodologies in biomedical research laboratories: a public research network's case study, Accred Qual. Assur. 20 (2015) 203–213.

[8] F.A. Digilio, A. Lanati, A. Bongiovanni, A. Mascia, M. Di Carlo, A. Barra, A.M. Cirafici, G. Colotti, A. Kisslinger, G. Lacerra, G.L. Liguori, Quality-based model for Life Sciences research guidelines, Accred Qual. Assur. 21 (2016), https://doi.org/10.1007/s00769-016-1205-0.

[9] V. Molinéro-Demilly, A. Charki, C. Jeoffrion, B. Lyonnet, S. O'Brien, L. Martin, An overview of Quality Management System implementation in a research laboratory, Int. J. Metrol. Qual. Eng. 9 (2018), https://doi.org/10.1051/ijmqe/2017025.

[10] V. Biasini, Implementation of a quality management system in a public research centre, Accred Qual. Assur. 17 (2012) 621–626.

[11] M. Poli, S. Pardini, I. Passarelli, I. Citti, D. Cornolti, E. Picano, The 4A's improvement approach: a case study based on UNI EN ISO 9001:2008, Total Qual. Manag. Bus. Excel. 26 (11) (2015) 1113–1130.

[12] Guidelines on Fair Data Management in Horizon 2020, European Commission (2016).

[13] M.D. Wilkinson, M. Dumontier, I.J. Aalbersberg, G. Appleton, M. Axton, A. Baak, N. Blomberg, J.W. Boiten, L.B. da Silva Santos, P.E. Bourne, J. Bouwman, A.J. Brookes, T. Clark, M. Crosas, I. Dillo, O. Dumon, S. Edmunds, C.T. Evelo, R. Finkers, A. Gonzalez-Beltran, A.J.G. Gray, P. Groth, C. Goble, J.S. Grethe, J. Heringa, P.A.C. t Hoen, R. Hooft, T. Kuhn, R. Kok, J. Kok, S.J. Lusher, M.E. Martone, A. Mons, A.L. Packer, B. Persson, P. Rocca-Serra, M. Roos, R. van

Schaik, S.A. Sansone, E. Schultes, T. Sengstag, T. Slater, G. Strawn, M.A. Swertz, M. Thompson, J. Van Der Lei, E. Van Mulligen, J. Velterop, A. Waagmeester, P. Wittenburg, K. Wolstencroft, J. Zhao, B. Mons, Comment: the FAIR Guiding Principles for scientific data management and stewardship, Sci. Data. 3 (2016) 1–9.

[14] J. Lötvall, A.F. Hill, F. Hochberg, E.I. Buzás, D. Di Vizio, C. Gardiner, Y.S. Gho, I.V. Kurochkin, S. Mathivanan, P. Quesenberry, S. Sahoo, H. Tahara, M.H. Wauben, K.W. Witwer, C. Théry, Minimal experimental requirements for definition of extracellular vesicles and their functions: a position statement from the International Society for Extracellular Vesicles, J. Extracell. Vesicles 3 (2014), https://doi.org/10.3402/jev.v3.26913.

[15] C. Théry, K.W. Witwer, E. Aikawa, M.J. Alcaraz, J.D. Anderson, R. Andriantsitohaina, A. Antoniou, T. Arab, F. Archer, G.K. Atkin-Smith, D.C. Ayre, J.M. Bach, D. Bachurski, H. Baharvand, L. Balaj, S. Baldacchino, N.N. Bauer, A.A. Baxter, M. Bebawy, C. Beckham, A. Bedina Zavec, A. Benmoussa, A.C. Berardi, P. Bergese, E. Bielska, C. Blenkiron, S. Bobis-Wozowicz, E. Boilard, W. Boireau, A. Bongiovanni, F.E. Borràs, S. Bosch, C.M. Boulanger, X. Breakefield, A.M. Breglio, M. Brennan, D.R. Brigstock, A. Brisson, M.L.D. Broekman, J.F. Bromberg, P. Bryl-Górecka, S. Buch, A.H. Buck, D. Burger, S. Busatto, D. Buschmann, B. Bussolati, E.I. Buzás, J.B. Byrd, G. Camussi, D.R.F. Carter, S. Caruso, L.W. Chamley, Y.T. Chang, A.D. Chaudhuri, C. Chen, S. Chen, L. Cheng, A.R. Chin, A. Clayton, S.P. Clerici, A. Cocks, E. Cocucci, R.J. Coffey, A. Cordeiro-da-Silva, Y. Couch, F.A.W. Coumans, B. Coyle, R. Crescitelli, M.F. Criado, C. D'Souza-Schorey, S. Das, P. de Candia, E.F. De Santana, O. De Wever, H.A. del Portillo, T. Demaret, S. Deville, A. Devitt, B. Dhondt, D. Di Vizio, L.C. Dieterich, V. Dolo, A.P. Dominguez Rubio, M. Dominici, M.R. Dourado, T.A.P. Driedonks, F.V. Duarte, H.M. Duncan, R.M. Eichenberger, K. Ekström, S. EL Andaloussi, C. Elie-Caille, U. Erdbrügger, J.M. Falcón-Pérez, F. Fatima, J.E. Fish, M. Flores-Bellver, A. Försönits, A. Frelet-Barrand, F. Fricke, G. Fuhrmann, S. Gabrielsson, A. Gámez-Valero, C. Gardiner, K. Gärtner, R. Gaudin, Y.S. Gho, B. Giebel, C. Gilbert, M. Gimona, I. Giusti, D.C.I. Goberdhan, A. Görgens, S.M. Gorski, D.W. Greening, J.C. Gross, A. Gualerzi, G.N. Gupta, D. Gustafson, A. Handberg, R.A. Haraszti, P. Harrison, H. Hegyesi, A. Hendrix, A.F. Hill, F.H. Hochberg, K.F. Hoffmann, B. Holder, H. Holthofer, B. Hosseinkhani, G. Hu, Y. Huang, V. Huber, S. Hunt, A.G.E. Ibrahim, T. Ikezu, J.M. Inal, M. Isin, A. Ivanova, H.K. Jackson, S. Jacobsen, S.M. Jay, M. Jayachandran, G. Jenster, L. Jiang, S.M. Johnson, J.C. Jones, A. Jong, T. Jovanovic-Talisman, S. Jung, R. Kalluri, S. ichi Kano, S. Kaur, Y. Kawamura, E.T. Keller, D. Khamari, E. Khomyakova, A. Khvorova, P. Kierulf, K.P. Kim, T. Kislinger, M. Klingeborn, D.J. Klinke, M. Kornek, M.M. Kosanović, Á.F. Kovács, E.M. Krämer-Albers, S. Krasemann, M. Krause, I.V. Kurochkin, G.D. Kusuma, S. Kuypers, S. Laitinen, S.M. Langevin, L.R. Languino, J. Lannigan, C. Lässer, L.C. Laurent, G. Lavieu, E. Lázaro-Ibáñez, S. Le Lay, M.S. Lee, Y.X.F. Lee, D.S. Lemos, M. Lenassi, A. Leszczynska, I.T.S. Li, K. Liao, S.F. Libregts, E. Ligeti, R. Lim, S.K. Lim, A. Linē, K. Linnemannstöns, A. Llorente, C.A. Lombard, M.J. Lorenowicz, Á.M. Lörincz, J. Lötvall, J. Lovett, M.C. Lowry, X. Loyer, Q. Lu, B. Lukomska, T.R. Lunavat, S.L.N. Maas, H. Malhi, A. Marcilla, J. Mariani, J. Mariscal, E.S. Martens-Uzunova, L. Martin-Jaular, M.C. Martinez, V.R. Martins, M. Mathieu, S. Mathivanan, M. Maugeri, L.K. McGinnis, M.J. McVey, D.G. Meckes, K.L. Meehan, I. Mertens, V.R. Minciacchi, A. Möller, M. Møller Jørgensen, A. Morales-Kastresana, J. Morhayim, F. Mullier, M. Muraca, L. Musante, V. Mussack, D.C. Muth, K.H. Myburgh, T. Najrana, M. Nawaz, I. Nazarenko, P. Nejsum, C. Neri, T. Neri,

R. Nieuwland, L. Nimrichter, J.P. Nolan, E.N.M. Nolte-'t Hoen, N. Noren Hooten, L. O'Driscoll, T. O'Grady, A. O'Loghlen, T. Ochiya, M. Olivier, A. Ortiz, L.A. Ortiz, X. Osteikoetxea, O. Ostegaard, M. Ostrowski, J. Park, D.M. Pegtel, H. Peinado, F. Perut, M.W. Pfaffl, D.G. Phinney, B.C.H. Pieters, R.C. Pink, D.S. Pisetsky, E. Pogge von Strandmann, I. Polakovicova, I.K.H. Poon, B.H. Powell, I. Prada, L. Pulliam, P. Quesenberry, A. Radeghieri, R.L. Raffai, S. Raimondo, J. Rak, M.I. Ramirez, G. Raposo, M.S. Rayyan, N. Regev-Rudzki, F.L. Ricklefs, P.D. Robbins, D.D. Roberts, S.C. Rodrigues, E. Rohde, S. Rome, K.M.A. Rouschop, A. Rughetti, A.E. Russell, P. Saá, S. Sahoo, E. Salas-Huenuleo, C. Sánchez, J.A. Saugstad, M.J. Saul, R.M. Schiffelers, R. Schneider, T.H. Schøyen, A. Scott, E. Shahaj, S. Sharma, O. Shatnyeva, F. Shekari, G.V. Shelke, A.K. Shetty, K. Shiba, P.R.M. Siljander, A.M. Silva, A. Skowronek, O.L. Snyder, R.P. Soares, B.W. Sódar, C. Soekmadji, J. Sotillo, P.D. Stahl, W. Stoorvogel, S.L. Stott, E.F. Strasser, S. Swift, H. Tahara, M. Tewari, K. Timms, S. Tiwari, R. Tixeira, M. Tkach, W.S. Toh, R. Tomasini, A.C. Torrecilhas, J.P. Tosar, V. Toxavidis, L. Urbanelli, P. Vader, B.W.M. van Balkom, S.G. van der Grein, J. Van Deun, M.J.C. van Herwijnen, K. Van Keuren-Jensen, G. van Niel, M.E. van Royen, A.J. van Wijnen, M.H. Vasconcelos, I.J. Vechetti, T.D. Veit, L.J. Vella, É. Velot, F.J. Verweij, B. Vestad, J.L. Viñas, T. Visnovitz, K.V. Vukman, J. Wahlgren, D.C. Watson, M.H.M. Wauben, A. Weaver, J.P. Webber, V. Weber, A.M. Wehman, D.J. Weiss, J.A. Welsh, S. Wendt, A.M. Wheelock, Z. Wiener, L. Witte, J. Wolfram, A. Xagorari, P. Xander, J. Xu, X. Yan, M. Yáñez-Mó, H. Yin, Y. Yuana, V. Zappulli, J. Zarubova, V. Žėkas, J. ye Zhang, Z. Zhao, L. Zheng, A.R. Zheutlin, A.M. Zickler, P. Zimmermann, A.M. Zivkovic, D. Zocco, E.K. Zuba-Surma, Minimal information for studies of extracellular vesicles 2018 (MISEV2018): a position statement of the International Society for Extracellular Vesicles and update of the MISEV2014 guidelines, J. Extracell. Vesicles 7 (1) (2018) 1–43.

[16] A. Clayton, E. Boilard, E.I. Buzas, L. Cheng, J.M. Falcón-Perez, C. Gardiner, D. Gustafson, A. Gualerzi, A. Hendrix, A. Hoffman, J. Jones, C. Lässer, C. Lawson, M. Lenassi, I. Nazarenko, L. O'Driscoll, R. Pink, P.R.M. Siljander, C. Soekmadji, M. Wauben, J.A. Welsh, K. Witwer, L. Zheng, R. Nieuwland, Considerations towards a roadmap for collection, handling and storage of blood extracellular vesicles, J. Extracell. Vesicles 8 (2019), https://doi.org/10.1080/20013078.2019.1647027.

[17] R.R. Kolb, M.L. Hoover, The history of quality in industry, Sandia Natl. Lab. (2012), https://doi.org/10.1080/14783363.2019.1576516.

[18] P. Hellman, Y. Liu, Development of quality management systems: how have disruptive technological innovations in quality management affected organizations? Qual. Innov. Prosper. 17 (2013) 104–119.

[19] F.W. Taylor, The Principles of Scientific Management, Dover Publications (2003).

[20] W.A. Shewhart, Statistical method from the viewpoint of quality control, New Foreword by W. Edwards Deming (1940).

[21] W.E. Deming, Out of the crisis: quality, productivity and competitive position, Massachusetts Inst. Technol. Cent. Adv. Eng. Study. (1986).

[22] J.M. Juran, B. Godfrey, R.E. Hoogstoel, E.G. Schilling, Juran ' S Quality Handbook, Train. Qual., 1999.

[23] B. Karlof, F. Loevingsson, The A-Z of Management Concepts and Models., Thorogood Publishing, 2005.

[24] A.R. Martínez-Lorente, F. Dewhurst, B.G. Dale, Total quality management: origins and evolution of the term, TQM Mag. 10 (5) (1998) 378–386.

[25] T. Redman, A. Wilkinson, E. Snape, M. Marchington, Managing with total quality management: theory and practice, Palgrave (2000).

[26] B. Lahidji, W. Tucker, Continuous quality improvement as a central tenet of TQM: history and current status, Qual. Innov. Prosper. 20 (2016) 157–168.
[27] OECD/ISO, International Regulatory Co-operation and International Organisations: The Case of the International Organization for Standardization (ISO), 2016.
[28] E. Ziegel, J. Lamprecht, ISO 9000. Technometrics 35 (2) (1993) 240, https://doi.org/10.2307/1269696.
[29] E. Hadjicostas, ISO 9000 quality management system, Qual. Assur. Anal. Chem. Train. Teach. (2010).
[30] ISO, ISO 9000 Family - Quality Management, 2019. ISO 9000 Fam. - Qual. Manag.
[31] D. Hoyle, J. Thompson, Quality management principles, QMS Convers. A Process Approach (2002), https://doi.org/10.1016/b978-0-08-049219-3.50008-6.
[32] H. Kivisaari, Risk Mitigation in Project Management: Case Horizon 2020, 2020.
[33] A.A. Baptista, F. Charrua-Santos, J.C. Páscoa, Prevention and risk management in RTD European projects developed by university-company's consortiums, Procedia Comput. Sci 64 (2015) 403–408.
[34] V. Valdivieso-Gómez, R. Aguilar-Quesada, Quality management systems for laboratories and external quality assurance programs, in: Qual. Control Lab., 2018.
[35] E. Directorate, C. Group, M. Committee, OECD series on principles of good laboratory practice and compliance monitoring, Ann. Ist. Super Sanita 33 (1997) 1–172.
[36] Cfr - Code of Federal Regulations title 21, Current good manufacturing practice for finished pharmaceuticals, part 211, Curr. Good Manuf. Pract. Finish. Pharm. (2013), https://www.accessdata.fda.gov/scripts/cdrh/cfdocs/cfcfr/CFRSearch.cfm?CFRPart=211.
[37] Quality assurance of pharmaceuticals, WHO Expert Comm. Specif. Pharm. Prep. Thirty-Sixth Report. Geneva. 2 (2007) 17–57.
[38] WHO, Guidelines for good clinical practice (GCP) for trials on pharmaceutical products, WHO Tech. Rep. Ser. (1995) 89–94.
[39] Handbook: Good Laboratory Practice (2nd), TDR pubblication WHO 2009
[40] M. Mendt, S. Kamerkar, H. Sugimoto, K.M. McAndrews, C.C. Wu, M. Gagea, S. Yang, E.V.R. Blanko, Q. Peng, X. Ma, J.R. Marszalek, A. Maitra, C. Yee, K. Rezvani, E. Shpall, V.S. LeBleu, R. Kalluri, Generation and testing of clinical-grade exosomes for pancreatic cancer, JCI Insight 3 (2018), https://doi.org/10.1172/jci.insight.99263.
[41] W. Whitford, P. Guterstam, Exosome manufacturing status, Future Med. Chem. 11 (2019) 1225–1236.
[42] Y.S. Chen, E.Y. Lin, T.W. Chiou, H.J. Harn, Exosomes in clinical trial and their production in compliance with good manufacturing practice, Tzu Chi Med. J. 32 (2020) 113–120.
[43] E. Rohde, K. Pachler, M. Gimona, Manufacturing and characterization of extracellular vesicles from umbilical cord–derived mesenchymal stromal cells for clinical testing, Cytotherapy 21 (2019) 581–592.
[44] ICH, ICH Official Web Site : ICH, Ich.Org., 2008.
[45] G. Andriolo, E. Provasi, V. Lo Cicero, A. Brambilla, S. Soncin, T. Torre, G. Milano, V. Biemmi, G. Vassalli, L. Turchetto, L. Barile, M. Radrizzani, Exosomes from human cardiac progenitor cells for therapeutic applications: development of a GMP-grade manufacturing method, Front. Physiol. 9 (2018), https://doi.org/10.3389/fphys.2018.01169.
[46] T. Lener, M. Gimona, L. Aigner, V. Börger, E. Buzas, G. Camussi, N. Chaput, D. Chatterjee, F.A. Court, H.A. del Portillo, L. O'Driscoll, S. Fais, J.M. Falcon-Perez, U. Felderhoff-Mueser, L. Fraile, Y.S. Gho, A. Görgens, R.C. Gupta, A. Hendrix, D.M. Hermann, A.F. Hill, F. Hochberg, P.A. Horn, D. de Kleijn, L. Kordelas, B.W. Kramer, E.M. Krämer-Albers, S. Laner-Plamberger, S. Laitinen,

T. Leonardi, M.J. Lorenowicz, S.K. Lim, J. Lötvall, C.A. Maguire, A. Marcilla, I. Nazarenko, T. Ochiya, T. Patel, S. Pedersen, G. Pocsfalvi, S. Pluchino, P. Quesenberry, I.G. Reischl, F.J. Rivera, R. Sanzenbacher, K. Schallmoser, I. Slaper-Cortenbach, D. Strunk, T. Tonn, P. Vader, B.W.M. van Balkom, M. Wauben, S. El Andaloussi, C. Théry, E. Rohde, B. Giebel, Applying extracellular vesicles based therapeutics in clinical trials - an ISEV position paper, J. Extracell. Vesicles 4 (2015), https://doi.org/10.3402/jev.v4.30087.

[47] S.G. Volsen, J.M. Kent, M. Masson, Quality: an old solution to new discovery dilemmas? Drug Discov. Today 9 (2004) 903–905.

[48] D.H. Riedl, M.K. Dunn, Quality assurance mechanisms for the unregulated research environment, Trends Biotechnol. 31 (2013) 552–554.

[49] R. Davies, Good Research Practice: It Is Time to Do what Others Think We Do, Quasar, 2013, pp. 21–23.

[50] S.G. Volsen, M.M. Masson, A novel audit model for assessing quality in non-regulated research, Qual. Assur. J. 12 (2) (2009) 57–63.

[51] Defra, Department for Environment, Food and Rural Affairs Joint Code of Practice for Research, JCoPR, 2015, pp. 1–8.

[52] Research Quality Association, Quality Systems Workbook, 2013, p. 44.

[53] RQA Working Party on Quality in Non-Regulated Research. Guidelines for Quality in Non-Regulated Scientific Research Booklet, RQA, 2008.

[54] WHO Handbook: Quality Practices in Basic Biomedical Research, WHO, 2013.

[55] C.G. Begley, A.M. Buchan, U. Dirnagl, Robust research: institutions must do their part for reproducibility, Nature 525 (2015) 25–27.

[56] F.S. Collins, L.A. Tabak, NIH plans to enhance reproducibility, Nature 505 (2014) 612–613.

[57] M. Poli, D. Cornolti, S. Pardini, G. Iervasi, How and why to implement a performance management system in public research institutions: the approach and the experience of a large multidisciplinary Italian centre, Int. J. Qual. Res. 12 (2018) 757–772.

[58] MRC Ethics Series Good Research Practice: Principles and Guidelines MRC Ethics Series Good Research Practice, 2012.

[59] O. Meagan Littrell, C. Stoeger, H. Maier, H. Fuchs, M. Hrabě de Angelis, L.A. Cassis, G.A. Gerhardt, R. Grondin, V. Gailus-Durner, Costs of implementing quality in research practice, Handb. Exp. Pharmacol. 257 (2019) 399–423.

[60] M. Poli, D. Petroni, S. Pardini, P.A. Salvadori, L. Menichetti, Implementation of a quality assurance system according to GMP and ISO 9001:2008 standard for radiopharmaceutical production in a public research centre, Accred Qual. Assur. 17 (2012) 341–348.

[61] S. Mancinelli, V. Zazzu, A. Turcato, G. Lacerra, F.A. Digilio, A. Mascia, M. Di Carlo, A.M. Cirafici, A. Bongiovanni, G. Colotti, A. Kisslinger, A. Lanati, G.L. Liguori, Applying design of experiments methodology to PEI toxicity assay on neural progenitor cells, Math. Model. Biol. (2015) 45–63.

[62] M. Yáñez-Mó, P.R.M. Siljander, Z. Andreu, A.B. Zavec, F.E. Borràs, E.I. Buzas, K. Buzas, E. Casal, F. Cappello, J. Carvalho, E. Colás, A. Cordeiro-Da Silva, S. Fais, J.M. Falcon-Perez, I.M. Ghobrial, B. Giebel, M. Gimona, M. Graner, I. Gursel, M. Gursel, N.H.H. Heegaard, A. Hendrix, P. Kierulf, K. Kokubun, M. Kosanovic, V. Kralj-Iglic, E.M. Krämer-Albers, S. Laitinen, C. Lässer, T. Lener, E. Ligeti, A. Line, G. Lipps, A. Llorente, J. Lötvall, M. Manček-Keber, A. Marcilla, M. Mittelbrunn, I. Nazarenko, E.N.M. Nolte-'t Hoen, T.A. Nyman, L. O'Driscoll, M. Olivan, C. Oliveira, É. Pállinger, H.A. Del Portillo, J. Reventós, M. Rigau, E. Rohde, M. Sammar, F. Sánchez-Madrid, N. Santarém, K. Schallmoser, M.S. Ostenfeld, W. Stoorvogel, R. Stukelj, S.G. Van Der Grein, M. Helena Vasconcelos, M.H.M. Wauben, O. De Wever, Biological properties of extracellular

vesicles and their physiological functions, J. Extracell. Vesicles (2015), https://doi.org/10.3402/jev.v4.27066.
[63] A. Becker, B.K. Thakur, J.M. Weiss, H.S. Kim, H. Peinado, D. Lyden, Extracellular vesicles in cancer: cell-to-cell mediators of metastasis, Canc. Cell 30 (2016) 836−848.
[64] J. Howitt, A.F. Hill, Exosomes in the pathology of neurodegenerative diseases, J. Biol. Chem. 52 (2016) 26589−26597.
[65] R. Xu, A. Rai, M. Chen, W. Suwakulsiri, D.W. Greening, R.J. Simpson, Extracellular vesicles in cancer — implications for future improvements in cancer care, Nat. Rev. Clin. Oncol. (2018), https://doi.org/10.1038/s4157.
[66] R. Kalluri, V.S. LeBleu, The biology, function, and biomedical applications of exosomes, Science 367 (2020), https://doi.org/10.1126/science.aau6977.
[67] V. Budnik, C. Ruiz-Cañada, F. Wendler, Extracellular vesicles round off communication in the nervous system, Nat. Rev. Neurosci. 17 (3) (2016) 160−172.
[68] S.V. Chitti, P. Fonseka, S. Mathivanan, Emerging role of extracellular vesicles in mediating cancer cachexia, Biochem. Soc. Trans. 46 (5) (2018) 1129−1136.
[69] C. Quek, A.F. Hill, The role of extracellular vesicles in neurodegenerative diseases, Biochem. Biophys. Res. Commun. 483 (4) (2017) 1178−1186.
[70] M. Colombo, G. Raposo, C. Théry, Biogenesis, secretion, and intercellular interactions of exosomes and other extracellular vesicles, Annu. Rev. Cell Dev. Biol. 30 (2014) 255−289.
[71] J. Meldolesi, Exosomes and ectosomes in intercellular communication, Curr. Biol. 28 (2018), https://doi.org/10.1016/j.cub.2018.01.059.
[72] M.W. Tibbitt, J.E. Dahlman, R. Langer, Emerging frontiers in drug delivery, J. Am. Chem. Soc. 138 (3) (2016) 704−717.
[73] D. Xitong, Z. Xiaorong, Targeted therapeutic delivery using engineered exosomes and its applications in cardiovascular diseases, Gene 575 (2016) 377−384.
[74] S. Manandhar, V.K. Kothandan, J. Oh, S.H. Yoo, J. Hwang, S.R. Hwang, A pharmaceutical investigation into exosomes, J. Pharm. Investig. 43 (2018) 32−45.
[75] C. Paganini, U. Capasso Palmiero, G. Pocsfalvi, N. Touzet, A. Bongiovanni, P. Arosio, Scalable production and isolation of extracellular vesicles: available sources and lessons from current industrial bioprocesses, Biotechnol. J. 14 (2019), https://doi.org/10.1002/biot.201800528.
[76] S.J. Gould, G. Raposo, As we wait: coping with an imperfect nomenclature for extracellular vesicles, J. Extracell. Vesicles 2 (2013) 1−3.
[77] K.W. Witwer, C. Théry, Extracellular vesicles or exosomes? On primacy, precision, and popularity influencing a choice of nomenclature, J. Extracell. Vesicles 8 (2019) 1−7.
[78] K.W. Witwer, C. Soekmadji, A.F. Hill, M.H. Wauben, E.I. Buzás, D. Di Vizio, J.M. Falcon-Perez, C. Gardiner, F. Hochberg, I.V. Kurochkin, J. Lötvall, S. Mathivanan, R. Nieuwland, S. Sahoo, H. Tahara, A.C. Torrecilhas, A.M. Weaver, H. Yin, L. Zheng, Y.S. Gho, P. Quesenberry, C. Théry, Updating the MISEV minimal requirements for extracellular vesicle studies: building bridges to reproducibility, J. Extracell. Vesicles 6 (1) (2017), https://doi.org/10.1080/20013078.2017.1396823.
[79] R. Lacroix, C. Judicone, M. Mooberry, M. Boucekine, N.S. Key, F. Dignat-George, A. Ambrozic, N. Bailly, C. Buffat, E. Buzas, A. Charpentier, B. Chatelain, J.M.D. Dogne, A. Falanga, E. Garcia Arias-Salgado, B. Gyorgy, M. Harrison, H. Kwaan, K. Lakota, V. Latger-Cannard, F. Massin, F. Mobarrez, F. Mojca, F. Mullier, J.P. Phelan, N. Snoj, S. Sodin-Semrl, E. Stezpień, S. Susen, A.D. Terrisse, V. Tintillier, A. Vignoli, Standardization of pre-analytical variables in plasma microparticle determination: results of the international society on Thrombosis

and haemostasis SSC collaborative workshop, J. Thromb. Haemostasis 11 (2013) 1190–1193.

[80] H. Kalra, R.J. Simpson, H. Ji, E. Aikawa, P. Altevogt, P. Askenase, V.C. Bond, F.E. Borràs, X. Breakefield, V. Budnik, E. Buzas, G. Camussi, A. Clayton, E. Cocucci, J.M. Falcon-Perez, S. Gabrielsson, Y.S. Gho, D. Gupta, H.C. Harsha, A. Hendrix, A.F. Hill, J.M. Inal, G. Jenster, E.M. Krämer-Albers, S.K. Lim, A. Llorente, J. Lötvall, A. Marcilla, L. Mincheva-Nilsson, I. Nazarenko, R. Nieuwland, E.N.M. Nolte-'t Hoen, A. Pandey, T. Patel, M.G. Piper, S. Pluchino, T.S.K. Prasad, L. Rajendran, G. Raposo, M. Record, G.E. Reid, F. Sánchez-Madrid, R.M. Schiffelers, P. Siljander, A. Stensballe, W. Stoorvogel, D. Taylor, C. Thery, H. Valadi, B.W.M. van Balkom, J. Vázquez, M. Vidal, M.H.M. Wauben, M. Yáñez-Mó, M. Zoeller, S. Mathivanan, Vesiclepedia: a compendium for extracellular vesicles with continuous community annotation, PLoS Biol. 10 (2012), https://doi.org/10.1371/journal.pbio.1001450.

[81] D.K. Kim, B. Kang, O.Y. Kim, D.S. Choi, J. Lee, S.R. Kim, G. Go, Y.J. Yoon, J.H. Kim, S.C. Jang, K.S. Park, E.J. Choi, K.P. Kim, D.M. Desiderio, Y.K. Kim, J. Lötvall, D. Hwang, Y.S. Gho, EVpedia: an integrated database of high-throughput data for systemic analyses of extracellular vesicles, J. Extracell. Vesicles (2013), https://doi.org/10.3402/jev.v2i0.20384.

[82] S.L. Subramanian, R.R. Kitchen, R. Alexander, B.S. Carter, K.H. Cheung, L.C. Laurent, A. Pico, L.R. Roberts, M.E. Roth, J.S. Rozowsky, A.I. Su, M.B. Gerstein, A. Milosavljevic, Integration of extracellular RNA profiling data using metadata, biomedical ontologies and Linked Data technologies, J. Extracell. Vesicles 4 (2015) 1–6.

[83] J. Van Deun, P. Mestdagh, P. Agostinis, Ö. Akay, S. Anand, J. Anckaert, Z.A. Martinez, T. Baetens, E. Beghein, L. Bertier, G. Berx, J. Boere, S. Boukouris, M. Bremer, D. Buschmann, J.B. Byrd, C. Casert, L. Cheng, A. Cmoch, D. Daveloose, E. De Smedt, S. Demirsoy, V. Depoorter, B. Dhondt, T.A.P. Driedonks, A. Dudek, A. Elsharawy, I. Floris, A.D. Foers, K. Gärtner, A.D. Garg, E. Geeurickx, J. Gettemans, F. Ghazavi, B. Giebel, T.G. Kormelink, G. Hancock, H. Helsmoortel, A.F. Hill, V. Hyenne, H. Kalra, D. Kim, J. Kowal, S. Kraemer, P. Leidinger, C. Leonelli, Y. Liang, L. Lippens, S. Liu, A. Lo Cicero, S. Martin, S. Mathivanan, P. Mathiyalagan, T. Matusek, G. Milani, M. Monguió-Tortajada, L.M. Mus, D.C. Muth, A. Németh, E.N.M. Nolte-'T Hoen, L. O'Driscoll, R. Palmulli, M.W. Pfaffl, B. Primdal-Bengtson, E. Romano, Q. Rousseau, S. Sahoo, N. Sampaio, M. Samuel, B. Scicluna, B. Soen, A. Steels, J.V. Swinnen, M. Takatalo, S. Thaminy, C. Théry, J. Tulkens, I. Van Audenhove, S. Van Der Grein, A. Van Goethem, M.J. Van Herwijnen, G. Van Niel, N. Van Roy, A.R. Van Vliet, N. Vandamme, S. Vanhauwaert, G. Vergauwen, F. Verweij, A. Wallaert, M. Wauben, K.W. Witwer, M.I. Zonneveld, O. De Wever, J. Vandesompele, A. Hendrix, EV-TRACK: transparent reporting and centralizing knowledge in extracellular vesicle research, Nat. Methods 14 (2017) 228–232.

[84] R. Hattemer-Apostel, Standard operating procedures - a novel perspective, Qual. Assur. J. 5 (2001) 207–219.

[85] J. Gough, M. Hamrell, Standard operating procedures (SOPs): how to write them to Be effective tools, Ther. Innov. Regul. Sci. 44 (2010) 463–468.

[86] A. Isak, Standard Operating Procedure (What Are They Good for), Intech. i, 2012.

[87] A. Blackmore, The myth of the SOP, Qual. Assur. J. 8 (2004) 3–12.

[88] J.G. Voelkel, K. Ishikawa, Guide to Quality Control, Technometrics (1989), 10.2307/1268827.

CHAPTER SIX

# Evolution of mechanical stability from lipid layers to complex bacterial envelope structures

**Žiga Pandur[a,b] and David Stopar[a,*]**
[a]University of Ljubljana, Biotechnical Faculty, Department of Food Science and Technology, Laboratory of Microbiology, Ljubljana, Slovenia
[b]University of Ljubljana, Faculty of Mechanical Engineering, Ljubljana, Slovenia
*Corresponding author: E-mail: david.stopar@bf.uni-lj.si

## Contents



*Advances in Biomembranes and Lipid Self-Assembly,* Volume 33
ISSN 2451-9634
https://doi.org/10.1016/bs.abl.2020.09.005

## Abstract

To survive in environment, bacteria must constantly maintain structural integrity of their envelope. All bacteria are separated from a surrounding media with a membrane, however only a few have just a membrane envelope layer. Most bacteria have evolved many additional layers to endure harsh conditions in the environment. The focus of this chapter is on different bacterial envelope structures, important for cell mechanical stability. We are not giving a chronological timeline leading to the present-day bacterial envelope structures, as these are largely unknown, but provide mechanistic explanations for the observed molecular assemblies in a given environment. Structural integrity of lipid monolayers and bilayers are reviewed, the reasons are given why they fail mechanically and why additional surface layers are needed. Though peptidoglycan is generally believed to be the sole structural element providing rigidity to the bacterial envelope, some recent findings imply that shear load can be equally distributed and shared between different layers. In this emerging new picture peptidoglycan remains an essential pillar in cell mechanical system but only as a part of a multi-composite cell barrier. The role of the outer membrane, S-layer, capsule, sheath and slime in cell mechanical stability is also discussed. Although the main emphasis is on canonical Gram-negative and Gram-positive bacteria, some fine examples of bacterial envelope adaptations to extreme environments are presented.

## 1. Introduction

Bacteria have no permanent cytoskeletal elements to support their structure and relay on cell envelope to maintain their shape. All bacterial cells are separated from a surrounding media with a membrane envelope but only a few have solely a membrane layer to provide structure, stability, and selectivity. Many additional envelope layers have evolved which help bacteria to maintain their integrity [1]. Cell envelope provides the intersection between two very different environments; the inner cellular lumen which is a highly regulated homeostatic biochemical system, and an unpredictable outer environment with large fluctuations of chemical and physical parameters [2]. Single cell organisms determine the ultimate physical and chemical boundaries of life on the planet. They can inhabit very low temperature environments down to −15 °C or grow at +121 °C, they can grow at very low or very high pH, low vacuum or pressures up to 1100 bar, they can thrive in hypoosmotic or hyperosmotic medium, at low and high redox potential, high electromagnetic radiation density as well as in radioactive environments [3–11].

Maintaining functional barrier in so diverse environments has resulted in different designs of bacterial envelope, from the most basic resembling a bag of enzymes protected by cytoplasmic membrane in some obligate intracellular bacterial parasites to the most sophisticated multilayer and multi-composite barrier designs found in endospores. Single cell organisms are found in all three domains of life: bacteria, archea and eukarya. They share similar basic cell plan with cytoplasm enveloped by cell membrane and cell wall. The details, however, differ immensely. In this chapter we will not discuss fungal membrane and semi striated cell wall layers of chitin, β-glucans, mannoproteins, and melanins [12]. We will also not discuss structure, composition and chemical bonding in archeal membranes and pseudopeptidoglycan that differs from bacterial peptidoglycan in chemical structure but resembles it in function and physical structure [13]. We will focus on bacterial structures important for cell mechanical stability that have evolved in different bacteria. We will not give a chronological timeline leading to the present-day envelope structures, as these are largely unknown, but provide where possible, mechanistic explanations for the observed bacterial barrier structures.

The literature on bacterial membrane and cell wall structures is very rich and covers in great details chemistry, biochemistry, genetics, and biophysics of different components [14–19]. Bacterial species display different membrane compositions and even the membrane composition of cells belonging to a single species is not constant but depends on the environment and growth condition to which the cells are exposed [20–22]. Bacterial membranes present a large diversity of amphiphilic lipids, including the common phospholipids phosphatidylglycerol (PG), phosphatidylethanolamine (PE) and cardiolipin, the less frequent phospholipids phosphatidylcholine (PC), and phosphatidylinositol (PI) and a variety of other membrane lipids, such as ornithine lipids, glycolipids, sphingolipids or hopanoids [23]. There are excellent reviews that give an overview about the bacterial membrane lipid structures, different metabolic pathways involved in their formation, and the distribution of membrane lipids and metabolic pathways across taxonomical groups and will not be discussed further [23–26]. Similar for peptidoglycan, there is a high diversity in the composition, sequence of the peptides and cross linkers that varies significantly from species to species and with the growth conditions [27]. We understand well the biogenesis of bacterial S-layers, a proteinaceous surface layer, which is formed by the self-assembly of monomeric proteins into a regularly spaced, two-dimensional array on the surface of bacteria [28]. Comprehensive description of structure, biosynthesis, transport, and assembly of bacterial capsular polysaccharides is known for some bacteria [29]. Although biosynthesis and assembly of structures embedded in cell wall and membrane such as pili, fimbria and flagella are vital for motility and adhesion, they will not be further discussed [30–34].

The main focus of this chapter will be on different surface layers and their contribution to the mechanical stability of bacteria in different environments. There are many known functions of bacterial envelope layers that include but are not limited to enzyme display, transport, motility, adsorption, biofilm formation, interactions with the host, or regulation of the immune system response [35–37]. However, it is often neglected that all surface layers also contribute to the maintenance of cell mechanical integrity. We will begin by reviewing structural integrity of lipid monolayers and bilayers and show how they fail mechanically and why additional surface layers are needed in bacteria. We will next describe the mechanical properties of peptidoglycan which is generally believed to be the sole structure providing rigidity to the bacterial envelope. However, some recent findings imply that shear load can be equally distributed and shared between different layers [38,39]. In this emerging new picture peptidoglycan remains an essential pillar in cell mechanical system but only as a part of a multi-composite cell barrier that needs to be protected as it is highly vulnerable to a variety of chemical attacks. Much like steel reinforced concrete, which provides tensile strength to the construction, but fails on corrosion. We will discuss the role of the outer membrane, S-layer, capsule, sheath and slime in cell mechanical stability. Although the main emphasis will be on canonical Gram-negative and Gram-positive bacteria, we will give some fine examples of surface layer adaptations to extreme environments.

# 2. Lipid monolayers

## 2.1 Lipid monolayer polymorphism

Lipid monolayers, monomolecular films at polar-apolar interface which reduce surface tension, are elemental building blocks of cell membranes and are used extensively as model systems to study adsorption, surface activity, wetting, ordering, lipid phase transitions, and membrane protein lipid interactions [40–44]. The most obvious geometric feature of lipid monolayers is spontaneous curvature, which is directly related to the lipid molecular geometry [45]. The common bacterial phospholipids such as PE, phosphatic acid, diacylglycerol or cardiolipin, have small polar headgroup to acyl chain ratio which makes them roughly conical in shape and impose a negative monolayer curvature. Monolayer with these lipids bends in such a way that the headgroups come closer together. Less frequently found PC and phosphatidylserine are cylindrical lipids that form a flat monolayer. Conversely, lipids with a large headgroup to acyl chain ratio, such as lysophosphatidylcholine or the large headgroups in PI phosphates confers an inverted conical shape to the lipids, thereby favoring the bending of the membrane into a positive curvature, bending the monolayer away from the headgroups (reviewed in Ref. [46–50]). Polymorphism in lipid monolayers is given in Fig. 1.

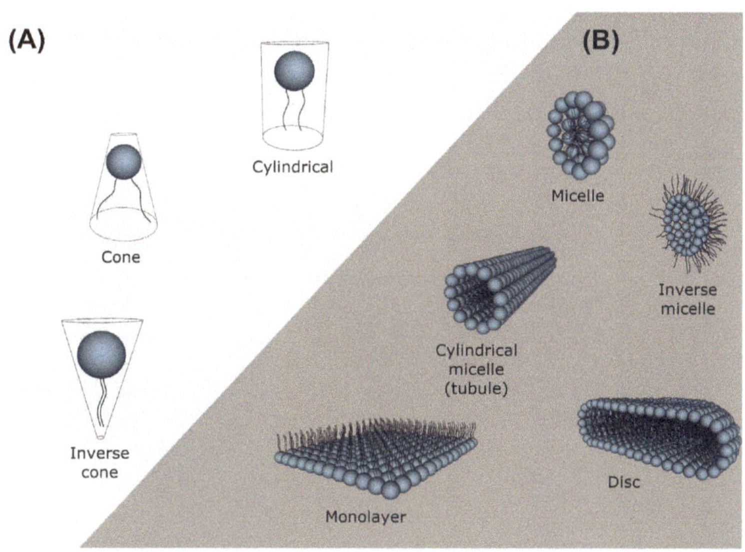


**Fig. 1** Different phospholipid molecular geometries: cylindrical, cone and inverse cone (A); different monolayer aggregates of phospholipid molecules: micelle, inverse micelle, cylindrical micelle (tubule), disc and monolayer (B). *Schematic drawings are adapted from Koynova et al. [51] and Patel et al. [52].*

## 2.2 Biological membrane is not a monolayer

Lipid monolayers provide a hydrophobic barrier and a good solvent for membrane proteins; however, they do not reflect the structure of biological membrane, which is composed of two monolayers. No known bacteria have their membrane composed of monolayers only. Lipid monolayer bending elasticity is not sufficient to damp the thermal fluctuations. In particular, a loose dynamic monolayer, such as an oil-water interface, has pronounced wrinkles associated with thermal fluctuations and may easily collapse [53]. The surface tension of lipid monolayers decreases with lipid density. At a certain very high surface density, further reduction of the surface tension is not possible, thus the monolayer become unstable at the interface and collapses. Calculations indicate that the energy to deform a monolayer by area compression is significantly greater than that needed to deform it by bending or chain tilting [54]. The collapse of lipid monolayers is dependent on temperature, lipid composition and may be reversible or irreversible [40]. Wrinkling or buckling of monolayer leads to the formation of different 3D lipid aggregates (e.g., bilayer folds, vesicles, tubes, micelles) [55]. The irreversible collapse of lipid monolayer to hexagonal phase or micelles would be detrimental to cell viability.

# 3. Lipid bilayers

Lipid bilayer is the single most important membrane structure found in bacteria (Fig. 2). Since lipid bilayers provide a universal biological membrane matrix, it is expected that many of the mechanical deficiencies of lipid

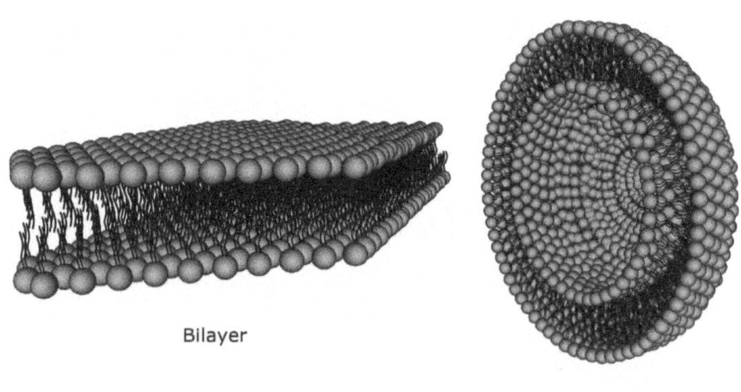


**Fig. 2** Models of phospholipid bilayer and vesicle. *Schematic drawings are adapted from Bozzuto et al. [57] and Khan et al. [56].*

monolayers are remedied in lipid bilayers. Two factors primarily govern whether a lipid will form a bilayer or not: solubility and shape. For a self-assembled structures such as a bilayers to form, the lipid should have a low solubility in water, which can also be described as a low critical micelle concentration [58,59]. Above the critical micelle concentration molecules will aggregate and form larger structures such as bilayers, micelles or inverted micelles (Fig. 1 and Fig. 2) [60]. Propensity for lipid bilayer depends on shape of the lipid [49,50]. If a particular lipid has too large a deviation from zero intrinsic curvature it will not form a lipid bilayer [61].

## 3.1 Lipid bilayer mechanical properties

The critical mechanical properties of fluid lipid bilayers are bilayer area compressibility/expansion modulus and bilayer bending constant. They determine the bilayer ability to compress, expand, or bend, and its propensity to curve [62]. Lipid bilayer is elastic by nature, giving it the ability to compress and expand and to protect the structural integrity of the cell even in times of mechanical deformation. Compressibility modulus quantifies the response of membrane area to tension, which under physiological conditions may arise from various perturbations, such as changes in osmolarity across the membrane or the addition of lipids or other molecules to one of the membrane's leaflets [63]. Surface flow properties of lipids are directly related to the structural rigidity of membranes [64]. For fluid bilayers the free rearrangement of molecules within the plane means that the structure does not support shear stress, and the shear modulus is by definition zero [65]. Unlike a bulk material, where the resistance to expansion comes from the intermolecular bonds, the resistance to expansion in a bilayer is a result of the extra hydrophobic area exposed to water upon pulling the lipids apart [66]. To a good first approximation the expansion modulus is $2\gamma$, where gamma ($\gamma$) is the surface tension of the water-lipid interface. Typically, the expansion modulus is 80–200 mN/m. The values vary strongly with solution conditions [67], but only weakly with tail length and lipid unsaturation [68]. On the other hand, the size of the hydrophilic head group will strongly influence the packing density [69]. A related term to expansion energy is the edge energy, the energy per unit length of a free edge contacting water. This can be thought of as the work needed to create a hole in the bilayer of a unit length. Usually, lipid bilayers begin to rupture at a relative surface expansion of the 2%–4%, which corresponds to an applied tension of 1–25 mN/m [70]. The edge energy is of significant importance in biology as it contributes to the auto-assembly and auto-healing of lipid bilayer structures [71]. The edge energy enables compartmentalization and resealing after sonoporation or electroporation. Due to the pronounced hydrophobic nature of lipid bilayers spontaneous pore formation is significantly suppressed which enables lipid bilayers to act as a barrier in biological membranes [72]. The mechanical properties of lipid bilayers are further dependent on lipid distribution and composition [68,73–75]. Adding new bilayers (making multilamellar structures) improves mechanical stability of membrane systems in proportion to the degree of lamellarity [76]. However, multilamellar membrane structures are usually not found in bacteria. Most likely due to the transport problems and decreased permeability.

## 3.2 Lipid bilayer bending elastic modulus

Arguably the most important lipid bilayer mechanical property is its bending elastic modulus. Lipid vesicles, for example, are strongly curved. Unlike compressive strain, it is very easy to sustain curvature strains (torques) in a lipid bilayer because of the smallness of its bending elastic modulus, which is typically a few $k_BT$ up to 20 $k_BT$ [77–79]. Nevertheless, compared to lipid monolayer the elastic bending modulus of the bilayer is doubled [80]. This provides much more stability for the biological membrane. The bending elastic constant of lipid bilayer determines membrane flexibility and a variety of membrane-associated physical and functional properties as reviewed by Marsh [80]. It can control membrane vesicle shape, affect the structure and formation of interlamellar attachments and non-lamellar lipid phases, play an important role in membrane fusion processes, determine the scale of thermally induced elastic bending fluctuations, modulate membrane-membrane interactions, rescales the lateral compressibility that might have functional consequences for membrane insertion of proteins [80]. Together with the spontaneous curvature the bending elastic constants can modulate or control the activity of membrane-associated enzymes and ion channels [81].

Lipid bilayer bending is dependent on its composition. Changes in membrane thickness can affect the bending rigidity of the lipid membrane [82]. Bending rigidity increases with the square of the bilayer thickness. In addition, bending rigidity is dependent on lipid head group charge. For instance, negatively charged dioleoylphosphatidylglycerol (DOPG) liposomes have tendency to become spherical compared to liposomes composed only of neutral lipids [83,84]. The repulsive forces between negative DOPG maximizes their distance and therefore stabilize spherical morphology of lipid bilayer and require larger forces to deform [69]. For example, vesicles composed of an *Escherichia coli* lipid extract are ~50% less stiff than those containing only DOPG [85]. Similarly, lipids composed of only zwitterionic dimyristoylphosphatidylcholine require approximately half the force to deform compared to DOPG [69]. In *Staphylococcus aureus* and *Caulobacter crescentus* the fraction of DOPG can be as high as 80% and the presence of surface charge stiffens the membrane because the repulsion between the charged lipids effectively resists membrane bending [86]. On the other hand, several peripheral and integral membrane proteins can increase bending and therefore reduce membrane stiffness [87] causing membrane invaginations and evaginations which results in the formation of tubules, vesicles or stacked membrane arrays [88].

## 3.3 Lipid bilayer phase transition

Lipid bilayer mechanical properties are strongly dependent on thermodynamics. The motion of individual lipid molecules at various temperatures is different, which results in different packing of acyl chains and hydrogen bonding between head groups, yielding different lipid phases [89]. Upon heating, lipids undergo a gel-to-liquid crystalline phase transition, which is accompanied by a large entropic change [90]. Changes in the internal energy of lipids with temperature are related to both lipid volume and area changes [91]. For instance, the average volume of a dipalmitoylphosphatidylcholine lipid bilayer changes by about 4% upon melting with the simultaneous lipid area increase of 25% [92,93]. Close to the chain melting transition of lipid bilayers, volume and enthalpy fluctuations increase whereby compressibility display pronounced maxima whereas the sound velocity has its minima [91]. It has been reported that at the transition temperature the bending rigidity significantly drops, bilayer gets thinner and the permeability for water increases [94,95]. In multicomponent lipid bilayers, the presence of saturated and unsaturated lipids with distinct transition temperatures complicates phase behavior. It is interesting to note that bacteria adapt their membrane composition so that they always grow at temperatures above the phase transition of their lipids. For example, if *E. coli* grows at 37 °C it has membrane lipid phase transition at 21 °C, whereas if it grows at 15 °C it has membrane lipid phase transition at 10 °C [96]. It is obvious that bacteria prefer liquid over gel phase membrane.

## 3.4 Failure of lipid bilayer stability

Although relatively stable lipid bilayers can be disrupted by a variety of physico-chemical stressors. The major challenges to lipid vesicle stability are: mechanical stress [97], oscillating electric field [98,99], static pressure [100], shock waves [101]; hydrolysis [102,103] and chemical assaults by different charged and surface active agents. Vesicles can become transiently porous and leak with no substantial volume change during the process of electroporation [104]. On a longer timescale (e.g. during shear stress) pores may grow to a critical size that leads to lipid bilayer disintegration into several fragments that subsequently reassemble into smaller vesicles. Similar in case of large elongations, the lipid vesicle may deform and fragment to smaller vesicles [97]. The integrity of lipid bilayer may be compromised by lipid hydrolysis. The rate of lipid hydrolysis is dependent on pH [105,106], temperature [105], buffer species [69,107], and ionic strength [108–111]. The ultimate disruption of lipid vesicles is solubilization with surface active molecules (e.g. detergents like Triton X-100). Amphiphilic compounds increase permeability of lipid vesicles by intercalating into the lipid bilayer, which at concentrations higher than critical micellar concentrations, leads to disruption and solubilization of the lipid bilayer [112,113].

Although there are many ways how lipid bilayers can be compromised, from biological point of view, lipid bilayer vulnerability to solubilization by surface-active compounds and failure to sustain osmotic pressure induced expansion are the leading causes of bacterial cell deaths. To curb this, bacteria need to reinforce the lipid bilayer structure or use additional envelope layers.

## 4. Cytoplasmic membrane layer

Simple homogenous membrane systems such as lipid monolayers or bilayers describe the fundamental physico-chemical properties of biological membranes but fail to describe the complexity of cytoplasmic membrane structures and functions. This is mainly due to a large number of lipid-lipid and lipid-protein interactions found in bacteria. For instance, *E. coli* has more than 100 different head group-acyl chain phospholipid combinations [114]. Such mixtures often exhibit properties intermediate to their components but are also capable of nonlinear phenomena which are not seen in single component systems such as phase separations [115,116]. If some of the lipid components are liquid at a given temperature while others are in the gel phase, the two phases may coexist in spatially separated populations [117]. The phase separation plays a critical role for proteins as they partition into one or the other phase and become locally concentrated or activated.

For example, in *Streptococcus pneumonia* the coexistence of liquid and gel phases minimize bending stress in the ellipsoid geometry defined by the cell wall. Gel phase lipids with high bending rigidity are spontaneously organized at the equator where curvature is minimal, thus marking the future division site, whereas liquid phase membranes are positioned onto the oblong hemispheres [118].

## 4.1 Material properties of cytoplasmic membranes

Cytoplasmic membrane material properties such as membrane stretching elasticity, its mean curvature, elastic modulus, edge energy, and flexoelectricity are different from simple bilayer model systems [16,79,119]. Biological membranes also have a degree of trans-bilayer asymmetry which increases mechanical stability compared to symmetric bilayers [120]. In sharp contrast to lipid bilayers, cytoplasmic membranes actively maintain, sense, and generate local membrane curvature. This active process is mediated and controlled by specialized proteins that can generate force by polymerization or molecular motors [121]. Maintaining membrane curvature is important because cell has control over processes like membrane scission, septal ring formation, nanotube formation, membrane invaginations, budding, trafficking, fusion, microdomain formation as well as local protein concentration and enzyme activation on membranes [121–130]. For example, four distinct types of membrane topologies can be found in rod-shaped bacteria: the cylindrical shape with zero Gaussian curvature; the curved tip of the septum with negative Gaussian curvature; the concave membrane shape found at the inner surface of the cell poles with positive Gaussian curvature; and the convex membranes found at the outer surface of the cell poles and endospores [131]. Related to membrane curvature is membrane flexoelectricity, a mechanoelectric phenomenon known from liquid crystal physics. In cytoplasmic membrane flexoelectricity manifests itself as the development of polarization upon changes in curvature [132]. Bioflexoelectricity enables membrane structures to function like soft micro and nano machines, sensors and actuators, thus providing important inputs to vital cellular phenomena such as membrane contact, electromotility, and mechanosensitivity [133].

## 4.2 Membrane failure energy

All the above functions of cytoplasmic membrane are possible only if membrane cohesion force is not exceeded. When cell membrane expands it can reach critical size and ruptures. In hypoosmotic environments water diffuses into the cell in an attempt to equalize chemical potential inside and outside of the cell resulting in cell volume increase. If critical elastic area compressibility modulus and membrane tensile strength are exceeded the failure energy is reached and cell membrane yields. Membrane failure energy can be increased significantly by cholesterol inclusion [134]. The cholesterol modulates the biological membrane in multiple ways. It changes the fluidity, thickness, compressibility, water penetration and intrinsic curvature of lipid bilayers [135]. The inclusion of cholesterol in lipid bilayers is important factor in increasing lipid bilayer cohesion, but only for lipids where both acyl chains are saturated, or mono or di-unsaturated. Polyunsaturated fatty acids induce greater fluidity because of their higher degree of conformational flexibility [136]. Multiple unsaturation in both lipid chains inhibits the condensing effect of cholesterol in bilayers. The failure energy in sphingomyelin/cholesterol mixture at approximately 40 mol% cholesterol could reach up to 173 mN/m, which is approximately 30-fold larger compared to diarachidonylphosphatidylcholine bilayers [134].

In bacterial membranes, however, cholesterol is absent. Thus, this kind of membrane protection is not an option. Several bacteria have hopanoids as minor membrane components [137]. It has been proposed that hopanoids may have similar function for modulation of molecular order of membranes as sterols in eukaryotic cells [138]. Although inclusion of hopanoids in bacterial membrane may protect bacterial cytoplasmic membrane from excessive expansion, it is not a general mechanism protecting bacteria from hypoosmotic shock. In contrast to eukaryotic cytoplasmic membrane, where membrane is fortified by physical connections to cytoskeleton, bacteria also have only limited temporal connections to the cytoskeleton (i.e. to FtsZ ring during the cell division) [139,140]. Therefore, in order to preserve membrane integrity a completely new design is needed in bacteria to protect cytoplasmic membrane from the adverse environmental effects.

## 5. Peptidoglycan layer

Most of bacteria rely on peptidoglycan in hypoosmotic environments to protect their cytoplasmic membrane [141]. In the bacterial world, one of the most important forces changing cell size and shape is water. Bacterial cytoplasm is a highly salty environment suspended in a less salty surrounding. As bacteria do not have active water transporters, they rely on osmosis for water transport. During water influx cell volume increases, thereby stretching lipid bilayer until it eventually bursts. Water flux response may occur within milliseconds of osmolarity change [142]. The role of peptidoglycan is to act as a barrier that physically blocks cell to take on too much water. The situation is similar to an inflation of a balloon inside a small box. Once expanding balloon hits the wall it pushes back on the balloon and no more air can be pressurized into the balloon. In contrast to the rigid box, peptidoglycan can reversibly expand providing a fitting fabric for the cell volume under osmotic pressure.

## 5.1 Matching the turgor pressure

In hypoosmotic environments inflow of the water increases the hydrodynamic (turgor) pressure inside the cell which results in cytoplasmic swelling. Force is required to counterbalance turgor pressure. The greater the difference between the osmolyte concentration outside and inside of the cell, the more force is necessary to withstand the pressure. Turgor pressure is defined by the ideal gas law-like, $P = RT\ (C_{int} - C_{ext})$, where R is the gas constant, T is the temperature, and $C_{int/ext}$ are the internal/external osmolarity, respectively. The turgor pressure in *E. coli* cells can vary in the range from 0.3 to 5 atm. Usually, it is higher in Gram-positive compared to Gram-negative bacteria [38,143]. To protect cell membrane from bursting during hypoosmotic stress bacteria synthesize peptidoglycan with a strength that matches the turgor pressure. For example, in low turgor environment a single layer of peptidoglycan can be enough to control the turgor pressure in *E. coli* [144]. On the other hand, much higher turgor pressure in some Gram-positive bacteria require substantially thicker peptidoglycan (Fig. 3) [27]. The turgor pressure is one of the most important growth parameters for bacteria. For example, during hypoosmotic shock in *B. subtilis* cytoplasmic membrane expands and electrically depolarize. This slows the motion of energy driven bacterial actin homolog MreB, which signals the rate and location of cell-wall synthesis [146]. Consequently, growth arrests, which is a safety measure preventing build-up of osmolytes inside the cell that would cause membrane to burst. If excess membrane is synthesized the growth resumes, which suggests that increased membrane tension is likely responsible for the growth inhibition during hypoosmotic shock [147].

## 5.2 Peptidoglycan composition

The peptidoglycan is composed of glycan strand and peptide stem. The glycan strands are made up of alternating N-acetylglucosamine (GlcNAc) and N-acetylmuramic acid (MurNAc) residues linked by β-1-4 bonds.

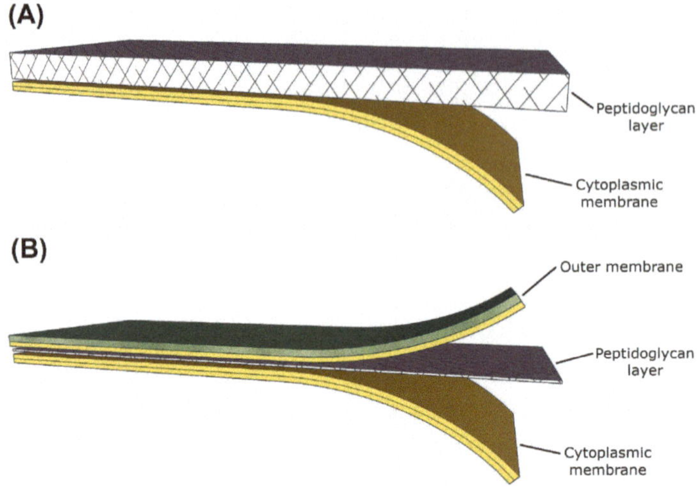


**Fig. 3** Schematic representation of bacterial cell envelope structure in Gram-positive bacteria with a cytoplasmic membrane which for simplicity is depicted to be composed of two identical leaflets and a thick peptidoglycan layer (A), cell envelope in Gram-negative bacteria with cytoplasmic membrane, thin peptidoglycan layer, and asymmetric outer membrane composed of lipopolysaccharide outer leaflet (green) and phospholipid inner leaflet (yellow) (B). The thickness of different layers is presented relative to the cytoplasmic membrane. *Schematic drawings are adapted from Madigan et al. [145].*

The basic disaccharide unit can be repeated several times (e.g. from 2 to 250) [148,149]. The D-lactoyl group of each MurNAc residue is substituted by a penta-peptide stem (usual composition is L-Ala-γ-D-Glu-meso-$A_2$pm-D-Ala-D-Ala). Cross-linking of the glycan strands generally occurs through neighboring peptide stems between the carboxyl group of D-Ala at position 4 on one peptide stem and the amino group of the diamino acid at position 3 of the other stem, either directly or through a short peptide bridge. Both the glycan and peptide part can be modified. In different bacteria glycan strands can be N-deacetylated, O-acetylated, N-glycosylated [150,151]. Amino acid composition of the peptide stem can vary at all five residues as well as in length and composition of the cross-linker. In addition, peptide stem and cross-linker can be further amidated, hydroxylated, acetylated, or decorated with amino acids or other groups including proteins [27,152]. The fine structure of peptidoglycan is not constant and varies with growth phase and bacterial development [22,153].

Although peptidoglycan is usually considered as a single component material it is best viewed as a composite material. Cytoskeletal filaments such as

MreB are interlinked to peptidoglycan via transmembrane protein MreD, periplasmic protein MreC and penicillin-binding proteins (PBPs) thus interlinking cell cytoskeleton with exoskeleton [154]. MreB is an actin homolog in *E. coli* that forms filaments and is implicated in cell shape maintenance. MreB filaments change peptidoglycan curvature and can exert an inward mechanical force [155,156]. If the intrinsic radius of curvature of MreB is less than the cell radius (R, ~500 nm) then the cell wall will experience an inward force and will be reshaped. The result is that the steady-state shape of the cell is a cylinder if the MreB filaments are of sufficient stiffness [157]. If MreB is deleted from the cell and the turgor pressure remains cell become spherical. The MreB filament reinforced peptidoglycan layer can be regarded as a fiber-reinforced composite material [156].

## 5.3 Peptidoglycan architecture

Peptidoglycan can form quasi 2D single layer structures in Gram-negative bacteria or thick 3D structures in Gram-positive bacteria. From an architecture point of view the most important peptidoglycan parameter is the degree of cross-linking. The degree of cross-linking varies a lot, from around 20% in *E. coli* to over 90% in *S. aureus* [158]. If we translate this to muropeptide content in *E. coli* most peptidoglycan units appear as monomers or dimers (50% and 40%, respectively), while higher oligomers being a minority [159]. Conversely, in *S. aureus*, the percentage of monomers is low (<10%) and most peptidoglycan units being present as oligomers (lengths up to nonamers were detected by HPLC and up to eicosamers deduced from calculation) [160].

Peptidoglycan architecture can be easily remodeled during growth [159,161]. For example, the transition of *E. coli* from an exponential growth into a resting stationary phase drastically modifies both the composition and architecture of peptidoglycan. The relative abundance of ld-$A_2$pm-$A_2$pm cross-linked muropeptides increases from approximately 5% to 12% of the total muropeptides [161], the degree of cross-linkage increases from 28% to 38% cross-linked muropeptides, the mean glycan chain length decreases from roughly 33 down to 17 disaccharides per chain, and the lipoprotein-bound muropeptides increases from 9% to 14% [159,161,162]. When peptidoglycan remodels there are two opposing contributions to the Gibbs free energy. During the addition of peptidoglycan basic unit, energy is stored in the activated MurNAc and GlcNAc units which drives the synthesis. This is opposed by the mechanical strain energy change which depends on the shape and size of the cell which increases with cell growth.

When assembly and disassembly reactions exactly balance the peptidoglycan stops growing [156]. The assembly and disassembly process is mediated by transpeptidases and hydrolases and is reversible. The most important enzymes for peptidoglycan architecture remodeling are DD-carboxypeptidase, which belong to the low-molecular weight penicillin-binding proteins (PBP) family. In contrast to the high-molecular weight PBP that synthesize the peptidoglycan, the DD-carboxypeptidase are thought to regulate the degree of cross-linking between glycan chains. By removing the terminal D-alanine from the ends of the peptide side chains, they prevent the transpeptidation reaction. Bacteria often possess several DD-carboxypeptidase and this multiplicity could be important to fine-tune architecture in response to environmental conditions [163–166]. The peptidoglycan synthesis is controlled by cytoskeletal elements from the inside of the cell, whereas peptidoglycan hydrolysis and remodeling are controlled from the enzymes incorporated in the outer membrane. However, recent data indicate that at least in some Gram-negative bacteria, the enlargement of the peptidoglycan layer requires control or activation of peptidoglycan synthases not only from inside the cell but also by synthases associated with the outer membrane, which provides new tools for peptidoglycan remodeling [141].

## 5.4 Biophysical properties of peptidoglycan responsible for stress bearing

Peptidoglycan is a very efficient biomaterial which provides mechanical strength to bacterial cells. It is light, porous, dynamic, strong and yet in constant flux. The peptidoglycan is viscoelastic solid that allows reversible expansion under pressure and gives cell its shape. Normally peptidoglycan is under dynamic stress in the living cell due to the cell turgor pressure. The pre-stressed state of peptidoglycan can relax if the cytoplasmic membrane is destroyed. For example, in *E. coli* upon membrane dissolution with detergent cell surface area decreased for 45% due to the relaxation of the peptidoglycan [167]. For a spherical cell with radius R and turgor pressure P the tensile stress in the cell wall is isotropic, $\sigma = PR/h$, where h is the peptidoglycan thickness. For a cylinder under pressure, $\sigma_{11}$ and $\sigma_{22}$ are defined as stresses in the circumferential and axial directions, respectively [156]. Peptidoglycan is more deformable in the direction of the long axis of the cell (elastic modulus $2.5 \times 10^7$ N m$^{-2}$) than in the direction perpendicular to the long axis (elastic modulus, $4.5 \times 10^7$ N m$^{-2}$) [168]. This is consistent with the observation that changes in the volume of osmotically shocked *E. coli* cells are mainly due to changes in the cell length, whereas cell diameter is virtually constant [169]. It was implied that the anisotropy in elasticity is the consequence of the predominant alignment of the more flexible peptides stems in the direction of the long axis of the cell and of the more rigid glycan strands perpendicular to the direction of the long axis [170]. Peptidoglycan has relatively homogenous pore size distribution: the mean radius of the pores is 2.06 nm for *E. coli* peptidoglycan and 2.12 nm for *B. subtilis* peptidoglycan [171]. From these values it was calculated that globular, uncharged proteins with molecular weights of up 22–24 kDa should be able to diffuse through the relaxed peptidoglycan. If peptidoglycan is stretched, globular proteins of up to 50 kDa or more are able to diffuse through peptidoglycan layer [172].

### 5.5 Obligate intracellular parasites lack peptidoglycan layer

Some obligate intracellular parasitic bacteria, such as *Ehrlichia* and *Anaplasma phagocytophilum*, have only cytoplasmic membrane without peptidoglycan [173]. Obligate intracellular bacteria replicate exclusively within the interior of the living host cells in an osmotically protected niche. Under these conditions peptidoglycan is dispensable. Moreover, the presence of peptidoglycan puts bacteria at risk of detection and destruction by host peptidoglycan recognition factors and downstream effectors [174]. This has resulted in a selective pressure to reduce the levels of peptidoglycan in obligate intracellular parasites. For example, in *Ehrlichia* and *A. phagocytophilum* no genes for peptidoglycan synthesis exist. In some other obligate intracellular parasites genes for peptidoglycan exist but peptidoglycan synthesis has not been detected or peptidoglycan is only transiently present during the bacterial cycle development. Although peptidoglycan is absent in some bacteria (i.e. *Mycoplasmas, Planctomyces, Chlamydiae*) all the other bacteria have peptidoglycan [27]. It appears that relaying solely on cytoplasmic membrane to maintain cell integrity is a very limited strategy in the bacterial world reserved only for the intracellular parasites.

## 6. Bacterial outer membrane layer

Peptidoglycan is a formidable nanomaterial, but it is not immune to destruction and needs protection from adverse environmental factors. Several agents, such as ß-lactams (i.e. penicillin, cephalosporin, carbapanems), lipopeptides (daptomycin), peptide antibiotics (i.e. colistin, polymyxin B) and ionophores (i.e. monensin, alinomycin) target cell wall [175]. It is interesting to note that Gram-negative bacteria have thinner peptidoglycan layer, yet they are more resistant to antibiotics than Gram-positive bacteria [176]. Some Gram-negative bacteria, such as Pseudomonas, are notorious in this regard [177]. There are several strategies how to protect peptidoglycan from destruction (i.e. synthesis of the outer membrane, S-layer, capsule, sheaths, slime) which we will now discuss.

The outer membrane is found mainly in Gram-negative bacteria (Fig. 4). The function of the outer membrane is to serve as a protective barrier that controls the influx and efflux of molecules. It is not immediately obvious why this layer is essential for bacteria. The outer membrane separates the periplasm from the external environment and function as a selective barrier that keeps hydrophobic or amphiphilic compounds that may destroy peptidoglycan from the cell [179,180]. For example, *E. coli* that inhabits the mammalian gut must have cell envelope that is particularly effective at excluding detergents such as bile salts that solubilize the cytoplasmic membrane [181]. This need not be a pressing issue for other Gram-negative bacteria, and their envelopes may differ in species and environment specific ways [14].

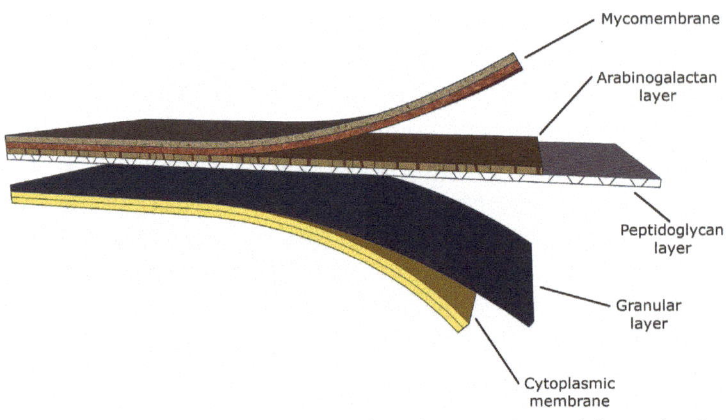


**Fig. 4** Schematic drawing of cell envelope found in members of the order *Corynebacteriales* adapted from Vincent et al. [178]. The inner leaflet of the outer mycomembrane is homogenous and composed of mycolic acid, whereas the outer leaflet is highly heterogeneous and consists of lipids, lipoglycans, and proteins. The thickness of different layers has been estimated to be: cytoplasmatic membrane 6.3 nm; granular layer 3.8 nm, periplasmic space 14.1 nm; peptidoglycan-arabinogalactan layer 6.3 nm; the asymmetric outer membrane 7.5 nm.

## 6.1 Outer membrane composition

The outer membrane is a unique asymmetrical bilayer structure with an inner leaflet composed of phospholipids and an outer leaflet composed mostly of lipopolysaccharides molecules (LPS). The LPS molecule is made of three parts: lipid A, a core oligosaccharide that is usually further subdivided into an inner and outer region; and an O-antigen polysaccharide [182]. Lipid A contains two β-(1−6)-linked glucosamine residues that are phosphorylated and have acylated tails anchoring the LPS molecule into the outer membrane. The inner core is proximal to lipid A and contains a high proportion of negatively charged bacteria-specific sugars, such as 2-keto-3-deoxyoctulosonate and phosphorylated L-glycero-D-manno-heptose. The outer core consists of common sugars, such as hexoses and hexosamines. Negatively charged inner core and lipid A form a dense, divalent cation-stabilized layer [183]. The strong electrostatic bridging interactions of $Ca^{2+}$ are essential for the outer membrane structure and function. Without charge screening by divalent cations, the LPS molecule is forced to overcome the thermodynamically unfavorable energy barrier and flips across the hydrophobic bilayer to minimize the repulsive electrostatic forces [184]. Consequently, addition of chelating agents to the outer membrane is a well-established method to permeabilize Gram-negative bacteria and to disrupt its structure in *E. coli* [185,186].

## 6.2 Attachment of the outer membrane to other surface layers

The outer membrane is not a free-floating envelope layer surrounding the bacterial cell, but it is covalently attached to the peptidoglycan. In *E. coli* and other related Gram-negative bacteria, Braun's lipoprotein (LPP) is the only lipoprotein that is responsible for the covalent linkage of the outer membrane to the peptidoglycan [187,188]. LPP is attached to the α-carboxyl group of meso-$A_2$pm residue of peptidoglycan through the ϵ-amino functional group of its C-terminal lysine residue while its N-terminal glyceryl-cysteine residue is modified by the addition of three fatty acids that allow incorporation of LPP to the inner leaflet of the outer membrane. LPP contributes to the stress bearing of bacterial cell by covalently linking OM to peptidoglycan layer and by regulating the width of periplasmic space [189]. Another peptidoglycan-associated lipoprotein (Pal) of Gram-negative bacteria is important in attachment of the outer membrane to the peptidoglycan. Pal possesses a region that forms a noncovalent binding pocket for the m-Dap residue (meso-diaminopimelate) of peptidoglycan [190]. Pal is tethered to the outer membrane via its N-terminal lipid moiety. Pal is transcribed as part of an operon that also encodes two proteins TolB and TolA, with which it interacts [191]. This makes Pal lipoprotein one of the most important mechanical structural elements in the bacterial envelope. It interacts noncovalently with the peptidoglycan layer, outer membrane protein OmpA, LPP, periplasmic protein TolB and the inner membrane protein TolA [192–198]. The Tol-Pal complex, an assembly of lipoproteins, integral membrane proteins and periplasmic proteins interacting with each other and the peptidoglycan, forms a web of covalent and noncovalent contacts between the outer membrane, peptidoglycan layer and the cytoplasmic membrane [199]. During bacterial cell division there are additional interactions in the divisome between the cytoplasmic membrane, peptidoglycan layer and the outer membrane which are guided by FtsK protein [200]. Similarly, FtsN and Tol-pal complex are required for proper invagination of cytoplasmic and outer membranes during constriction in cell division [201]. These examples clearly illustrate the interconnected nature of multilayer bacterial cell envelope fabric.

## 6.3 Outer membrane stress bearing

The covalently cross-linked peptidoglycan had been thought for decades to be the only envelope layer that matters for bacterial mechanics [202,203]. However, some recent findings indicate that the outer membrane can bear stress loads through large molecular assemblies interlinking the outer and the inner membrane. The outer membrane is actually even more stiff than the cell peptidoglycan and makes a large contribution to cell mechanics [39]. It has been suggested that the stiffness and strength of *E. coli* cells during hyperosmotic shock are largely due to the outer membrane [38,39]. The outer membrane is under negligible or slightly compressive load during steady-state growth. However, steady-state growth is unlikely to be common for bacteria in the environment and bacteria frequently encounter hyperosmotic fluctuations and dynamic mechanical forces, which will engage the load-bearing nature of the outer membrane. Both LPS and proteins contribute to the stiffness of the outer membrane. Compromising the outer membrane, either chemically or genetically, greatly increases deformation of the cell envelope in response to stretching, bending and indentation forces, and induces increased levels of cell lysis upon mechanical perturbation [39]. The outer membrane thus contributes to the bacteria's ability to survive and to tune its mechanical properties based on its environment.

## 6.4 *Mycobacterium tuberculosis* outer membrane

*Mycobacterium tuberculosis* ability to resist, and persist in the moderately acid environment of the phagosome or phagolysosome has been linked to complex cell envelope with double membrane which acts as a formidable permeability barrier for antibacterial effectors, including protons [204]. The cell envelope of *M. tuberculosis* has an unusual, waxy coating on its surface due to the presence of mycolic acid (Fig. 4). This coating makes cells impervious to Gram-staining, and as a result, *M. tuberculosis* can appear either Gram-negative or Gram-positive. In mycobacteria and corynebacteria, the permeability barrier consists of an outer mycomembrane bilayer composed of mycolic acids and other lipids. Mycolic acids are very-long-chain (C30—C90) α-alkyl, β-hydroxy fatty acids, which contribute to extremely hydrophobic bacterial surface. The inner leaflet of the outer membrane is homogeneous and mainly composed of mycolic acids. The mycoloyl residues are covalently linked to the cell wall arabinogalactan, which connects peptidoglycan to the outer membrane [178]. The outer monolayer of the outer membrane is heterogenous and composed of various glycolipids, including trehalose monomycolate and trehalose dimycolate which accumulates in a cord-like fashion on the surface of the cells [205]. This provides a thick layer of lipid on the outer part of the cell and protects the tubercle bacillus from noxious chemicals and the host's immune system [206,207]. The composition and amounts of mycolic acids have been shown to affect the virulence, growth rate, colony morphology and permeability of *M. tuberculosis* [208—211]. Mycolic acids in the outer membrane also play a mechanical role [212—214]. The cyclopropane rings in mycolic acids of *M. tuberculosis* contribute to the structural integrity of the cell wall complex and protect cell from oxidative stress [215,216]. Between cytoplasmic membrane and peptidoglycan, there is an extra granular layer which is composed of penicillin-binding proteins, lipoproteins and lipoteihoic acids. The function and precise composition of this layer is still poorly understood [178].

## 6.5 *Deinococcus radiodurans* outer membrane

The *Deinococcus radiodurans* is well known for its extreme resistance to gamma radiation, UV rays, oxidative stress and desiccation stress [217]. *D. radiodurans* is a Gram-positive bacterium possessing the ability to synthesize carotenoids which bind to the cell envelope and protect against UV and oxidative stress [218]. The structure of the cell envelope of *D. radiodurans* is different from that of the other Gram-positive bacteria (Fig. 5). The complex envelope of

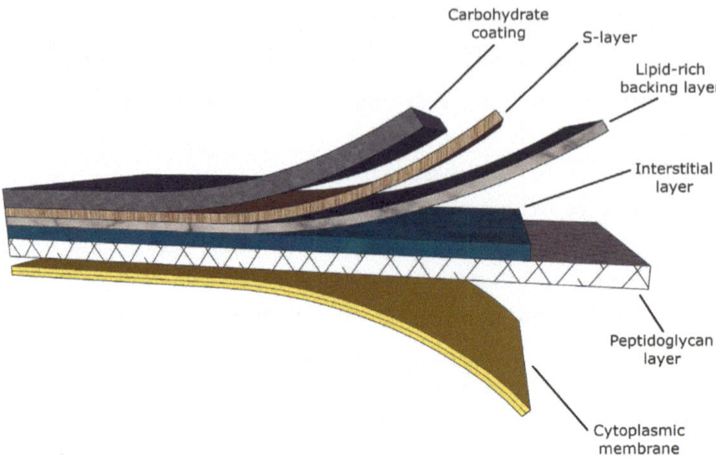


**Fig. 5** *D. radiodurans* envelope structure. The complex structure of cell envelope has been implicated in bacterium extreme resistance to gamma radiation, UV rays, oxidative stress and desiccation stress. *Schematic drawing is adapted from Yu et al. [102], Yu et al. [219] and Rothfuss et al. [220].*

this organism constitutes a cytoplasmic membrane, a thick peptidoglycan layer, an interstitial layer which is only known to be composed of water soluble proteins, a lipid rich backing layer followed by a hexagonally packed S-surface layer (Hpi) and finally the outermost carbohydrate coat. The composition of the *D. radiodurans* outer membrane is very different from that of Gram-negative bacteria. In contrast to Gram-negative bacteria the outer membrane of *D. radiodurans* has no LPS or usual bacterial phospholipids instead, there are high proportions of glucose-containing phospholipids [221].

## 7. S-layers

The S-layers are frequently observed surface layers in bacteria [222]. It is therefore an exception to the rule that in *E. coli* and *B. subtilis*, the canonical bacterial model systems, the S-layers are absent. Most S-layers are formed by noncovalent, entropy-driven assembly of a single (glyco)protein protomer on the bacterial surface, giving rise to proteinaceous paracrystalline layers (Fig. 6). Generally, a single S-layer is present, constituting 5 to 10% of the total cell protein which is a substantial energy burden for a cell. When the other cell surface components (e.g. capsules, glycocalyx) are absent, S-layer is the outermost cell envelope component. Because many

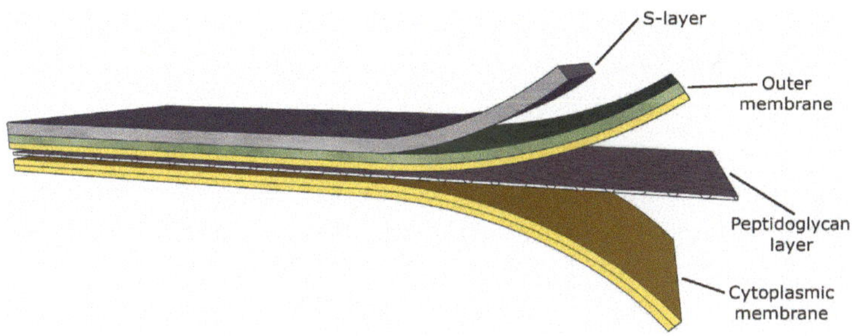


**Fig. 6** S-layer can form as an outermost envelope layer on both Gram-positive and Gram-negative bacteria. Only Gram-negative cell envelope is depicted. The LPS and phospholipid leaflets in the outer membrane are indicated with different colors. The S-layer usually consists of a crystal-like single layer of proteins. The S-layer lattices can have oblique, square, or hexagonal symmetry [223].

bacteria have S-layers it has been speculated that they play important roles in the interaction between the cell and its environment. Various functions have been proposed for S-layers, including shape maintenance and molecular sieving. The S-layer may be a virulence factor, protecting pathogenic bacteria against complement killing, facilitating binding of bacteria to host molecules, or enhancing their ability to associate with macrophages (reviewed in Ref. [224–227]). The S-layer can mechanically reinforce cell envelope by initiating mineral precipitation [228].

## 7.1 S-layer as a primordial membrane

S-layers are the simplest biological (glyco)protein membranes which have developed during the evolution. The S-layer is assembled from identical protein macromolecules. Theoretically, it is not possible to form a closed protein membrane with less redundancy of information. Therefore, it is tempting to speculate that a simple protein membrane capable of dynamic growth could have initiated formation of a barrier membrane envelope in an early stage of biological evolution [229–232]. The S-layers are recognized as versatile patterning elements for the generation of complex supramolecular structures involving other molecules such as lipids, proteins, glycans, and nucleic acids as well as inorganic materials (e.g. nanoparticles). For example, S-layers have been shown to interact specifically with a great variety of amphiphilic molecules (e.g. phosphor- and ether-lipids) generating more complex supramolecular membrane structures with potential for high transmembrane selectivity [233]. When S-layer proteins of *Bacillus stearothermophilus* were recrystallized on positively charged unilamellar liposomes an inversion of the zeta-potential was observed from +29.1 mV to -27.1 mV. In addition, covalent crosslinked recrystallized S-layer proteins stabilized lipid vesicles. S-layer coated liposomes released only half the amount of liposome enclosed cargo upon exposure to shear forces or ultrasonication or temperature shifts from 25 to 55 °C compared to non-coated liposomes [234] suggesting that S-layer can have a considerable mechanical effect on bacterial cell.

## 7.2 Antifouling properties of S-layer membranes

It was demonstrated that S-layer lattices allow free passage for molecules with a molecular weight of up to 30 kDa and have sharp exclusion limits between molecular weights of 30 and 45 kDa proposing a limiting pore diameter in the range of 3—4.5 nm [235,236]. It appears that molecules capable of penetrating the S-layer have also free passage through the peptidoglycan layer. It is therefore unlikely that S-layers have the potential to function as an effective barrier against lysogenic enzymes and are thus unable to protect peptidoglycan degradation [237]. To allow free passage of molecules S-layers should have antifouling properties [238—240]. It has been suggested that the ultra-low fouling properties of S-layers may be due to their zwitterionic surface properties [241]. In zwitterionic antifouling surfaces ions alternate perfectly in the sub-nanometer scale between positive and negative charges preventing adsorption of naturally occurring molecules, particularly proteins. Alternatively, it has been demonstrated that S-layer could be shedded from the surface to maintain the surface clean [242].

## 7.3 S-layer and mineralization

A unique ecological role was be demonstrated for the cyanobacterial S-layer of *Synechococcus* strain GL24. This bacterium was found to induce mineralization of fine-grain gypsum ($CaSO_4 \cdot 2H_2O$) and calcite ($CaCO_3$) in a fresh water lake [243]. The S-layer has a hexagonal monomer arrangement and provides regularly spaced, chemically identical nucleation sites for mineral growth [228,244,245]. Mineral formation begins within the large holes of the protein array when $Ca^{2+}$ binds to negatively charged sites on the S-layer protein and is joined by $SO_4^{2-}$, initiating the formation of a mineral aggregate. Eventually, the S-layer becomes encrusted with mineral and is shed so that cells have a patchy appearance with respect to the location of mineralized portions of their surface. Shedding of S-layer material could be a common process of bacteria to get rid of mineral depositions on their cell surface thereby maintaining basic vital processes such as growth and division as well as nutrient transport.

## 8. Capsular layers

Capsule polymers are highly flexible hydrocolloids that extend radially from the cell and are rarely cross-linked to one another by covalent bonds (Fig. 7). The most effective linkages between capsule polymers seem to be electrostatic where highly acidic residues can be inter-bridged by divalent metal ions such as $Mg^{2+}$ or $Ca^{2+}$. However, these metals are highly exchangeable and can be displaced by other multivalent species [246]. Capsules are one of the most frequently encountered surface layers [247]. Many species of Gram-positive and Gram-negative organisms are enveloped by a capsule.

### 8.1 Capsule composition

Capsules are comprised of high-molecular-weight capsular polysaccharide chains that are exported across the bacterial cell envelope and retained at the cell surface. The polymers of capsules are branched or unbranched homo- or hetero-polysaccharides but can occasionally be proteinaceous as well [246,248,249]. The chemical composition of capsule displays

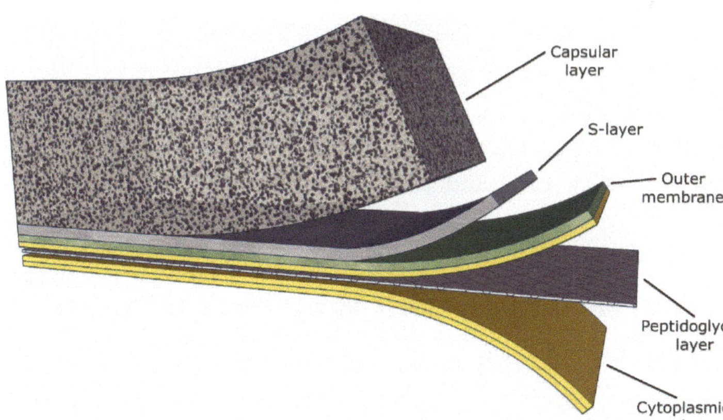


**Fig. 7** Both Gram-positive and Gram-negative bacteria may have capsular layer. Only Gram-negative envelope structure is depicted with a relatively thick gel like capsular layer. The capsular layer can have various thickness and can extend up to several μm into the extracellular space [145].

remarkable diversity even within a single species. For instance, *E. coli* has over 80 different capsular (K antigen) serotypes [250,251]. *E. coli* capsules have been subdivided into four groups based on structural, biochemical, and genetic criteria (reviewed in Whitfield [252]). Capsule polysaccharides in groups 1 and 4 have shorter carbohydrate chains, which are covalently linked to the lipid A core of the outer membrane producing $K_{LPS}$ molecules [253,254]. Longer chain polysaccharides form capsules independently of covalent attachment to LPS. For example, $K30_{CPS}$ is anchored noncovalently to the bacterial surface by an outer membrane lectin, Wzi [255]. Capsular polysaccharides in groups 2 and 3 are thought to be linked to the surface via a diacylglycerolphosphate moiety [256].

## 8.2 Multifunctional role of capsule

In Gram-negative bacteria capsule is linked covalently and noncovalently to the outer membrane, which in turn is connected to the peptidoglycan and the cytoplasmic membrane. The viscoelastic nature of capsule strengthens the envelope [257]. Extending great distances from the cell surface (often up to several micrometers), capsules are used as cementing substances, to bind bacteria together into microcolony formats, to bind the cells to substrata, or to form flocs, eventually developing large biofilms which cover surfaces [258,259]. Capsular polysaccharides offer physical barrier to antibiotics or bind to cationic antibiotics, thus lowering peptidoglycan degradation. Capsule prevents bacterial cell desiccation due to hydrated nature of polysaccharides keeping all envelope layers moist and lubricated, capsule facilitate bacteria to adhere onto host cell surface, help the bacteria to escape from host immune response by inhibiting complement cascade, prevent antibody opsonization, and preclude the recognition by macrophage [260]. Capsular proteins represent a major virulence factor together with toxins, adhesins [261], LPS, and exopolysaccharides [262].

# 9. Extracellular polymeric substances (EPS)

Like capsules, bacterial extracellular polymeric substances or slimes include a wide range of extracellular homo and heteropolymers [248,263]. They are abundant in nature and are weakly adsorbed to the cell or float freely in the aqueous milieu. They are frequently the most abundant components of biofilms. Sometimes bacteria overproduce their capsular materials but for some reason, fail to anchor them securely to their surfaces.

Such polymers are sloughed off into the surroundings and float freely until they become associated with other solid surfaces or bacteria. Electron microscopy of natural biofilms or suspended flocs frequently reveals envelope and other cellular debris from lyzed bacteria intermixed with intact cells throughout a polymeric matrix; this matrix was initially part of a capsule but is now more appropriately classified as extracellular slime or EPS since it is no longer entirely associated with the cell [249].

## 10. Sheath layers

Sheaths are less frequently encountered envelope structures than capsules or EPS in the bacterial world [247]. Sheats are tubes enclosing linear chains of cells composed of an amorphous polymeric network. Sheaths are found in *Thiothrix spp., Thioploca spp.* or some cyanobacteria [264,265]. In *Thiploca* sheaths, for example, cells can even glide within the sheath [264]. The sheaths are an integral peripheral structure of these genera and have far-reaching design purposes over and above those of more usual bacteria [266]. Sheaths precipitate iron (*Sphaerotilus*) or manganese (*Leptothrix*) oxides [267]. For *Leptothrix discophora*, the manganese-oxidizing capability is determined by proteins within the sheath [268,269]. *Leptothrix* spp. sheaths comprises a metal-impregnated organic matrix that may be synthesized and excreted as globules from the surface of the bacterial cell envelope. Transfer to the immature sheath skeleton by globule, threads or pili like secretions contributes to the thickening of the sheath layer in which aqueous-phase inorganic elements began to deposit, resulting in high electron density fibrous structure [270]. Sheaths play important roles in ecosystems by accumulating biologically important elements and induce the crystallization of unexpected mineral phases [271]. This bacteria may also possess capsule which is structurally differentiated from the sheath, the outer membrane, as well as S-layer (Fig. 8) [272,273].

## 11. Endospores

In terrestrial (soil) environments where many bacteria found potentially lethal extreme conditions which includes cycles of heat and cold, freezing and thawing, physical abrasion, extreme desiccation, exposure to corrosive chemicals, attack by other organisms and their extracellular degradative enzymes, or prolonged exposure to solar radiation there are many

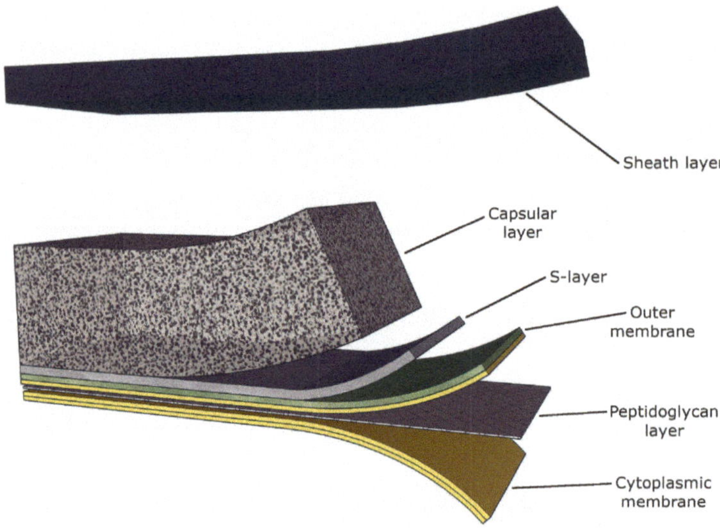


**Fig. 8** Sheath layer is a dislocated structural layer of some bacteria that encloses chains of bacterial cells. Sheath layers are usually metal-impregnated organic matrices. *Schematic drawing of Gram-negative envelope structure with additional sheath layer was adapted from Madigan et al. [145].*

spore forming bacteria [274]. Endospores are the toughest living biological structure known that can survive extreme conditions for years. A sporulating bacterium cannot predict beforehand how long or in what environment it will spend its dormant state. Therefore, the sporulating cell must prepare for the worst each time it undergoes differentiation. There are several reasons why endospores can survive harsh environments as reviewed by Nicholson et al. [274]. These factors include the genetic makeup of the sporulating species, the precise sporulation conditions, the spore coats, the relative impermeability of the spore core, the low water content of the hydrated spore's core, the high level of minerals in the spore core, the saturation of spore DNA with $\alpha/\beta$-type small acid-soluble proteins (SASP), and repair of damage to macromolecules during spore germination and outgrowth.

## 11.1 Endospore envelope structure

The distinct morphological feature of endospore is its layered envelope structure. It is reminiscent of, but much more sophisticated version of the vegetative bacterial envelope structure. The core of the endospore is surrounded, by the inner spore membrane, germ cell wall, cortex, outer spore

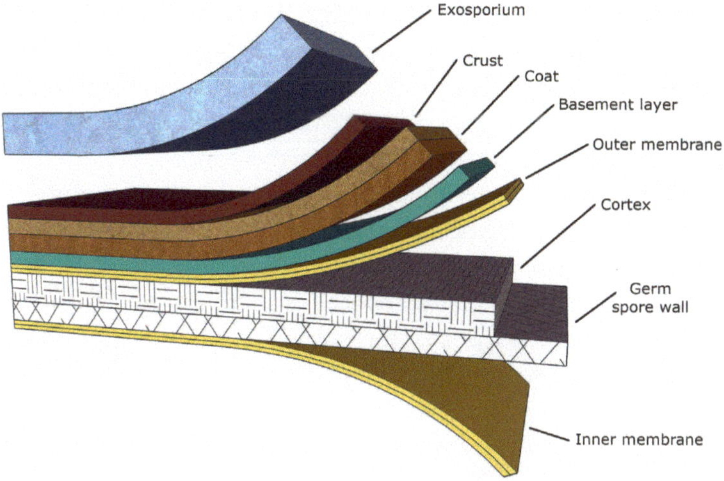


**Fig. 9** Bacterial endospore is the most complex biological cell envelope structure known. The multiple layer structure of the spore envelope protects genome, which is housed in partially dehydrated core. Both germ cell wall and cortex layer are composed of peptidoglycan but with different architecture. The outer membrane is derived from mother forespore phospholipid membrane. The coat has four distinct sublayers: basement layer, inner coat, outer coat, and crust. The outermost protective layer exosporium may not be present in all endospores. *Schematic drawing is adapted from literature by McKenney et al. [275], Stewart et al. [277] and Cogliati et al. [276].*

membrane, coat, and exosporium (Fig. 9). There are four layers in spore coat observed in thin-section electron microscopy of the *B. subtilis*: a basement layer, a lamellar inner coat, a more coarsely layered outer coat and a recently identified layer named the crust [275]. The coat has an important role in limiting access of hydrolyzing enzymes to the peptidoglycan-containing cortex. In some species, such as *B. anthracis*, a different type of outermost layer exists, called the exosporium [278]. The exosporium is a distinct glycoprotein layer that surrounds the spore and is separated by a gap called the interspace. In *Bacillus cereus* and its close relatives *B. anthracis* and *Bacillus thuringensis*, the exosporium consists of a crystalline two-dimensional basal layer adorned with hair-like projections [279].

## 11.2 Dissolution of endospore envelope

Although an endospore is manly a dormant structure fit for the long-term bacterial survival, spores must be able to germinate rapidly when nutrients become available. Spores continually monitor their surroundings using an array of germination receptors embedded in the inner spore membrane. Germinants of *B. subtilis* include sugars, amino acids, peptidoglycan fragments and ions [280,281]. As germinants will reach their receptors traversing the spore coat and cortex, the coat must ultimately be both protective and permeable. The endospre coat can therefore be conceptualized as a molecular sieve that excludes large molecules, such as lysozyme, while allowing the passage of small-molecule germinants [282,283]. Most of the layered structures of the endospore are rapidly lost upon germination: the coat is shed, and the cortex is degraded by cell wall hydrolases. The germ cell wall is retained to serve as a primer for vegetative cell wall assembly. The inner spore membrane becomes the plasma membrane, whereas the fate of the outer spore membrane is less clear. It may be lost during spore germination, but there is at least one organism (*Acetonema longum*) in which it persists after germination, thus generating a double membrane vegetative cell akin to Gram-negative bacteria [284].

## 12. How did bacterial envelope evolve?

The bacterial envelope is an extremely complex and diverse layered structure of vital importance for bacteria. Although it is very important to know how the envelope evolved, we do not have at present a definite answer. Helpful suggestions are coming from whole-genome sequencing, molecular biology, biochemistry, biophysics and bioinformatics studies [285–287]. For example, we have vague ideas how the outer membrane might have evolved. During endospore formation Gram-positive bacteria have temporally two membranes. Contrary to other endospore formers *A. longum* keeps the outer membrane even after germination [284]. Hence, it is possible that Gram-negative bacteria evolved from an ancestral endospore-forming Gram-positive bacteria [286]. During the process, the thickness of the peptidoglycan was reduced, and teichoic acids were lost, presumably before the acquisition of the LPS molecules. Currently however, there are too many unknowns in this scenario. One thing is to keep the second membrane, the other is to make it functional (i.e. connect it to the peptidoglycan, target protein to the other membrane, change physiology to allow for lower turgor pressures, transport nutrient across two membranes, and many other modifications we find in modern Gram-negative bacteria). Nevertheless, with the new research tools at our disposal we may be in a position to provide some meaningful answers to these questions.

## 13. Conclusions

Bacterial cell envelope is a multicomponent multilayer composite material of exquisite complexity. Bacteria take advantage of physical interactions between envelope molecular components to achieve mechanical stability and selectivity unmatched in biology. These interactions are typically complex and cannot be easily and quantitatively deciphered. Progress is being made in developing a quantitative understanding of the interactions important for cell envelope biogenesis and integrity. We have a general idea how mechanical forces influence molecular chemistry of the envelope and how mechanical stress shape and reshape the cell. We are beginning to appreciate that cell mechanics is not revolving solely around peptidoglycan layer but is an integrated response of all cell envelope layers. The cell envelope structure is interconnected on many different levels. Envelope molecular constituents foremost participate in building a particular layer structure (i.e. lipid bilayer, peptidoglycan, outer membrane, capsule, sheath) where they have distinct molecular interactions within the layer. However, there are also numerous covalent and non-covalent interactions between adjacent layers (i.e. LPP connecting the outer membrane to peptidoglycan) or trans-layer connections between distant layers (i.e. Pal interconnecting the cytoplasmic membrane to the outer membrane via peptidoglycan) that we are beginning to unravel. Multiple interlayer connections reinforce envelope structure. Being light, strong, flexible, porous, and selective cell envelope will continue to inspire engineers to make exciting new material composite structures. However, many molecular details of bacterial cell envelope endurance are still missing and await further theoretical explanation.

## References

[1] H.T. Tien, A. Ottova, The lipid bilayer concept: experimental realization and current applications, Memb. Sci. Technol. (2003) 1–73.

[2] J.J. Thwaites, N.H. Mendelson, Mechanical behaviour of bacterial cell walls, Adv. Micro. Physiol. (1991) 173–222.

[3] E.M. Rivkina, E.I. Friedmann, C.P. McKay, D.A. Gilichinsky, Metabolic activity of permafrost bacteria below the freezing point, Appl. Environ. Microbiol. 66 (2000) 3230–3233.

[4] K. Takai, K. Nakamura, T. Toki, U. Tsunogai, M. Miyazaki, J. Miyazaki, H. Hirayama, S. Nakagawa, T. Nunoura, K. Horikoshi, Cell proliferation at 122 C and isotopically heavy $CH_4$ production by a hyperthermophilic methanogen under high-pressure cultivation, Proc. Natl. Acad. Sci. U.S.A. 105 (2008) 10949–10954.

[5] C. Schleper, G. Püher, B. Kuhlmorgen, W. Zillig, Life at extremely low pH, Nature 375 (1995) 741–742.

[6] H. Liu, Y. Zhou, R. Liu, K.-Y. Zhang, R. Lai, *Bacillus solisalsi* sp. nov., a halotolerant, alkaliphilic bacterium isolated from soil around a salt lake, Int. J. Syst. Evol. Microbiol. 59 (2009) 1460–1464.
[7] G. Horneck, Survival of microorganisms in space: a review, Adv. Space Res. 1 (1981) 39–48.
[8] R.D. Sleator, C. Hill, Bacterial osmoadaptation: the role of osmolytes in bacterial stress and virulence, FEMS Microbiol. Rev. 26 (2002) 49–71.
[9] M.G. Liebensteiner, N. Tsesmetzis, A.J.M. Stams, B.P. Lomans, Microbial redox processes in deep subsurface environments and the potential application of (per)chlorate in oil reservoirs, Front. Microbiol. 5 (2014).
[10] G. Beretta, A.F. Mastorgio, L. Pedrali, S. Saponaro, E. Sezenna, The effects of electric, magnetic and electromagnetic fields on microorganisms in the perspective of bioremediation, Rev. Environ. Sci. Bio/Technology. 18 (2019) 29–75.
[11] A. Krisko, M. Radman, Biology of extreme radiation resistance: the way of Deinococcus radiodurans, Cold Spring Harb. Perspect. Biol. 5 (2013) a012765–a012765.
[12] N.A.R. Gow, J.-P. Latge, C.A. Munro, The fungal cell wall: structure, biosynthesis, and function, Microbiol. Spectr. 5 (2017).
[13] S.-V. Albers, B.H. Meyer, The archaeal cell envelope, Nat. Rev. Microbiol. 9 (2011) 414–426.
[14] T.J. Silhavy, D. Kahne, S. Walker, The bacterial cell envelope, Cold Spring Harb. Perspect. Biol. 2 (2010) a000414–a000414.
[15] D.-J. Scheffers, M.G. Pinho, Bacterial cell wall synthesis: new insights from localization studies, Microbiol. Mol. Biol. Rev. 69 (2005) 585–607.
[16] G.K. Auer, D.B. Weibel, Bacterial cell mechanics, Biochemistry 56 (2017) 3710–3724.
[17] J.A.H. Romaniuk, L. Cegelski, Bacterial cell wall composition and the influence of antibiotics by cell-wall and whole-cell NMR, Philos. Trans. R. Soc. B Biol. Sci. 370 (2015) 20150024.
[18] A.J.F. Egan, R.M. Cleverley, K. Peters, R.J. Lewis, W. Vollmer, Regulation of bacterial cell wall growth, FEBS J. 284 (2017) 851–867.
[19] C. Paradis-Bleau, G. Kritikos, K. Orlova, A. Typas, T.G. Bernhardt, A genome-wide screen for bacterial envelope biogenesis mutants identifies a novel factor involved in cell wall precursor metabolism, PLoS Genet. 10 (2014) e1004056.
[20] M.F. Brown, Lipid polymorphism and membrane properties, in: R. Epand (Ed.), Lipid Polymorph. Membr. Prop., Academic press, San Diego, CA, 1997, pp. 285–356.
[21] W. Dowhan, Molecular basis for membrane phospholipid diversity:why are there so many lipids? Annu. Rev. Biochem. 66 (1997) 199–232.
[22] G.W. Huisman, D.A. Siegele, M.M. Zambrano, R. Kolter, Morphological and physiological changes during stationary phase, in: F.C. Neidhardt (Ed.), *Escherichia Coli* Salmonella Typhimurium Cell. Mol. Biol., American Society for Microbiology, Washington DC, 1996, pp. 1672–1682.
[23] C. Sohlenkamp, O. Geiger, Bacterial membrane lipids: diversity in structures and pathways, FEMS Microbiol. Rev. 40 (2016) 133–159.
[24] C. Ratledge, S.G. Wilkinson, Microbial Lipids LK, Academic press, London, 1988 [etc.] SE - .. dl.; 25 cm. https://wur.on.worldcat.org/oclc/898843218.
[25] H. Goldfine, Bacterial membranes and lipid packing theory, J. Lipid Res. 25 (1984) 1501–1507.
[26] J.E. Cronan, Bacterial membrane lipids: where do we stand? Annu. Rev. Microbiol. 57 (2003) 203–224.
[27] W. Vollmer, Structural variation in the glycan strands of bacterial peptidoglycan, FEMS Microbiol. Rev. 32 (2008) 287–306.

[28] R.P. Fagan, N.F. Fairweather, Biogenesis and functions of bacterial S-layers, Nat. Rev. Microbiol. 12 (2014) 211–222.
[29] L.M. Willis, C. Whitfield, Structure, biosynthesis, and function of bacterial capsular polysaccharides synthesized by ABC transporter-dependent pathways, Carbohydr. Res. 378 (2013) 35–44.
[30] M. Lukaszczyk, B. Pradhan, H. Remaut, The biosynthesis and structures of bacterial pili, Subcell Biochem. (2019) 369–413.
[31] O. Mol, B. Oudega, Molecular and structural aspects of fimbriae biosynthesis and assembly in *Escherichia coli*, FEMS Microbiol. Rev. 19 (1996) 25–52.
[32] R. Liu, H. Ochman, Stepwise formation of the bacterial flagellar system, Proc. Natl. Acad. Sci. U.S.A. 104 (2007) 7116–7121.
[33] K. Dimmitt, S. Bradford, M. Simon, Synthesis of bacterial flagella. I. Requirement for protein and ribonucleic acid synthesis during flagellar regeneration in Bacillus subtilis, J. Bacteriol. 95 (1968) 801–810.
[34] O.A. Soutourina, P.N. Bertin, Regulation cascade of flagellar expression in Gram-negative bacteria, FEMS Microbiol. Rev. 27 (2003) 505–523.
[35] A. Cockayne, Bacterial cell walls, in: Encycl. Immunol., Elsevier, 1998, pp. 320–323.
[36] M.-P. Chapot-Chartier, S. Kulakauskas, Cell wall structure and function in lactic acid bacteria, Microb. Cell Factories 13 (2014) S9.
[37] X. Liu, H.M. Grabherr, R. Willmann, D. Kolb, F. Brunner, U. Bertsche, D. Kühner, M. Franz-Wachtel, B. Amin, G. Felix, M. Ongena, T. Nürnberger, A.A. Gust, Host-induced bacterial cell wall decomposition mediates pattern-triggered immunity in Arabidopsis, Elife 3 (2014).
[38] H. Hwang, N. Paracini, J.M. Parks, J.H. Lakey, J.C. Gumbart, Distribution of mechanical stress in the *Escherichia coli* cell envelope, Biochim. Biophys. Acta Biomembr. 1860 (2018) 2566–2575.
[39] E.R. Rojas, G. Billings, P.D. Odermatt, G.K. Auer, L. Zhu, A. Miguel, F. Chang, D.B. Weibel, J.A. Theriot, K.C. Huang, The outer membrane is an essential load-bearing element in Gram-negative bacteria, Nature 559 (2018) 617–621.
[40] S. Baoukina, S.J. Marrink, D.P. Tieleman, Structure and dynamics of lipid monolayers: theory and applications, in: Biomembr. Front., Humana Press, Totowa, NJ, 2009, pp. 75–99.
[41] D. Risović, A. Penezić, V. Čadež, S. Šegota, B. Gašparović, Surface free energy tuning of supported mixed lipid layers, RSC Adv. 6 (2016) 52475–52484.
[42] S.A. Akimov, V.A.J. Frolov, P.I. Kuzmin, J. Zimmerberg, Y.A. Chizmadzhev, F.S. Cohen, Domain formation in membranes caused by lipid wetting of protein, Phys. Rev. E. 77 (2008) 051901.
[43] B.L. Stottrup, S.L. Keller, Phase behavior of lipid monolayers containing DPPC and cholesterol analogs, Biophys. J. 90 (2006) 3176–3183.
[44] M. Elderdfi, A.F. Sikorski, Langmuir-monolayer methodologies for characterizing protein-lipid interactions, Chem. Phys. Lipids 212 (2018) 61–72.
[45] D. Koller, K. Lohner, The role of spontaneous lipid curvature in the interaction of interfacially active peptides with membranes, Biochim. Biophys. Acta Biomembr. 1838 (2014) 2250–2259.
[46] L.V. Chernomordik, M.M. Kozlov, Protein-lipid interplay in fusion and fission of biological membranes, Annu. Rev. Biochem. 72 (2003) 175–207.
[47] G. Di Paolo, P. De Camilli, Phosphoinositides in cell regulation and membrane dynamics, Nature 443 (2006) 651–657.
[48] J. Zimmerberg, M.M. Kozlov, How proteins produce cellular membrane curvature, Nat. Rev. Mol. Cell Biol. 7 (2006) 9–19.

[49] T.J. McIntosh, S.A. Simon, Roles of bilayer material properties in function and distribution of membrane proteins, Annu. Rev. Biophys. Biomol. Struct. 35 (2006) 177–198.
[50] O.S. Andersen, R.E. Koeppe, Bilayer thickness and membrane protein function: an energetic perspective, Annu. Rev. Biophys. Biomol. Struct. 36 (2007) 107–130.
[51] R. Koynova, B. Tenchov, Transitions between lamellar and nonlamellar phases in membrane lipids and their physiological roles, OA Biochem. 1 (2013).
[52] V. Patel, R. Lalani, D. Bardoliwala, S. Ghosh, A. Misra, Lipid-based oral formulation strategies for lipophilic drugs, AAPS PharmSciTech 19 (2018) 3609–3630.
[53] A.R. Thiam, B. Antonny, J. Wang, J. Delacotte, F. Wilfling, T.C. Walther, R. Beck, J.E. Rothman, F. Pincet, COPI buds 60-nm lipid droplets from reconstituted water-phospholipid-triacylglyceride interfaces, suggesting a tension clamp function, Proc. Natl. Acad. Sci. U.S.A. 110 (2013) 13244–13249.
[54] P.I. Kuzmin, S.A. Akimov, Y.A. Chizmadzhev, J. Zimmerberg, F.S. Cohen, Line tension and interaction energies of membrane rafts calculated from lipid splay and tilt, Biophys. J. 88 (2005) 1120–1133.
[55] S. Baoukina, L. Monticelli, H.J. Risselada, S.J. Marrink, D.P. Tieleman, The molecular mechanism of lipid monolayer collapse, Proc. Natl. Acad. Sci. U.S.A. 105 (2008) 10803–10808.
[56] M. Khan, N. Dosoky, J. Williams, Engineering lipid bilayer membranes for protein studies, Int. J. Mol. Sci. 14 (2013) 21561–21597.
[57] G. Bozzuto, A. Molinari, Liposomes as nanomedical devices, Int. J. Nanomed. 10 (2015) 975–999.
[58] J.N. Israelachvili, Interactions of biological membranes and structures, in: Intermol. Surf. Forces, Elsevier, 2011, pp. 577–616.
[59] H. Heerklotz, R.M. Epand, The enthalpy of acyl chain packing and the apparent water-accessible apolar surface area of phospholipids, Biophys. J. 80 (2001) 271–279.
[60] D. Lombardo, M.A. Kiselev, S. Magazù, P. Calandra, Amphiphiles self-assembly: basic concepts and future perspectives of supramolecular approaches, Adv. Condens. Matter Phys. 2015 (2015) 1–22.
[61] N. Fuller, R.P. Rand, The influence of lysolipids on the spontaneous curvature and bending elasticity of phospholipid membranes, Biophys. J. 81 (2001) 243–254.
[62] R.M. Venable, F.L.H. Brown, R.W. Pastor, Mechanical properties of lipid bilayers from molecular dynamics simulation, Chem. Phys. Lipids 192 (2015) 60–74.
[63] M. Doktorova, M.V. LeVine, G. Khelashvili, H. Weinstein, A new computational method for membrane compressibility: bilayer mechanical thickness revisited, Biophys. J. 116 (2019) 487–502.
[64] G. Espinosa, I. Lopez-Montero, F. Monroy, D. Langevin, Shear rheology of lipid monolayers and insights on membrane fluidity, Proc. Natl. Acad. Sci. U.S.A. 108 (2011) 6008–6013.
[65] A.V. Granato, The shear modulus of liquids, J. Phys. IV 06 (1996) C8-C1–C8-C9.
[66] D. Boal, Mechanics of the Cell, second ed., Mech. Cell, 2012, pp. 1–608.
[67] C.A. Rutkowski, L.M. Williams, T.H. Haines, H.Z. Cummins, The elasticity of synthetic phospholipid vesicles obtained by photon correlation spectroscopy, Biochemistry 30 (1991) 5688–5696.
[68] W. Rawicz, K.C. Olbrich, T. McIntosh, D. Needham, E. Evans, Effect of chain length and unsaturation on elasticity of lipid bilayers, Biophys. J. 79 (2000) 328–339.
[69] N. Kato, A. Ishijima, T. Inaba, F. Nomura, S. Takeda, K. Takiguchi, Effects of lipid composition and solution conditions on the mechanical properties of membrane vesicles, Membranes (Basel) (2015) 22–47.

[70] P.-A. Boucher, B. Joós, M.J. Zuckermann, L. Fournier, pore formation in a lipid bilayer under a tension ramp: modeling the distribution of rupture tensions, Biophys. J. 92 (2007) 4344–4355.
[71] M. Asgari, A. Biria, Free energy of the edge of an open lipid bilayer based on the interactions of its constituent molecules, Int. J. Non Lin. Mech. 76 (2015) 135–143.
[72] T. Portet, R. Dimova, A new method for measuring edge tensions and stability of lipid bilayers: effect of membrane composition, Biophys. J. 99 (2010) 3264–3273.
[73] S. Ali, J.M. Smaby, M.M. Momsen, H.L. Brockman, R.E. Brown, Acyl chain-length asymmetry alters the interfacial elastic interactions of phosphatidylcholines, Biophys. J. 74 (1998) 338–348.
[74] B. Brüning, R. Stehle, P. Falus, B. Farago, Influence of charge density on bilayer bending rigidity in lipid vesicles: a combined dynamic light scattering and neutron spin-echo study, Eur. Phys. J. E. 36 (2013) 77.
[75] P. Girard, J. Prost, P. Bassereau, Passive or active fluctuations in membranes containing proteins, Phys. Rev. Lett. 94 (2005) 088102.
[76] D. Vorselen, F.C. MacKintosh, W.H. Roos, G.J.L. Wuite, Competition between bending and internal pressure governs the mechanics of fluid nanovesicles, ACS Nano 11 (2017) 2628–2636.
[77] H. Bermúdez, D.A. Hammer, D.E. Discher, Effect of bilayer thickness on membrane bending rigidity, Langmuir 20 (2004) 540–543.
[78] R. Phillips, T. Ursell, P. Wiggins, P. Sens, Emerging roles for lipids in shaping membrane-protein function, Nature 459 (2009) 379–385.
[79] A.G. Petrov, Flexoelectricity of model and living membranes, Biochim. Biophys. Acta Biomembr. 1561 (2002) 1–25.
[80] D. Marsh, Elastic curvature constants of lipid monolayers and bilayers, Chem. Phys. Lipids 144 (2006) 146–159.
[81] J.C. Bozelli, W. Jennings, S. Black, Y.H. Hou, D. Lameire, P. Chatha, T. Kimura, B. Berno, A. Khondker, M.C. Rheinstädter, R.M. Epand, Membrane curvature allosterically regulates the phosphatidylinositol cycle, controlling its rate and acyl-chain composition of its lipid intermediates, J. Biol. Chem. 293 (2018) 17780–17791.
[82] S. Purushothaman, P. Cicuta, O. Ces, N.J. Brooks, Influence of high pressure on the bending rigidity of model membranes, J. Phys. Chem. B 119 (2015) 9805–9810.
[83] F. Nomura, M. Nagata, T. Inaba, H. Hiramatsu, H. Hotani, K. Takiguchi, Capabilities of liposomes for topological transformation, Proc. Natl. Acad. Sci. U.S.A. 98 (2001) 2340–2345.
[84] F. Nomura, T. Inaba, S. Ishikawa, M. Nagata, S. Takahashi, H. Hotani, K. Takiguchi, Microscopic observations reveal that fusogenic peptides induce liposome shrinkage prior to membrane fusion, Proc. Natl. Acad. Sci. U.S.A. 101 (2004) 3420–3425.
[85] G.F. White, K.I. Racher, A. Lipski, F.R. Hallett, J.M. Wood, Physical properties of liposomes and proteoliposomes prepared from *Escherichia coli* polar lipids, Biochim. Biophys. Acta Biomembr. 1468 (2000) 175–186.
[86] H.A. Faizi, S.L. Frey, J. Steinkühler, R. Dimova, P.M. Vlahovska, Bending rigidity of charged lipid bilayer membranes, Soft Matter 15 (2019) 6006–6013.
[87] P.W. Fowler, J. Hélie, A. Duncan, M. Chavent, H. Koldsø, M.S.P. Sansom, Membrane stiffness is modified by integral membrane proteins, Soft Matter 12 (2016) 7792–7803.
[88] N. Jamin, M. Garrigos, C. Jaxel, A. Frelet-Barrand, S. Orlowski, Ectopic neo-formed intracellular membranes in *Escherichia coli*: a response to membrane protein-induced stress involving membrane curvature and domains, Biomolecules 8 (2018) 88.
[89] K. Simons, J.L. Sampaio, Membrane organization and lipid rafts, Cold Spring Harb. Perspect. Biol. 3 (2011) a004697–a004697.

[90] D. Balleza, Mechanical properties of lipid bilayers and regulation of mechanosensitive function, Channels 6 (2012) 220–233.
[91] T. Heimburg, Mechanical aspects of membrane thermodynamics. Estimation of the mechanical properties of lipid membranes close to the chain melting transition from calorimetry, Biochim. Biophys. Acta Biomembr. 1415 (1998) 147–162.
[92] J.F. Nagle, D.A. Wilkinson, Lecithin bilayers. Density measurement and molecular interactions, Biophys. J. 23 (1978) 159–175.
[93] D. Needham, E. Evans, Structure and mechanical properties of giant lipid (DMPC) vesicle bilayers from 20.degree.C below to 10.degree.C above the liquid crystal-crystalline phase transition at 24.degree, Clin. Biochem. 27 (1988) 8261–8269.
[94] R. Dimova, B. Pouligny, C. Dietrich, Pretransitional effects in dimyristoylphosphatidylcholine vesicle membranes: optical dynamometry study, Biophys. J. 79 (2000) 340–356.
[95] B. Gumí-Audenis, F. Sanz, M.I. Giannotti, Impact of galactosylceramides on the nanomechanical properties of lipid bilayer models: an AFM-force spectroscopy study, Soft Matter 11 (2015) 5447–5454.
[96] T. Heimburg, Phase Transitions in Biological Membranes, 2019, pp. 39–61.
[97] P. Marmottant, T. Biben, S. Hilgenfeldt, Deformation and rupture of lipid vesicles in the strong shear flow generated by ultrasound-driven microbubbles, Proc. R. Soc. A Math. Phys. Eng. Sci. 464 (2008) 1781–1800.
[98] K.A. Riske, R. Dimova, Electro-deformation and poration of giant vesicles viewed with high temporal resolution, Biophys. J. 88 (2005) 1143–1155.
[99] T. Portet, F. Camps I Febrer, J.M. Escoffre, C. Favard, M.P. Rols, D.S. Dean, Visualization of membrane loss during the shrinkage of giant vesicles under electropulsation, Biophys. J. 96 (2009) 4109–4121.
[100] W. Ding, M. Palaiokostas, G. Shahane, W. Wang, M. Orsi, Effects of high pressure on phospholipid bilayers, J. Phys. Chem. B 121 (2017) 9597–9606.
[101] Y.R. Sliozberg, Structural Changes in Lipid Vesicles Generated by the Shock Blast Waves: Coarse-Grained Molecular Dynamics Simulation, 2013.
[102] H. Wu, L. Yu, Y. Tong, A. Ge, S. Yau, M. Osawa, S. Ye, Enzyme-catalyzed hydrolysis of the supported phospholipid bilayers studied by atomic force microscopy, Biochim. Biophys. Acta Biomembr. 1828 (2013) 642–651.
[103] S.R. Tabaei, M. Rabe, H. Zetterberg, V.P. Zhdanov, F. Höök, Single lipid vesicle assay for characterizing single-enzyme kinetics of phospholipid hydrolysis in a complex biological fluid, J. Am. Chem. Soc. 135 (2013) 14151–14158.
[104] R.L. Knorr, M. Staykova, R.S. Gracià, R. Dimova, Wrinkling and electroporation of giant vesicles in the gel phase, Soft Matter 6 (2010) 1990–1996.
[105] W.W. Sułkowski, D. Pentak, K. Nowak, A. Sułkowska, M.P. Goertz, N. Goyal, G.A. Montano, B.C. Bunker, The influence of temperature, cholesterol content and pH on liposome stability, J. Mol. Struct. 744–747 (2005) 737–747.
[106] M.P. Goertz, N. Goyal, G.A. Montano, B.C. Bunker, Lipid bilayer reorganization under extreme pH conditions, Langmuir (2011) 5481–5491.
[107] M.M. Koerner, L.A. Palacio, J.W. Wright, K.S. Schweitzer, B.D. Ray, H.I. Petrache, Electrodynamics of lipid membrane interactions in the presence of zwitterionic buffers, Biophys. J. 101 (2011) 362–369.
[108] M.M.A.E. Claessens, B.F. Van Oort, F.A.M. Leermakers, F.A. Hoekstra, M.A.C. Stuart, Charged lipid vesicles: effects of salts on bending rigidity, stability, and size, Biophys. J. 87 (2004) 3882–3893.
[109] A. Kodama, M. Morandi, R. Ebihara, T. Jimbo, M. Toyoda, Y. Sakuma, M. Imai, N. Pu, M.I. Angelova, Migration of deformable vesicles induced by ionic stimuli, Langmuir (2018).

[110] R.A. Böckmann, A. Hac, T. Heimburg, H. Grubmüller, Effect of sodium chloride on a lipid bilayer, Biophys. J. 85 (2003) 1647–1655.
[111] R.A. Böckmann, H. Grubmüller, Multistep binding of divalent cations to phospholipid bilayers: a molecular dynamics study, Angew. Chem. Int. Ed. 43 (2004) 1021–1024.
[112] K. Inque, T. Kitagawa, Effect of lipid composition on sensitivity of lipid membranes to triton X-100, Biochim. Biophys. Acta Biomembr. 426 (1976) 1–16.
[113] B.R. Casadei, C.C. Domingues, E. De Paula, K.A. Riske, Direct visualization of the action of triton X-100 on giant vesicles of erythrocyte membrane lipids, Biophys. J. 106 (2014) 2417–2425.
[114] A. Jeucken, M.R. Molenaar, C.H.A. van de Lest, J.W.A. Jansen, J.B. Helms, J.F. Brouwers, A comprehensive functional characterization of *Escherichia coli* lipid genes, Cell Rep. 27 (2019) 1597–1606.e2.
[115] J.-W. Park, D.J. Ahn, Temperature effect on nanometer-scale physical properties of mixed phospholipid monolayers, Colloids Surf. B Biointerfaces 62 (2008) 157–161.
[116] H.M. Seeger, G. Marino, A. Alessandrini, P. Facci, Effect of physical parameters on the main phase transition of supported lipid bilayers, Biophys. J. 97 (2009) 1067–1076.
[117] M. Edidin, The state of lipid rafts: from model membranes to cells, Annu. Rev. Biophys. Biomol. Struct. 32 (2003) 257–283.
[118] P. Calvez, J. Jouhet, V. Vié, C. Durmort, A. Zapun, Lipid phases and cell geometry during the cell cycle of Streptococcus pneumoniae, Front. Microbiol. 10 (2019).
[119] Y. Sun, T.-L. Sun, H.W. Huang, Physical properties of *Escherichia coli* spheroplast membranes, Biophys. J. 107 (2014) 2082–2090.
[120] L. Lu, W.J. Doak, J.W. Schertzer, P.R. Chiarot, Membrane mechanical properties of synthetic asymmetric phospholipid vesicles, Soft Matter 12 (2016) 7521–7528.
[121] H.T. McMahon, E. Boucrot, Membrane curvature at a glance, J. Cell Sci. 128 (2015) 1065–1070.
[122] M. Pinot, S. Vanni, E. Ambroggio, D. Guet, B. Goud, J.-B. Manneville, Feedback Between Membrane Tension, Lipid Shape and Curvature in the Formation of Packing Defects, 2018.
[123] H.T. McMahon, J.L. Gallop, Membrane curvature and mechanisms of dynamic cell membrane remodelling, Nature 438 (2005) 590–596.
[124] Y. Shibata, J. Hu, M.M. Kozlov, T.A. Rapoport, Mechanisms shaping the membranes of cellular organelles, Annu. Rev. Cell Dev. Biol. 25 (2009) 329–354.
[125] A. Martyna, B. Bahsoun, M.D. Badham, S. Srinivasan, M.J. Howard, J.S. Rossman, Membrane remodeling by the M2 amphipathic helix drives influenza virus membrane scission, Sci. Rep. 7 (2017) 44695.
[126] H.P. Erickson, M. Osawa, FtsZ constriction force – curved protofilaments bending membranes, Subcell Biochem (2017) 139–160.
[127] T.N.V. Saaki, H. Strahl, L.W. Hamoen, Membrane curvature and the tol-pal complex determine polar localization of the chemoreceptor tar in *Escherichia coli*, J. Bacteriol. 200 (2018).
[128] H. Alimohamadi, B. Ovryn, P. Rangamani, Modeling membrane nanotube morphology: the role of heterogeneity in composition and material properties, Sci. Rep. 10 (2020) 2527.
[129] W. Römer, L. Berland, V. Chambon, K. Gaus, B. Windschiegl, D. Tenza, M.R.E. Aly, V. Fraisier, J.-C. Florent, D. Perrais, C. Lamaze, G. Raposo, C. Steinem, P. Sens, P. Bassereau, L. Johannes, Shiga toxin induces tubular membrane invaginations for its uptake into cells, Nature 450 (2007) 670–675.
[130] I.K. Jarsch, F. Daste, J.L. Gallop, Membrane curvature in cell biology: an integration of molecular mechanisms, J. Cell Biol. 214 (2016) 375–387.

[131] H. Strahl, J. Errington, Bacterial membranes: structure, domains, and function, Annu. Rev. Microbiol. 71 (2017) 519–538.
[132] Q. Deng, L. Liu, P. Sharma, Flexoelectricity in soft materials and biological membranes, J. Mech. Phys. Solid. 62 (2014) 209–227.
[133] V. Vitkova, A.G. Petrov, Lipid Bilayers and Membranes, 2013, pp. 89–138.
[134] D. Needham, R.S. Nunn, Elastic deformation and failure of lipid bilayer membranes containing cholesterol, Biophys. J. 58 (1990) 997–1009.
[135] S.-T. Yang, A.J.B. Kreutzberger, J. Lee, V. Kiessling, L.K. Tamm, The role of cholesterol in membrane fusion, Chem. Phys. Lipids 199 (2016) 136–143.
[136] S.E. Feller, K. Gawrisch, A.D. MacKerell, Polyunsaturated fatty acids in lipid bilayers: intrinsic and environmental contributions to their unique physical properties, J. Am. Chem. Soc. 124 (2002) 318–326.
[137] G. Ourisson, M. Rohmer, K. Poralla, Prokaryotic hopanoids and other polyterpenoid sterol surrogates, Annu. Rev. Microbiol. 41 (1987) 301–333.
[138] J.P. Sáenz, D. Grosser, A.S. Bradley, T.J. Lagny, O. Lavrynenko, M. Broda, K. Simons, Hopanoids as functional analogues of cholesterol in bacterial membranes, Proc. Natl. Acad. Sci. U.S.A. 112 (2015) 11971–11976.
[139] Y.-L. Shih, L. Rothfield, The bacterial cytoskeleton, Microbiol. Mol. Biol. Rev. 70 (2006) 729–754.
[140] C. Di Ciano-Oliveira, A.C.P. Thirone, K. Szaszi, A. Kapus, Osmotic stress and the cytoskeleton: the R(h)ole of Rho GTPases, Acta Physiol. 187 (2006) 257–272.
[141] A. Typas, M. Banzhaf, C.A. Gross, W. Vollmer, From the regulation of peptidoglycan synthesis to bacterial growth and morphology, Nat. Rev. Microbiol. 10 (2012) 123–136.
[142] M. Jansen, A. Blume, A comparative study of diffusive and osmotic water permeation across bilayers composed of phospholipids with different head groups and fatty acyl chains, Biophys. J. 68 (1995) 997–1008.
[143] J.T. Mika, P.E. Schavemaker, V. Krasnikov, B. Poolman, Impact of osmotic stress on protein diffusion in L actococcus lactis, Mol. Microbiol. 94 (2014) 857–870.
[144] J.C. Gumbart, M. Beeby, G.J. Jensen, B. Roux, *Escherichia coli* peptidoglycan structure and mechanics as predicted by atomic-scale simulations, PLoS Comput. Biol. 10 (2014) e1003475.
[145] M.T. Madigan, J.M. Martinko, P.V. Dunlap, D.P. Clark, Brock Biology of Microorganisms, twelfth ed., Pearson Education, San Francisco, 2009.
[146] E.R. Rojas, K.C. Huang, J.A. Theriot, Homeostatic cell growth is accomplished mechanically through membrane tension inhibition of cell-wall synthesis, Cell Syst. 5 (2017) 578–590.e6.
[147] E.R. Rojas, K.C. Huang, Regulation of microbial growth by turgor pressure, Curr. Opin. Microbiol. 42 (2018) 62–70.
[148] R.C. Hughes, Autolysis of Bacillus cereus cell walls and isolation of structural components, Biochem. J. 121 (1971) 791–802.
[149] I.G. Boneca, Z.-H. Huang, D.A. Gage, A. Tomasz, Characterization of *Staphylococcus aureus* cell wall glycan strands, evidence for a new β- N- acetylglucosaminidase activity, J. Biol. Chem. 275 (2000) 9910–9918.
[150] D. Sychantha, A.S. Brott, C.S. Jones, A.J. Clarke, Mechanistic pathways for peptidoglycan O-acetylation and de-O-acetylation, Front. Microbiol. 9 (2018).
[151] M.-H. Laaberki, J. Pfeffer, A.J. Clarke, J. Dworkin, O -acetylation of peptidoglycan is required for proper cell separation and S-layer anchoring in Bacillus anthracis, J. Biol. Chem. 286 (2011) 5278–5288.
[152] A.K. Yadav, A. Espaillat, F. Cava, Bacterial strategies to preserve cell wall integrity against environmental threats, Front. Microbiol. 9 (2018).

[153] M.A. de Pedro, F. Cava, Structural constraints and dynamics of bacterial cell wall architecture, Front. Microbiol. 6 (2015).
[154] T. Kruse, J. Bork-Jensen, K. Gerdes, The morphogenetic MreBCD proteins of *Escherichia coli* form an essential membrane-bound complex, Mol. Microbiol. 55 (2005) 78–89.
[155] F. van den Ent, C.M. Johnson, L. Persons, P. de Boer, J. Löwe, Bacterial actin MreB assembles in complex with cell shape protein RodZ, EMBO J. 29 (2010) 1081–1090.
[156] S.X. Sun, H. Jiang, Physics of bacterial morphogenesis, Microbiol. Mol. Biol. Rev. 75 (2011) 543–565.
[157] H. Jiang, F. Si, W. Margolin, S.X. Sun, Mechanical control of bacterial cell shape, Biophys. J. 101 (2011) 327–335.
[158] F. Chattaway, Microbial cell walls and membranes, Biochem. Educ. 10 (1982) 33.
[159] B. Glauner, J. V Höltje, U. Schwarz, The composition of the murein of *Escherichia coli*, J. Biol. Chem. 263 (1988) 10088–10095.
[160] M.A. Snowden, H.R. Perkins, Peptidoglycan cross-linking in *Staphylococcus aureus*. An apparent random polymerisation process, Eur. J. Biochem. 191 (1990) 373–377.
[161] A.G. Pisabarro, M.A. de Pedro, D. Vázquez, Structural modifications in the peptidoglycan of *Escherichia coli* associated with changes in the state of growth of the culture, J. Bacteriol. 161 (1985) 238–242.
[162] B. Blasco, A.G. Pisabarro, M.A. de Pedro, Peptidoglycan biosynthesis in stationary-phase cells of *Escherichia coli*, J. Bacteriol. 170 (1988) 5224–5228.
[163] D. Gully, D. Gargani, K. Bonaldi, C. Grangeteau, C. Chaintreuil, J. Fardoux, P. Nguyen, R. Marchetti, N. Nouwen, A. Molinaro, P. Mergaert, E. Giraud, A peptidoglycan-remodeling enzyme is critical for bacteroid differentiation in bradyrhizobium spp. during legume symbiosis, Mol. Plant Microbe Interact. 29 (2016) 447–457.
[164] A.S. Ghosh, C. Chowdhury, D.E. Nelson, Physiological functions of D-alanine carboxypeptidases in *Escherichia coli*, Trends Microbiol. 16 (2008) 309–317.
[165] D.E. Nelson, K.D. Young, Contributions of PBP 5 and DD-carboxypeptidase penicillin binding proteins to maintenance of cell shape in *Escherichia coli*, J. Bacteriol. 183 (2001) 3055–3064.
[166] K.D. Young, Bacterial shape, Mol. Microbiol. 49 (2004) 571–580.
[167] A.L. Koch, S.L. Lane, J.A. Miller, D.G. Nickens, Contraction of filaments of *Escherichia coli* after disruption of cell membrane by detergent, J. Bacteriol. 169 (1987) 1979–1984.
[168] X. Yao, M. Jericho, D. Pink, T. Beveridge, Thickness and elasticity of gram-negative murein sacculi measured by atomic force microscopy, J. Bacteriol. 181 (1999) 6865–6875.
[169] G. Van Den Bogaart, N. Hermans, V. Krasnikov, B. Poolman, Protein mobility and diffusive barriers in *Escherichia coli*: consequences of osmotic stress, Mol. Microbiol. 64 (2007) 858–871.
[170] W. Vollmer, P. Born, Bacterial cell envelope peptidoglycan, in: Microb. Glycobiol., Elsevier, 2010, pp. 15–28.
[171] R.C. Hughes, P.F. Thurman, E. Stokes, Estimates of the porosity of Bacillus licheniformis and Bacillus subtilis cell walls, Z. für Immunitatsforsch. Exp. Klin. Immunol. 149 (1975) 126–135.
[172] W. Vollmer, J.-V. Holtje, The architecture of the murein (peptidoglycan) in gram-negative bacteria: vertical scaffold or horizontal layer(s)? J. Bacteriol. 186 (2004) 5978–5987.
[173] M. Lin, Y. Rikihisa, Ehrlichia chaffeensis and Anaplasma phagocytophilum lack genes for lipid A biosynthesis and incorporate cholesterol for their survival, Infect. Immun. 71 (2003) 5324–5331.

[174] C. Otten, M. Brilli, W. Vollmer, P.H. Viollier, J. Salje, Peptidoglycan in obligate intracellular bacteria, Mol. Microbiol. 107 (2018) 142–163.
[175] K. Bush, Antimicrobial agents targeting bacterial cell walls and cell membranes, Rev. Sci. Tech. l'OIE. 31 (2012) 43–56.
[176] H. Nikaido, Bacterial resistance to antibiotics as a function of outer membrane permeability, J. Antimicrob. Chemother. 22 (1988) 17–22.
[177] Z. Pang, R. Raudonis, B.R. Glick, T.-J. Lin, Z. Cheng, Antibiotic resistance in *Pseudomonas aeruginosa*: mechanisms and alternative therapeutic strategies, Biotechnol. Adv. 37 (2019) 177–192.
[178] A.T. Vincent, S. Nyongesa, I. Morneau, M.B. Reed, E.I. Tocheva, F.J. Veyrier, The mycobacterial cell envelope: a relict from the past or the result of recent evolution? Front. Microbiol. 9 (2018).
[179] K.L. May, M. Grabowicz, The bacterial outer membrane is an evolving antibiotic barrier, Proc. Natl. Acad. Sci. U.S.A. 115 (2018) 8852–8854.
[180] H. Nikaido, Molecular basis of bacterial outer membrane permeability revisited, Microbiol. Mol. Biol. Rev. 67 (2003) 593–656.
[181] D.G. Thanassi, L.W. Cheng, H. Nikaido, Active efflux of bile salts by *Escherichia coli*, J. Bacteriol. 179 (1997) 2512–2518.
[182] B. Bertani, N. Ruiz, Function and biogenesis of lipopolysaccharides, EcoSal Plus 8 (2018).
[183] A.V. Hughes, D.S. Patel, G. Widmalm, J.B. Klauda, L.A. Clifton, W. Im, Physical properties of bacterial outer membrane models: neutron reflectometry & molecular simulation, Biophys. J. 116 (2019) 1095–1104.
[184] L.A. Clifton, M.W.A. Skoda, A.P. Le Brun, F. Ciesielski, I. Kuzmenko, S.A. Holt, J.H. Lakey, Effect of divalent cation removal on the structure of gram-negative bacterial outer membrane models, Langmuir 31 (2015) 404–412.
[185] H.-L. Alakomi, A. Paananen, M.-L. Suihko, I.M. Helander, M. Saarela, Weakening effect of cell permeabilizers on gram-negative bacteria causing biodeterioration, Appl. Environ. Microbiol. 72 (2006) 4695–4703.
[186] L. Leive, Release of lipopolysaccharide by EDTA treatment of *E. coli*, Biochem. Biophys. Res. Commun. 21 (1965) 290–296.
[187] V. Braun, K. Rehn, Chemical characterization, spatial distribution and function of a lipoprotein (Murein-Lipoprotein) of the *E. coli* cell wall. The specific effect of trypsin on the membrane structure, Eur. J. Biochem. 10 (1969) 426–438.
[188] V. Braun, U. Sieglin, The covalent murein-lipoprotin structure of the *Escherichia coli* cell wall. The attachment site of the lipoprotein on the murein, Eur. J. Biochem. 13 (1970) 336–346.
[189] M. Mathelié-Guinlet, A.T. Asmar, J.-F. Collet, Y.F. Dufrêne, Lipoprotein Lpp regulates the mechanical properties of the *E. coli* cell envelope, Nat. Commun. 11 (2020).
[190] R. Godlewska, K. WiÅ›niewska, Z. Pietras, E.K. Jagusztyn-Krynicka, Peptidoglycan-associated lipoprotein (Pal) of Gram-negative bacteria: function, structure, role in pathogenesis and potential application in immunoprophylaxis, FEMS Microbiol. Lett. 298 (2009) 1–11.
[191] A. Kovacs-Simon, R.W. Titball, S.L. Michell, Lipoproteins of bacterial pathogens, Infect. Immun. 79 (2011) 548–561.
[192] M. Leduc, K. Ishidate, N. Shakibai, L. Rothfield, Interactions of *Escherichia coli* membrane lipoproteins with the murein sacculus, J. Bacteriol. 174 (1992) 7982–7988.
[193] E. Bouveret, R. Derouiche, A. Rigal, R. Lloubes, C. Lazdunski, H. Benedetti, Peptidoglycan-associated lipoprotein-TolB interaction. A possible key to explaining the formation of contact sites between the inner and outer membranes of *Escherichia coli*, J. Biol. Chem. 270 (1995) 11071–11077.

[194] T. Clavel, P. Germon, A. Vianney, R. Portalier, J.C. Lazzaroni, TolB protein of *Escherichia coli* K-12 interacts with the outer membrane peptidoglycan-associated proteins Pal, Lpp and OmpA, Mol. Microbiol. 29 (1998) 359–367.
[195] E. Cascales, A. Bernadac, M. Gavioli, J.-C. Lazzaroni, R. Lloubes, Pal lipoprotein of *Escherichia coli* plays a major role in outer membrane integrity, J. Bacteriol. 184 (2002) 754–759.
[196] E. Bouveret, H. Bénédetti, A. Rigal, E. Loret, C. Lazdunski, In vitro characterization of peptidoglycan-associated lipoprotein (PAL)-peptidoglycan and PAL-TolB interactions, J. Bacteriol. 181 (1999) 6306–6311.
[197] E. Cascales, M. Gavioli, J.N. Sturgis, R. Lloubes, Proton motive force drives the interaction of the inner membrane TolA and outer membrane Pal proteins in *Escherichia coli*, Mol. Microbiol. 38 (2000) 904–915.
[198] C. Deprez, R. Lloubès, M. Gavioli, D. Marion, F. Guerlesquin, L. Blanchard, Solution structure of the *E.coli* TolA C-terminal domain reveals conformational changes upon binding to the phage g3p N-terminal domain, J. Mol. Biol. 346 (2005) 1047–1057.
[199] Y.-C. Yeh, L.R. Comolli, K.H. Downing, L. Shapiro, H.H. McAdams, The Caulobacter tol-pal complex is essential for outer membrane integrity and the positioning of a polar localization factor, J. Bacteriol. 192 (2010) 4847–4858.
[200] A.M. Berezuk, S. Glavota, E.J. Roach, M.C. Goodyear, J.R. Krieger, C.M. Khursigara, Outer membrane lipoprotein RlpA is a novel periplasmic interaction partner of the cell division protein FtsK in *Escherichia coli*, Sci. Rep. 8 (2018) 1–14.
[201] M.A. Gerding, Y. Ogata, N.D. Pecora, H. Niki, P.A.J. de Boer, The trans -envelope Tol-Pal complex is part of the cell division machinery and required for proper outer-membrane invagination during cell constriction in *E. coli*, Mol. Microbiol. 63 (2007) 1008–1025.
[202] A.L. Koch, Orientation of the peptidoglycan chains in the sacculus of *Escherichia coli*, Res. Microbiol. 149 (1998) 689–701.
[203] J. V Höltje, Growth of the stress-bearing and shape-maintaining murein sacculus of *Escherichia coli*, Microbiol. Mol. Biol. Rev. 62 (1998) 181–203.
[204] O.H. Vandal, C.F. Nathan, S. Ehrt, Acid resistance in *Mycobacterium tuberculosis*, J. Bacteriol. 191 (2009) 4714–4721.
[205] K. Takayama, C. Wang, G.S. Besra, Pathway to synthesis and processing of mycolic acids in *Mycobacterium tuberculosis*, Clin. Microbiol. Rev. 18 (2005) 81–101.
[206] P.J. Brennan, H. Nikaido, The envelope of mycobacteria, Annu. Rev. Biochem. 64 (1995) 29–63.
[207] M. Daffé, P. Draper, The envelope layers of mycobacteria with reference to their pathogenicity, Adv. Microb. Physiol. (1997) 131–203.
[208] E. Dubnau, J. Chan, C. Raynaud, V.P. Mohan, M.-A. Lanéelle, K. Yu, A. Quémard, I. Smith, M. Daffé, Oxygenated mycolic acids are necessary for virulence of *Mycobacterium tuberculosis* in mice, Mol. Microbiol. 36 (2002) 630–637.
[209] M.S. Glickman, J.S. Cox, W.R. Jacobs, A novel mycolic acid cyclopropane synthetase is required for cording, persistence, and virulence of *Mycobacterium tuberculosis*, Mol. Cell. 5 (2000) 717–727.
[210] J. Liu, C.E. Barry, G.S. Besra, H. Nikaido, Mycolic acid structure determines the fluidity of the mycobacterial cell wall, J. Biol. Chem. 271 (1996) 29545–29551.
[211] Y. Yuan, Y. Zhu, D.D. Crane, C.E. Barry III, The effect of oxygenated mycolic acid composition on cell wall function and macrophage growth in *Mycobacterium tuberculosis*, Mol. Microbiol. 29 (1998) 1449–1458.
[212] C. Asselineau, J. Asselineau, Trehalose-containing glycolipids, Prog. Chem. Fats Other Lipids 16 (1978) 59–99.

[213] C.E. Barry, R.E. Lee, K. Mdluli, A.E. Sampson, B.G. Schroeder, R.A. Slayden, Y. Yuan, Mycolic acids: structure, biosynthesis and physiological functions, Prog. Lipid Res. 37 (1998) 143–179.
[214] A. Zureck, G. P. Kubica, L. G. Wayne (editors), the mycobacteria — a sourcebook, Part A and Part B (microbiology series volume 15). 1553 S., 225 abb., 122 tab. New York-basel 1984. Marcel dekker. $ 295.00. ISBN: 0-8247-7009-9 (Part A), 0-8247-1917-4 (Part B), J. Basic Microbiol. 25 (1985) 662–662.
[215] K.M. George, Y. Yuan, D.R. Sherman, C.E. Barry, The biosynthesis of cyclopropanated mycolic acids in *Mycobacterium tuberculosis*, J. Biol. Chem. 270 (1995) 27292–27298.
[216] Y. Yuan, R.E. Lee, G.S. Besra, J.T. Belisle, C.E. Barry, Identification of a gene involved in the biosynthesis of cyclopropanated mycolic acids in *Mycobacterium tuberculosis*, Proc. Natl. Acad. Sci. U.S.A. 92 (1995) 6630–6634.
[217] J.R. Battista, Against all odds:the survival strategies of *Deinococcus radiodurans*, Annu. Rev. Microbiol. 51 (1997) 203–224.
[218] D. Slade, M. Radman, Oxidative stress resistance in Deinococcus radiodurans, Microbiol. Mol. Biol. Rev. 75 (2011) 133–191.
[219] J. Yu, T. Li, S. Dai, Y. Weng, J. Li, Q. Li, H. Xu, Y. Hua, B. Tian, A tamB homolog is involved in maintenance of cell envelope integrity and stress resistance of *Deinococcus radiodurans*, Sci. Rep. 7 (2017) 45929.
[220] H. Rothfuss, J.C. Lara, A.K. Schmid, M.E. Lidstrom, Involvement of the S-layer proteins Hpi and SlpA in the maintenance of cell envelope integrity in *Deinococcus radiodurans* R1, Microbiology 152 (2006) 2779–2787.
[221] B.G. Thompson, R. Anderson, R.G.E. Murray, Unusual polar lipids of Micrococcus radiodurans strain Sark, Can. J. Microbiol. 26 (1980) 1408–1411.
[222] U.B. Sleytr, P. Messner, D. Pum, M. Sára, Occurrence, location, ultrastructure and morphogenesis of S-layers, in: Cryst. Bact. Cell Surf. Proteins, Elsevier, 1996, pp. 5–33.
[223] U.B. Sleytr, B. Schuster, E.-M. Egelseer, D. Pum, S-layers: principles and applications, FEMS Microbiol. Rev. 38 (2014) 823–864.
[224] M. Sara, E.M. Egelseer, Functional aspects of S-layers, in: U.W. Sleytr, P. Messner, D. Pum, M. Sara (Eds.), Cryst. Bact. Cell Surf. Proteins, Academic press, Austin, Tex, 1996, pp. 103–131.
[225] U.B. Sleytr, P. Messner, Crystalline surface layers on bacteria, Annu. Rev. Microbiol. 37 (1983) 311–339.
[226] P. Messner, K. Steiner, K. Zarschler, C. Schäffer, S-layer nanoglycobiology of bacteria, Carbohydr. Res. 343 (2008) 1934–1951.
[227] P. Messner, E.M. Egelseer, U.B. Sleytr, C. Schäffer, Bacterial surface layer glycoproteins and "non-classical" secondary cell wall polymers, in: Microb. Glycobiol., Elsevier, 2010, pp. 109–128.
[228] J. Šmarda, D. Šmajs, J. Komrska, V. Krzyžánek, S-layers on cell walls of cyanobacteria, Micron 33 (2002) 257–277.
[229] U.B. Sleytr, R. Plohberger, The dynamic process of assembly of two-dimensional arrays of macromolecules on bacteria cell walls, Electron Microscopy Mol. Dim. (1980) 36–47.
[230] D. Pum, P. Messner, U.B. Sleytr, Role of the S layer in morphogenesis and cell division of the archaebacterium Methanocorpusculum sinense, J. Bacteriol. 173 (1991) 6865–6873.
[231] U. Sleytr, Bacterial S-layers, Trends Microbiol. 7 (1999) 253–260.
[232] U.B. Sleytr, M. Sára, D. Pum, B. Schuster, P. Messner, C. Schäffer, Self-assembling protein systems: microbial S-layers, in: S.R. Fahnestock, A. Steinbüchel (Eds.),

Biopolym, Wiley-VCH Verlag GmbH & Co. KGaA, Weinheim, Germany, 2005. Online.
[233] U.B. Sleytr, P. D, E. E.M, I. N, S. N, S-layer proteins, in: Knoll. Wolfgang (Ed.), Handb. Biofunctional Surfaces, Pan Stanford Publishing, 2013, pp. 507–568.
[234] C. Mader, S. Küpcü, M. Sára, U. Sleytr, Stabilizing effect of an S-layer on liposomes towards thermal or mechanical stress, Biochim. Biophys. Acta Biomembr. 1418 (1999) 106–116.
[235] P. Messner, C. Schäffer, E.-M. Egelseer, U.B. Sleytr, Occurrence, structure, chemistry, genetics, morphogenesis, and functions of S-layers, in: Prokaryotic Cell Wall Compd, Springer Berlin Heidelberg, Berlin, Heidelberg, 2010, pp. 53–109.
[236] T. Pavkov-Keller, S. Howorka, W. Keller, The structure of bacterial S-layer proteins, Prog. Mol. Biol. Transl. Sci. (2011) 73–130.
[237] M. Sára, U.B. Sleytr, Molecular sieving through S layers of Bacillus stearothermophilus strains, J. Bacteriol. 169 (1987) 4092–4098.
[238] M. Sára, U.B. Sleytr, Charge distribution on the S layer of Bacillus stearothermophilus NRS 1536/3c and importance of charged groups for morphogenesis and function, J. Bacteriol. 169 (1987) 2804–2809.
[239] K. Gruber, U.B. Sleytr, Influence of an S-layer on surface properties of Bacillus stearothermophilus, Arch. Microbiol. 156 (1991) 181–185.
[240] S. Weigert, M. Sára, Surface modification of an ultrafiltration membrane with crystalline structure and studies on interactions with selected protein molecules, J. Membr. Sci. 106 (1995) 147–159.
[241] Y.-L. Ji, Q.-F. An, Q. Zhao, W.-D. Sun, K.-R. Lee, H.-L. Chen, C.-J. Gao, Novel composite nanofiltration membranes containing zwitterions with high permeate flux and improved anti-fouling performance, J. Membr. Sci. 390–391 (2012) 243–253.
[242] A. Chandramohan, E. Duprat, L. Remusat, S. Zirah, C. Lombard, A. Kish, Novel mechanism for surface layer shedding and regenerating in bacteria exposed to metal-contaminated conditions, Front. Microbiol. 9 (2019).
[243] J.B. Thompson, F.G. Ferris, Cyanobacterial precipitation of gypsum, calcite, and magnesite from natural alkaline lake water, Geology 18 (1990) 995.
[244] S. Schultze-Lam, G. Harauz, T.J. Beveridge, Participation of a cyanobacterial S layer in fine-grain mineral formation, J. Bacteriol. 174 (1992) 7971–7981.
[245] S. Schultze-Lam, T.J. Beveridge, Physicochemical characteristics of the mineral-forming S-layer from the cyanobacterium Synechococcus strain GL24, Can. J. Microbiol. 40 (1994) 216–223.
[246] R.J.C. McLean, T.J. Beveridge, Metal-binding capacity of bacterial surfaces and their ability to form mineralized aggregates, in: H.L. Ehrlich, C.L. Brierley (Eds.), Microb. Miner. Recover., McGraw-Hill Publishing Co., New York, 1990, pp. 185–222.
[247] T.J. Beveridge, L.L. Graham, Surface layers of bacteria, Microbiol. Rev. 55 (1991) 684–705.
[248] T.J. Beveridge, Ultrastructure, chemistry, and function of the bacterial wall, Int. Rev. Cytol. (1981) 229–317.
[249] C. Whitfield, Bacterial extracellular polysaccharides, Can. J. Microbiol. 34 (1988) 415–420.
[250] K. Jann, B. Jann, Polysaccharide antigens of *Escherichia coli*, Clin. Infect. Dis. 9 (1987) S517–S526.
[251] I. Orskov, F. Orskov, B. Jann, K. Jann, Serology, chemistry, and genetics of O and K antigens of *Escherichia coli*, Bacteriol. Rev. 41 (1977) 667–710.
[252] C. Whitfield, Biosynthesis and assembly of capsular polysaccharides in *Escherichia coli*, Annu. Rev. Biochem. 75 (2006) 39–68.
[253] P.A. Amor, C. Whitfield, Molecular and functional analysis of genes required for expression of group IB K antigens in *Escherichia coli*: characterization of the his- region

containing gene clusters for multiple cell-surface polysaccharides, Mol. Microbiol. 26 (1997) 145–161.
[254] P.R. MacLachlan, W.J. Keenleyside, C. Dodgson, C. Whitfield, Formation of the K30 (group I) capsule in *Escherichia coli* O9:K30 does not require attachment to lipopolysaccharide lipid A-core, J. Bacteriol. 175 (1993) 7515–7522.
[255] S. Sachdeva, N. Kolimi, S.A. Nair, T. Rathinavelan, Key diffusion mechanisms involved in regulating bidirectional water permeation across *E. coli* outer membrane lectin, Sci. Rep. 6 (2016) 28157.
[256] E.C. Gotschlich, B.A. Fraser, O. Nishimura, J.B. Robbins, T.Y. Liu, Lipid on capsular polysaccharides of gram-negative bacteria, J. Biol. Chem. 256 (1981) 8915–8921.
[257] G.R. de S. Araújo, N.B. Viana, F. Gómez, B. Pontes, S. Frases, The mechanical properties of microbial surfaces and biofilms, Cell Surf. 5 (2019) 100028.
[258] J.W. Costerton, R.T. Irvin, K.J. Cheng, The bacterial glycocalyx in nature and disease, Annu. Rev. Microbiol. 35 (1981) 299–324.
[259] M.E. Bayer, H. Thurow, Polysaccharide capsule of *Escherichia coli*: microscope study of its size, structure, and sites of synthesis, J. Bacteriol. 130 (1977) 911–936.
[260] I.S. Roberts, The biochemistry and genetics of capsular polysaccharide production IN bacteria, Annu. Rev. Microbiol. 50 (1996) 285–315.
[261] D.A. Rasko, V. Sperandio, Anti-virulence strategies to combat bacteria-mediated disease, Nat. Rev. Drug Discov. 9 (2010) 117–128.
[262] H.-J. Wu, A.H.-J. Wang, M.P. Jennings, Discovery of virulence factors of pathogenic bacteria, Curr. Opin. Chem. Biol. 12 (2008) 93–101.
[263] T.J. Beveridge, The structure of bacteria, in: J.S. Poindexter, E.R. Leadbetter (Eds.), Bact. Nat., Plenum Publishing Corp., New York, 1989, pp. 1–65.
[264] J.M. Larkin, W.R. Strohl, Beggiatoa, Thiothrix, and Thioploca, Annu. Rev. Microbiol. 37 (1983) 341–367.
[265] R.Y. Stanier, G.C. Bazine, Phototrophic prokaryotes: the cyanobacteria, Annu. Rev. Microbiol. 31 (1977) 225–274.
[266] T.J. Beveridge, The bacterial surface: general considerations towards design and function, Can. J. Microbiol. 34 (1988) 363–372.
[267] W.C. Ghiorse, Biology of iron-and manganese-depositing bacteria, Annu. Rev. Microbiol. 38 (1984) 515–550.
[268] L.F. Adams, W.C. Ghiorse, Characterization of extracellular $Mn^{2+}$-oxidizing activity and isolation of an $Mn^{2+}$-oxidizing protein from Leptothrix discophora SS-1, J. Bacteriol. 169 (1987) 1279–1285.
[269] F.C. Boogerd, J.P. de Vrind, Manganese oxidation by Leptothrix discophora, J. Bacteriol. 169 (1987) 489–494.
[270] M. Furutani, Assemblage of bacterial saccharic microfibrils in sheath skeleton formed by cultured leptothrix sp. strain OUMS1, J. Mar. Sci. Res. Dev. 3 (2013).
[271] C.S. Chan, Microbial polysaccharides template Assembly of nanocrystal fibers, Science 303 (2004) 1656–1658 (80-. ).
[272] T.J. Beveridge, R.J. Doyle, Metal Ions and Bacteria, Wiley, 1989.
[273] L.L. Graham, R. Harris, W. Villiger, T.J. Beveridge, Freeze-substitution of gram-negative eubacteria: general cell morphology and envelope profiles, J. Bacteriol. 173 (1991) 1623–1633.
[274] W.L. Nicholson, N. Munakata, G. Horneck, H.J. Melosh, P. Setlow, Resistance of Bacillus endospores to extreme terrestrial and extraterrestrial environments, Microbiol. Mol. Biol. Rev. 64 (2000) 548–572.
[275] P.T. McKenney, A. Driks, P. Eichenberger, The Bacillus subtilis endospore: assembly and functions of the multilayered coat, Nat. Rev. Microbiol. 11 (2013) 33–44.
[276] S. Cogliati, J. Gabriel Costa, Bacterial spores and its relatives as agents of mass destruction, J. Bioterrorism Biodefense 07 (2016).

[277] G.C. Stewart, The exosporium layer of bacterial spores: a connection to the environment and the infected host, Microbiol. Mol. Biol. Rev. 79 (2015) 437–457.
[278] C. Steichen, P. Chen, J.F. Kearney, C.L. Turnbough, Identification of the immunodominant protein and other proteins of the Bacillus anthracis exosporium, J. Bacteriol. 185 (2003) 1903–1910.
[279] L. Kailas, C. Terry, N. Abbott, R. Taylor, N. Mullin, S.B. Tzokov, S.J. Todd, B.A. Wallace, J.K. Hobbs, A. Moir, P.A. Bullough, Surface architecture of endospores of the Bacillus cereus/anthracis/thuringiensis family at the subnanometer scale, Proc. Natl. Acad. Sci. U.S.A. 108 (2011) 16014–16019.
[280] D. Paredes-Sabja, P. Setlow, M.R. Sarker, Germination of spores of Bacillales and Clostridiales species: mechanisms and proteins involved, Trends Microbiol. 19 (2011) 85–94.
[281] I.M. Shah, M.-H. Laaberki, D.L. Popham, J. Dworkin, A eukaryotic-like ser/thr kinase signals bacteria to exit dormancy in response to peptidoglycan fragments, Cell 135 (2008) 486–496.
[282] A.I. Aronson, P. Fitz-James, Structure and morphogenesis of the bacterial spore coat, Bacteriol. Rev. 40 (1976) 360–402.
[283] A. Driks, Bacillus subtilis spore coat, Microbiol. Mol. Biol. Rev. 63 (1999) 1–20.
[284] E.I. Tocheva, E.G. Matson, D.M. Morris, F. Moussavi, J.R. Leadbetter, G.J. Jensen, Peptidoglycan remodeling and conversion of an inner membrane into an outer membrane during sporulation, Cell 146 (2011) 799–812.
[285] M. Hegreness, R. Kishony, Analysis of genetic systems using experimental evolution and whole-genome sequencing, Genome Biol. 8 (2007) 201.
[286] W. Vollmer, Bacterial outer membrane evolution via sporulation? Nat. Chem. Biol. 8 (2012) 14–18.
[287] D. Zerbib, Bacterial cell envelopes: composition, architecture, and origin, in: D. Miklavčič (Ed.), Handb. Electroporation, Springer International Publishing, Cham, 2017, pp. 417–436.

FSC
www.fsc.org
MIX
Papier | Fördert
gute Waldnutzung
FSC® C083411

Kolibri 360 GmbH
Ferdinand-Jühlke-Straße 7
99095 Erfurt, Deutschland
produktsicherheit@kolibri360.de